FUNCTIONAL PAVEMENTS

Functional Pavements

Editors

Xianhua Chen & Jun Yang
Southeast University, China

Markus Oeser
RWTH Aachen University, Germany

Haopeng Wang
Delft University of Technology, The Netherlands

CRC Press
Taylor & Francis Group
Boca Raton   London   New York

CRC Press is an imprint of the
Taylor & Francis Group, an **informa** business

A BALKEMA BOOK

Published by:
CRC Press/Balkema
Schipholweg 107C, 2316 XC Leiden, The Netherlands

© 2021 by Taylor & Francis Group, LLC
CRC Press/Balkema is an imprint of the Taylor & Francis Group, an informa business

No claim to original U.S. Government works

**Visit the Taylor & Francis Web site at
http://www.taylorandfrancis.com**

**and the CRC Press Web site at
http://www.crcpress.com**

Typeset by Integra Software Services Pvt. Ltd., Pondicherry, India

Although all care is taken to ensure integrity and the quality of this publication and the information herein, no responsibility is assumed by the publishers nor the author for any damage to the property or persons as a result of operation or use of this publication and/or the information contained herein.

Library of Congress Cataloging-in-Publication Data

ISBN: 978-0-367-72610-2 (hbk)
ISBN: 978-1-003-15622-2 (eBook)
DOI: 10.1201/9781003156222
https://doi.org/10.1201/9781003156222

Functional Pavements – Chen et al (eds)
© 2021 Taylor & Francis Group, London, ISBN 978-0-367-72610-2

Table of contents

Asphalt mixture evaluation and performance

Pavement geotechnics & cementitious materials

Preface

The Chinese-European Workshop (CEW) on Functional Pavement Design was founded to promote activities relating to experimental characterization, advanced analysis, material development and production, design and construction of functional pavements. The previous five workshops were successfully hosted by

- Southeast University, China, 2010
- Harbin Institute of Technology, China, 2012
- RWTH Aachen University, Germany, 2014
- Delft University of Technology, The Netherlands, 2016
- Changsha University of Science and Technology, China, 2018

Following the great success of the five earlier CEW events, the 6th Workshop was held at Southeast University, in Nanjing, China, on October 18-21, 2020. The theme of CEW2020 is **SMART⁺**, which represents ***Sustainable, Safe, Multi-functional, Advanced Road for Tomorrow***. Due to the pandemic at this extraordinary time, CEW 2020 was held online. Over the years, participation increased with delegates not only from China and Europe but also from other parts of the world. The Workshops are meant to create an international platform for exchanging innovative ideas on functional pavements, sharing the recent developments and bridging the gap between functional and structural requirements of pavements.

All submitted manuscripts were peer-reviewed by at least two reviewers and the Editors. On the basis of their recommendations, more than 70 papers were accepted and published in the conference proceedings. The presentations were arranged in two parallel tracks and grouped in themes related to: Asphalt binders for flexible pavements, Asphalt mixture evaluation and performance, Pavement construction and maintenance, Pavement surface properties and vehicle interaction, Cementitious materials for rigid pavements, Pavement geotechnics and environment. Five keynote speakers were invited to present at the conference. Three special sessions (China-Netherlands, China-Germany, China-Britain-South Africa) were held to address specific topics on functional pavements.

The Editors would like to thank the Organizing Committee for their efforts in organization of online sessions. We also thank all the colleagues who contributed to the success of the conference through reviewing the manuscripts and ensuring the excellent quality of the accepted papers. The success of this conference would have not been possible without the generous sponsorship of several organizations.

We hope that CEW2020 will contribute to the establishment of a new generation of pavement design methodologies in which functional and structural requirements of pavements are combined and balanced.

The Editors
October 2020
Nanjing, China

Functional Pavements – Chen et al (eds)
© 2021 Taylor & Francis Group, London, ISBN 978-0-367-72610-2

Committees

Executive Committee

Chair:
Xianhua Chen, *Southeast University, China*

Members:
Markus Oeser, *RWTH Aachen University, Germany*
Sandra Erkens, *Delft University of Technology, The Netherlands*
Xueyan Liu, *Delft University of Technology, The Netherlands*
Athanasios (Tom) Scarpas, *Khalifa University, United Arab Emirates*
Yiqiu Tan, *Harbin Institute of Technology, China*
Zhen Leng, *The Hong Kong Polytechnic University, China*
Dawei Wang, Professor, *Harbin Institute of Technology, China*
Jun Yang, *Southeast University, China*

Secretary:
Bin Yu, *Southeast University, China*
Haopeng Wang, *Delft University of Technology, The Netherlands*

International Advisory Committee

Honored Chair:
Wei Huang, Professor, CAE Academician, *Southeast University, China*

Members:
Aimin Sha, Professor, *Chang'an University, China*
Bernhard Steinauer, Professor, *RWTH-AACHEN University, Germany*
Boming Tang, Professor, *Chongqing Jiaotong University, China*
Changwen Miao, Professor, CAE Academician, *Southeast University, China*
Frohmut Wellner, Professor, *Dresden University of Technology, Germany*
Fuming Wang, Professor, CAE Academician, *Zhengzhou University, China*
Fwa Tien Fang, Professor, *National University of Singapore, Singapore*
Imad L. Al-Qadi, Professor, *University of Illinois at Urbana-Champaign, USA*
Jianlong Zheng, Professor, CAE Academician, *Changsha University of Technology, China*
Jingyun Chen, Professor, *Dalian University of Technology, China*
Lingbing Wang, Professor, *Virginia Technology University, USA*
Lijun Sun, Professor, *Tongji University, China*
Michael Kaliske, Professor, *Dresden University of Technology, Germany*
Shaopu Yang, Professor, *Shijiazhuang Tiedao University, China*
Shengyue Wang, Professor, *Southeast University, China*
Susan Tighe, Professor, Academician of Canadian NAE, *University of Waterloo, Canada*
Viktor Mechtcherine, Professor, Academician of the Russian NAE, *Dresden University of Technology, Germany*
Wolfram Ressel, Professor, *University of Stuttgart, Germany*
Xiaoming Huang, Professor, *Southeast University, China*
Xiaoning Zhang, Professor, *South China University of Technology, China*
Yubao Guo, Professor, *RWTH Aachen University, Germany*

Functional Pavements – Chen et al (eds)
© 2021 Taylor & Francis Group, London, ISBN 978-0-367-72610-2

Scientific Committee

Baoshan Huang, Professor, *University of Tennessee, USA*
Bernhard Hofko, Associate Professor, *Vienna Tech University, Austria*
Bjorn Birgisson, Professor, *Texas A&M University, USA*
Bin Yu, Professor, *Southeast University, China*
Bo Yao, Associate Professor, *Nanjing University of Science and Technology, China*
Changfa Ai, Professor, *Southwest Jiaotong University, China*
Chia-pei Chou, Professor, *National Taiwan University, China*
David Woodward, Dr., *Ulster University, Ireland*
Davide Lo Presti, Professor, *University of Palermo, Italy*
Dawei Wang, Professor, *Harbin Institute of Technology, China*
Degou Cai, Researcher, *China Academy of Railway Sciences, China*
Elias Kassa, Professor, *Norwegian University of Science and Technology, Norway*
Erol Tutumluer, Professor, *University of Illinois at Urbana-Champaign, USA*
Feng Li, Professor, *Beihang University, China*
Filippo G. Praticò, Professor, *University Mediterranea of Reggio Calabria, Italy*
Filippo Giustozzi, Senior Lecturer, *Royal Melbourne Institute of Technology, Australia*
Gary W. Chai, Senior Research Fellow, *Griffith University, Australia*
Gordon Airey, Professor, *The University of Nottingham, UK*
Guoyang Lu, Research Assistant Professor, *The Hong Kong Polytechnic University, China*
Guangji Xu, Phd, *Southeast University, China*
Hainian Wang, Professor, *Changan University, China*
Haizhu Lu, PhD, *Jiangsu Conservation Center, China*
Hancheng Dan, Professor, *Central South University, China*
Hao Wang, Associate Professor, *Rutgers University, USA*
Hervé Di Benedetto, Professor, *University of Lyon, France*
Hongduo Zhao, Professor, *Tongji University, China*
Hui Li, Professor, *Tongji University, China*
Huining Xu, Professor, *Harbin Institute of Technology, China*
Jianchuan Cheng, Professor, *Southeast University, China*
Jiangmiao Yu, Professor, *South China University of Technology, China*
Jian-shiuh Chen, Professor, *National Cheng-kung University, China*
Jianzhong Pei, Professor, *Chang'an University, China*
Jie Gao, Lecturer, *East China Jiaotong University, China*
Jinxiang Hong, Researcher, *Jiangsu Academy of Building Materials Science, China*
John T. Harvey, Professor, *University of California, Davis, USA*
Jorge A. Prozzi, Professor, *The University of Texas at Austin, USA*
Jun Yang, Professor, *Southeast University, China*
Katerina Varveri, Assistant Professor, *Delft University of Technology, The Netherlands*
Lei Zhang, Professor, *Southeast University, China*
Liang He, Associate Professor, *Chongqing Jiaotong University, China*
Michael P. Wistuba, Professor, *TU Braunschweig, Germany*
Mingliang Li, Phd, *Highway Research Institute of the Ministry of Communications, China*
Nicole Kringos, Professor, *KTH Royal Institute of Technology, Sweden*
Pengfei Liu, Senior Researcher, *RWTH Aachen University, Germany*
Peng Xiao, Professor, *Yangzhou University, China*

Asphalt binders for flexible pavements

Functional Pavements – Chen et al (eds)
© 2021 Taylor & Francis Group, London, ISBN 978-0-367-72610-2

Preliminary study of reinforced asphalt binder with functionalized fiber

J. Xiong
Guangxi Transportation Science and Technology Group Co., Ltd., Nanning, Guangxi, China

Z. Su
School of Chemistry and Chemical Engineering, Guangxi University, Nanning, Guangxi, China

S. Ma
School of Resources, Environment and Materials, Guangxi University, Nanning, Guangxi, China

Y. Ye, J. Li* & R. Lan
School of Chemistry and Chemical Engineering, Guangxi University, Nanning, Guangxi, China

ABSTRACT: In this study, a new set of functionalized polyacrylonitrile (PAN) fibers were utilized successfully in asphalt. HEMA-PDA/PAN (HPAN) fibers, which were surface grafted with polyadopamine and polyhydroxyethyl to promote the asphalt-fiber adhesion and to avoid the creation of weak interfaces in modified asphalt, were added in asphalt and the results showed significant increase of material modulus. Based on the microstructure morphological analyses performed in scanning electron and atomic force microscopes, the surface of polydopamine functionalized fibers was rough with nano-particles on their surface indicating the new substances formed.

1 INTRODUCTION

Fiber reinforcement of asphalt mixtures is a promising research direction in the field of pavement engineering (Ho et al., 2016, Kaloush et al., 2010). Synthetic or other fiber types mainly improve the fatigue cracking resistance of asphalt mixtures and together with the potential increase of strength can be selected as additives to produce durable pavements (Gibson and Li, 2015, Apostolidis et al., 2019). Dalhat (Dalhat et al., 2019) reported that PAN/SBS composite modified asphalt mixture possesses higher anti-rutting performance than a single modified asphalt mixture. Wang et al. (Wang et al., 2018) reported that adding an appropriate amount of PAN to the asphalt mixture can improve it's anti-fatigue properties. However, fiber reinforced asphalt mixtures are to meet the current demands in structural and functional performance of pavements. Some researchers have functionalized the synthetics fibers to improve the overall mechanical performance of asphalt mixtures. Xiang et al. (Xiang et al., 2018) found that the basalt fiber functionalized with KH550 had shown enhanced adhesion in asphalt mixtures. Also, polyacrylonitrile fiber has an inert surface. Therefore, researchers have discovered that polydopamine (PDA) can adhere to most substances as a functional coating (Ryu et al., 2018, Liebscher et al., 2013). Su et al. (Su et al., 2019) adhered PDA to the surface of polyacrylonitrile fiber to graft graphene to improve the mechanical properties of fiber reinforced asphalt.

In this paper, the functionalized PAN fibers, which have been surface grafted with polyhydroxyethyl methacrylate and PDA, were used to improve the adhesion with asphalt binders.

---

* Corresponding author

Scanning electron microscope (SEM) and atomic force microscope (AFM) were used to show the microstructure of the modified fibers. The mechanical properties of functionalized fiber reinforced asphalt binders were evaluated by a dynamic shear rheometer (DSR) and the preliminary results presented here

2 MATERIALS AND METHODS

2.1 *Materials*

The asphalt used in this experiment is ESSO AH-70. Dopamine hydrochloride and hydroxyethyl methacrylate were provided by Shanghai McLean Biochemical Technology Co., Ltd. China. Trimethylaminomethane was provided by Sinopharm group chemical reagent Co., Ltd. China. Anhydrous ethanol was provided by Guangdong Guanghua Technology Co., Ltd. China. Polyacrylonitrile fiber with a length of 6 mm was purchased from Changzhou Tianyi engineering fiber Co., Ltd. China.

2.2 *Experiment process of modified fiber of modified fiber*

First, prepare Tris-HCl buffer with pH= 8.5. 2 g/L of dopamine hydrochloride and 20 g/L of hydroxyethyl methacrylate were added, then the fibers were added and stirred under vacuum for 10h. The resulting product was named as HPAN.

2.3 *Preparation of fiber modified asphalt*

The fibers were added in 150 g of asphalt, dispersed in a high-speed disperser for 90 min, and maintained at 175 °C. 1%, 2%, 3% (based on the asphalt quality) fiber modified asphalt was prepared.

2.4 *Textural characterization*

The micromorphology of PAN and HPAN were analyzed by SEM and AFM. The fibers were glued on conductive glue and sprayed with gold to obtain its microscopic appearance under the electron microscope. The three-dimensional shapes of fibers were obtained through the tapping mode of AFM. The mechanical property of fiber modified asphalt was evaluated by DSR (Ahmed, 2016, Li et al., 2019). Frequency sweep was performed on the asphalt sample, and the test temperature is 46-64 °C.

3 RESULTS AND DISCUSSION

3.1 *SEM & AFM*

Figure 1(a and b) are 8000-fold micrographs of PAN and HPAN, respectively. Figure 1(a) shows the smooth surface of PAN while that of the modified PAN in Figure 1(b) is rough, and there are many obvious particles on the surface, indicating that PDA and polyhydroxyethyl methacrylate substances have been successfully deposited on the fiber surface.

Figure 2 shows the 3D diagrams of PAN and MPAN. It can be seen from the AFM diagram that the surface of the PAN is smooth, while the surface of the MPAN has obvious particles, and the surface is highly undulating. This indicates that PDA-PHEMA successfully adhered to the fiber surface, which is consistent with the SEM findings.

3.2 *Complex modulus*

Figure 3 shows that the complex modulus of HPAN modified asphalt is greater than complex modulus of PAN modified asphalt at the same content and temperature. At 46 °C, the

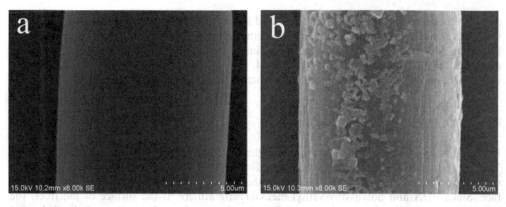

Figure 1.   SEM images of PAN (a) and HPAN (b).

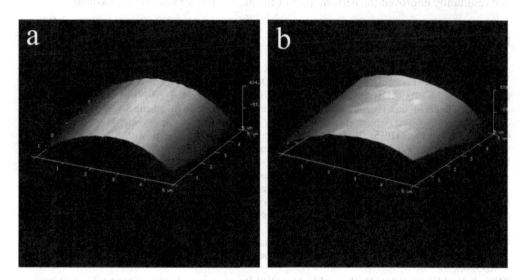

Figure 2.   AFM images of PAN (a) and HPAN (b).

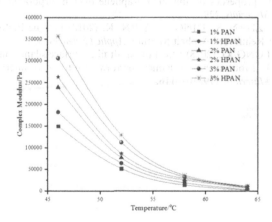

Figure 3.   Relationship between complex modulus and temperature of different content of PAN and HPAN modified asphalt.

complex modulus of 3% HPAN fiber modified asphalt is 15.6% higher than that of 3% PAN modified asphalt. This is because there are many hydroxyl groups and amine groups on the PDA-PHEMA coating, and from the SEM images below, it can be seen that the surface roughness of the fiber is significantly increased, thereby enhancing the adsorption of the fiber to asphalt and improving the deformation resistance of fiber modified asphalt.

4 CONCLUSIONS

DSR test showed that the deformation resistance of 3% HPAN modified asphalt is about 15.6% higher than that of pristine PAN modified asphalt. SEM and AFM results confirmed that polydopamine and polyhydroxyethyl methacrylate successfully adhered to the fiber surface. Since PDA and polyhydroxyethyl methacrylate adhere to the surface of the fiber, the introduction of amine groups, hydroxyl functional groups on the surface of the fiber and increase the surface roughness of the fiber which enhanced the adhesion of the fiber to asphalt and resultantly improved the performance of the modified fiber incorporated asphalt.

REFERENCES

APOSTOLIDIS, P., LIU, X., DANIEL, G. C., ERKENS, S. & SCARPAS, T. (2019) Effect of synthetic fibres on fracture performance of asphalt mortar. *Road Materials & Pavement Design*, 1–14.

DALHAT, M. A., AL-ADHAM, K. & AL-ABDUL WAHHAB, H. I. (2019) Multiple Stress–Creep–Recovery Behavior and High-Temperature Performance of Styrene Butadiene Styrene and Polyacrylonitrile Fiber–Modified Asphalt Binders. *Journal of Materials in Civil Engineering*, 31.

GIBSON, N. & LI, X. (2015) Characterizing Cracking of Asphalt Mixtures with Fiber Reinforcement. *Transportation Research Record: Journal of the Transportation Research Board*, 2507, 57–66.

HO, C.-H., SHAN, J., WANG, F., CHEN, Y. & ALMONNIEAY, A. (2016) Performance of Fiber-Reinforced Polymer-Modified Asphalt: Two-Year Review in Northern Arizona. *Transportation Research Record: Journal of the Transportation Research Board*, 2575, 138–149.

KALOUSH, K. E., BILIGIRI, K. P., ZEIADA, W. A., RODEZNO, M. C. & REED, J. X. (2010) Evaluation of Fiber-Reinforced Asphalt Mixtures Using Advanced Material Characterization Tests. *Journal of Testing & Evaluation*, 38, 400–411.

LIEBSCHER, J., MROWCZYNSKI, R., SCHEIDT, H. A., FILIP, C., HADADE, N. D., TURCU, R., BENDE, A. & BECK, S. (2013) Structure of polydopamine: a never-ending story? *Langmuir*, 29, 10539–48.

RYU, J. H., MESSERSMITH, P. & LEE, H. (2018) Polydopamine Surface Chemistry - A Decade of Discovery. *ACS Applied Materials & Interfaces*, 10.

SU, Z., MUHAMMAD, Y., SAHIBZADA, M., LI, J., MENG, F., WEI, Y., ZHAO, Z. & ZHANG, L. (2019) Preparation and properties of aminated graphene fiber incorporated modified asphalt. *Construction and Building Materials*, 229.

WANG, H., YANG, Z., ZHAN, S., DING, L. & JIN, K. (2018) Fatigue Performance and Model of Polyacrylonitrile Fiber Reinforced Asphalt Mixture. *Applied Sciences*, 8.

XIANG, Y., XIE, Y. & LONG, G. (2018) Effect of basalt fiber surface silane coupling agent coating on fiber-reinforced asphalt: From macro-mechanical performance to micro-interfacial mechanism. *Construction and Building Materials*, 179, 107–116.

Functional Pavements – Chen et al (eds)
© 2021 Taylor & Francis Group, London, ISBN 978-0-367-72610-2

Use of waste oil/styrene-butadiene-rubber blends as rejuvenators for aged bitumen

Shisong Ren, Xueyan Liu, Peng Lin & Sandra Erkens
Section of Pavement Engineering, Faculty of Civil Engineering & Geosciences, Delft University of Technology, Delft, The Netherlands

ABSTRACT: This study elaborates the influence of blends formulated by waste oil (WO) and styrene-butadiene rubber (SBR) on thermo-mechanical and chemical compositional characteristics of rejuvenated bitumen. The results show that both WEO (waste engine oil) and WCO (waste cooking oil) can compensate viscous component for reclaimed bitumen, and the effect of WCO is more significant. However, WO-rejuvenated bitumen has considerable drawbacks in terms of temperature susceptibility and flow resistance, but dramatic improvement after being added with SBR has been noticed. Furthermore, the rejuvenation mechanism of WO/SBR blends is revealed by using FTIR tests, which shows that the physical blending mainly determines the rejuvenation mechanism when the WO/SBR blends are used in bitumen. The combination of WO and SBR is beneficial for improving the temperature susceptibility and rutting resistance of rejuvenated bitumen.

1 INTRODUCTION

At the end-of-life of asphalt pavements, the aged bitumen in old asphalt, which is named reclaimed asphalt pavement (RAP), is a degraded material of high modulus [4] but with the potential to be re-used with the proper treatment [1–3]. Recently, although many highway agencies allow the incorporation of RAP in new pavements [4], the pavement materials consisting of RAP could show high thermal and fatigue cracking vulnerability [5].

Nowadays, the most commonly used rejuvenation method for aged RAP binders is the addition of softening additives of light fractions which can balance the chemical composition and thus the physio-mechanical properties of aged binders [3]. Among those additives, waste oil (WO) is commonly used for the aged bitumen, including WCO and WEO [6]. Although many studies have reported that adding WO could compromise the physio-mechanical properties for aged bitumen, there are inevitable issues of using WO into the rejuvenation of aged bitumen. The main challenge of WO-rejuvenated bitumen is on the determination of rejuvenator dosage, which can not only restore the physio-mechanical properties but strengthen the anti-rutting ability for aged bitumen as well [7, 8]. Hence, the motivation and objective of this study is to explore the potential of using waste oil and SBR to restore and improve the rheological and chemical properties of aged bitumen.

2 MATERIALS AND METHODS

2.1 *Materials and sample preparation*

Base bitumen ZH70$^{\#}$ is used to prepare the aged bitumen with air-blowing process at 260 °C for 48 hours. The conventional performance for neat bitumen and waste oil are displayed in Table 1 and 2, respectively. Meanwhile, the 60 °C viscosity and 25 °C penetration value of

Table 1. Conventional properties and chemical components of aged asphalt.

Items	Aged bitumen	Test methods
Penetration (25 °C, 0.1mm)	10	ASTM D5
Softening point (°C)	84.1	ASTM D36
Ductility (10 °C, cm)	0.2	ASTM D113
Ductility (15 °C, cm)	1.3	ASTM D113
Viscosity (60 °C, Pa·s)	32359	AASHTO T316
Saturate (S, wt%)	10.31	ASTM D4124
Aromatic (A, wt%)	8.19	
Resin (R, wt%)	42.83	
Asphaltene (At, wt%)	38.67	
Colloidal index CI *	0.96	

* Colloidal index CI = (Asphaltene + Saturate)/(Aromatic + Resin)

Table 2. Basic properties of WCO and WEO.

Items	WCO	WEO
Specific gravity (g·cm^{-3})	0.926	0.884
Viscosity (60°C, mPa·s)	28.75	55.00
Flash point (°C)	171	173

aged bitumen is 32359 Pa·s and 10-dmm, while the softening point and 5 °C ductility is 84.1 °C and 0.1-cm, respectively.

The aged bitumen is mixed with WO for about 15min with the rotation speed of 1000 r/min and mixing temperature of 150 °C. The dosage of waste oil is 5, 10 and 15wt%, which is referred as WCO5 (WEO5), WCO10 (WEO10) and WCO15 (WEO15), respectively. Further, both WCO10 and WEO10 rejuvenated bitumen are further modified by SBR with various content of 1, 2 and 3wt% to obtain WO/SBR rejuvenated bitumen, which are called as WCOSBR1 (WEOSBR1), WCOSBR2 (WEOSBR2) and WCOSBR3 (WEOSBR3).

2.2 Rheological and chemical properties characterization

Dynamic shear rheometer (DSR, TA-HR1) is performed to determine these viscoelastic behaviors for bitumen with 25-mm diameter and 1-mm gap. Both complex modulus (G^*) and phase angle (δ) of aged and rejuvenated bitumen are obtained through frequency sweep tests with the frequency increasing from 10^{-1} rad/s to 10^2 rad/s at the various temperature of 45, 50, 55, 60 and 65 °C, respectively.

FTIR is used to explore the rejuvenation and modification mechanism of rejuvenated bitumen with the WO/SBR blends. The molecular interaction of bitumen sample in IR ranges of 400-4000 cm^{-1} is observed.

3 RESULTS AND DISCUSSION

3.1 Temperature susceptibility

In this study, the complex number index (CNI) parameter is selected to evaluate the temperature susceptibility of aged and rejuvenated bitumen as follows:

$$CNI = G'TS + G''TS \cdot i \qquad (1)$$

The relationship between temperature (T) with storage modulus (G') and loss modulus (G") of aged, WCO- and WCO/SBR-rejuvenated bitumen is displayed in Figure 1. Obviously, the increasing of temperature remarkably reduces both storage and loss modulus of bitumen. There exists fairly good linear-relation between temperature with lgG' and lgG" values. The G'TS and G"TS values of bitumen can be obtained from the absolute slope of regression equations of lgG'-T and lgG"-T, severally.

From Table 3, WCO and WEO increases the G'TS as well as G"TS values of aged and rejuvenated bitumen, indicating that adding WCO and WEO can rise the temperature susceptibility for aged and rejuvenated binders. This phenomenon is attributed to the temperature-dependency of light components in WO. Meanwhile, the temperature susceptibility of WO-rejuvenated bitumen improves after being added with SBR copolymer. Moreover, WCO/SBR-rejuvenated bitumen has superior thermal susceptibility to that of WEO/SBR-rejuvenated bitumen.

3.2 Viscoelastic properties

Figure 2 presents the effects of WO type and concentration, SBR copolymer dosage as well as frequency on the complex modulus of aged and rejuvenated bitumen. This result indicates that the complex modulus value of bitumen increases linearly as the testing frequency increases. Over the whole frequency range, the aged bitumen has the highest modulus values because of the loss of light compounds. Meanwhile, adding WCO and WEO could remarkably reduce complex modulus of aged bitumen, which continues to decrease as the WO dosage increases. Obviously, adding WCO and WEO has adverse influence on anti-deformation ability for aged and rejuvenated bitumen. The modulus value of rejuvenated bitumen is improved by incorporating the SBR. With the increase of SBR dosage, the complex modulus of

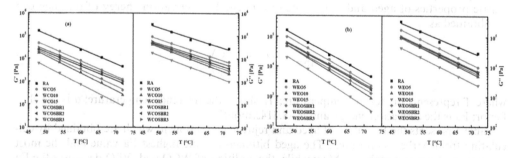

Figure 1. Storage modulus (G') and loss modulus (G") of aged bitumen, WCO and WCO/SBR rejuvenated bitumen (a), and WEO and WEO/SBR rejuvenated bitumen (b).

Table 3. CNI values of aged and rejuvenated bitumen.

Bitumen samples	CNI=G'TS+G"TS·i	Bitumen samples	CNI=G'TS+G"TS·i
RA	0.06427+0.04414i	RA	0.06427+0.04414i
WCO5	0.06630+0.04642i	WEO5	0.06775+0.05316i
WCO10	0.07564+0.05316i	WEO10	0.07259+0.05339i
WCO15	0.07752+0.05197i	WEO15	0.07520+0.05201i
WCOSBR1	0.06970+0.04688i	WEOSBR1	0.06628+0.05084i
WCOSBR2	0.06140+0.04488i	WEOSBR2	0.06417+0.04951i
WCOSBR3	0.06040+0.04359i	WEOSBR3	0.06298+0.04891i

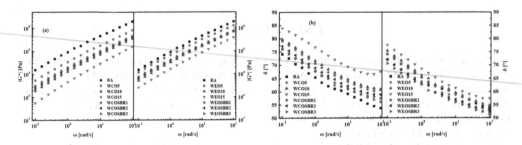

Figure 2. Complex modulus (a) and phase angle (b) of aged and rejuvenated bitumen.

rejuvenated bitumen increases gradually, which is lower than that of WCO5 and WEO5 rejuvenated bitumen.

Compared to aged bitumen, WCO and WEO rejuvenated bitumen has higher phase angle value, showing that WO can restore the viscous component and modulus of aged bitumen dramatically. However, the increase of WO dosage weakens the elastic properties of rejuvenated bitumen, which is importance of improving both high- and low-temperature performance for bitumen. Moreover, adding SBR can remarkably decline phase angle value and strengthen elastic property of WO-rejuvenated bitumen. Furthermore, the phase angle for WCO-rejuvenated binder shows larger than WEO-rejuvenated bitumen. This result indicates that WCO can recover the viscous characteristic for aged bitumen more easily than WEO. It also can be seen than the impact of SBR on improving elastic properties for WEO-rejuvenated binder is more obvious.

3.3 *Activation energy*

The relationship between temperature and shift factor α of aged and rejuvenated bitumen is analyzed by the Arrhenius formula to assess the influence of WO and SBR on the viscoelastic properties of aged and rejuvenated bitumen. The activation energy of bitumen can be obtained as

$$\ln\alpha = \frac{Ea}{R}\left(\frac{1}{T} - \frac{1}{T_0}\right) \qquad (2)$$

where T represents the testing temperature; T_0 shows the reference temperature; α is the shift factor; Ea is the activation energy and R=8.314J/mol·K.

Table 4 presents the Ea values of aged and rejuvenated bitumen, which are obtained by calculating from Arrhenius formula. The aged bitumen has the highest Ea value and the most excellent performance stability. Meanwhile, the addition of WCO and WEO decreases the Ea value of aged bitumen dramatically, which means WCO and WEO both make negative influence on temperature sensibility for rejuvenated bitumen. Moreover, Ea value of rejuvenated bitumen can be further enhanced by adding SBR, which can strengthen the temperature sensitivity for WO-rejuvenated binder obviously. Further, WEO rejuvenated bitumen has better temperature susceptibility performance than WCO rejuvenated bitumen.

Table 4. The Ea values of aged and rejuvenated bitumen.

Codes	RA	WCO5	WCO10	WCO15	WCOSBR1	WCOSBR2	WCOSBR3
Ea/KJ·mol-1	67.09	62.22	60.27	55.92	57.28	58.96	58.58
Codes	RA	WEO5	WEO10	WEO15	WEOSBR1	WEOSBR2	WEOSBR3
Ea/KJ·mol-1	67.09	66.50	65.96	63.37	60.56	61.56	61.54

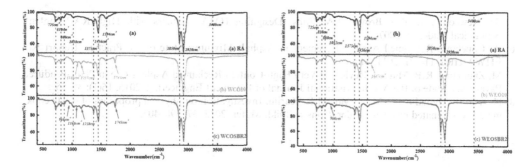

Figure 3. FTIR plots of aged and rejuvenated bitumen.

3.4 *Rejuvenation mechanism of WO/SBR blends in bitumen*

Aim to further characterize the rejuvenation mechanisms of WO and SBR on the microstructure of recycled bitumen, the main IR vibrational peaks of aged and rejuvenated bitumen are shown in Figure 3. It is well known that FTIR results can determine whether or not chemical reaction happens during the mixing of aged asphalt, WO and SBR. Compared with reclaimed asphalt, waste cooking oil rejuvenated asphalt has additional characteristic peaks at 1160, 1257 and 1745 cm^{-1}, which is attributed to the addition of waste oil. Besides, WO/SBR rejuvenated bitumen has the additional characteristic peak at 966 cm^{-1}, which are assigned as –CH=CH– out-of-plane deformation. Furthermore, new absorption peaks of WEO/SBR rejuvenated bitumen appear at 966 cm^{-1} and 1747 cm^{-1}, which could characterize the existence of WEO and SBR. However, compared with the absorption peaks of raw materials, there is no distinct formation or disappearance of characteristic peaks for WO and SBR rejuvenated bitumen. In a word, the rejuvenation mechanisms of WO and WO/ SBR rejuvenated bitumen mainly focus on the physical blending.

4 CONCLUSIONS

In this study, waste oil (WCO and WEO) and SBR were used as rejuvenators to restore and enhance the viscoelastic of rejuvenated bitumen. Blending WCO and WEO significantly decreases the complex modulus of aged bitumen, which can be enhanced by the addition of SBR. Meanwhile, WCO and WEO can restore the viscous components and modulus of aged bitumen dramatically but weakens the elastic behavior of rejuvenated bitumen, which can be remarkably enhanced by SBR. Based on master-curves, WCO and WEO both have adverse influence on the rutting resistance, activation energy as well as the temperature susceptibility of rejuvenated bitumen, which can be improved by incorporating the SBR copolymer in WO. Further, FTIR tests have shown that the rejuvenation mechanism occurs mostly due to the physical blending of WO/SBR blends in aged binders.

REFERENCES

1. J.C. Peterson. A Review of the Fundamentals of Asphalt Oxidation: Chemical, Physicochemical Property, and Durability Relationship. Transportation Research Circular, Number E-C140, Transportation Research Board of National Academies, Washington, D.C., 2009.
2. A. EI-Shorbagy, S. EI-Badawy, A. Gabr. Investigation of waste oils as rejuvenators of aged bitumen for sustainable pavement. Constr. Build. Mater. 2019, 220: 228–237.
3. P. Apostolidis, X. Liu, C. Kasbergen, A. Scarpas. Synthesis of Asphalt Binder Aging and the State of the Art of Antiaging Technologies. Transportation Research Record. 2017, 2633 (1): 147–153.
4. I. Al-Qadi, M. Elseifi, S. Carpenter. Reclaimed Asphalt Pavement - A Literature Review. FHWA-ICT -07-001, 2007.

5. D. Newcomb, E. Ray Brown, J.A. Epps. Designing HMA mixtures with High RAP content: A practical guide, 41, 2007.
6. A. Copeland. Reclaimed Asphalt Pavement in Asphalt Mixtures: State of the Practice. Report No. FHWA-HRT-11-021, 2011, McLean.
7. M. Zaumanis, R.B. Mallick. Review of Very High-Content Reclaimed Asphalt Use in Plant-Produced Pavements: State of the Art. International Journal of Pavement Engineering. 2015, 16(1): 39–55.
8. H. Asli, E. Ahmadiniam, M. Zargar, M.R. Karim. Investigation on physical properties of waste cooking oil-Rejuvenated bitumen binder. Constr. Build. Mater. 2012, 37: 398–405.

Functional Pavements – Chen et al (eds)
© 2021 Taylor & Francis Group, London, ISBN 978-0-367-72610-2

Smart bionic de-icing bituminous material based on autocrine microcapsules with a temperature responsive character

Xin-Ming Xie, Li-Qing Wang & Jun-Feng Su*
School of Material Science and Engineering, Tiangong University, Tianjin, China

ABSTRACT: bitumen is one of the most widely used pavement materials in road construction. Due to the low temperature in winter, the snow on the road is frozen due to snowfall. The close combination of ice and the road greatly reduces the adhesion coefficient of the road, which has a serious adverse impact on road stability and traffic safety. Intelligent de-icing material is an urgent need for bitumen pavement especially in winter to prevent traffic accidents. In this study, inspired by the biomimetic de-icing structure of bird feather auto-secreted oil, and based on an oil-controlled release microcapsule with temperature response characteristics, an intelligent de-icing bitumen material for long-life anti-ice pavement was synthesized.

1 PRESENT SITUATION AND DEVELOPMENT TREND OF ROAD DE-ICING RESEARCH

At present, de-icing technology can be divided into passive de-icing and active de-icing. Passive methods include three traditional methods: mechanical method, chemical method and thermal method. There are a lot of researches on passive de-icing, mainly using artificial methods to remove the ice from the road surface. Traditional passive de-icing methods have obvious limitations, such as low efficiency, environmental pollution, road aging and other disadvantages, which can neither prevent the formation of ice nor effectively de-icing [1]. Active de-icing method is to add a certain amount of special de-icing materials to the road surface to change the deformation characteristics of the road surface and the contact state between tires and the road surface, so that the road surface itself has a certain de-icing ability. The literature review shows that rubber particle addition technology and road antifreeze coating technology have been successfully applied to realize the purpose of road de-icing. However, rubber particles reduce the durability of bitumen pavement. Road antifreeze coating technology because of road wear, coating can not achieve long-term snow removal and de-icing function [2]. Microencapsulation technology is a miniature packaging technology with the microencapsulation shell as the container. By adding the microcapsule contained with deicing agent into the bitumen binder and triggering release under specific conditions, the deicing performance of bitumen pavement can be improved without affecting material properties. As a deicing agent, glycerin reduces the freezing point of water and the interface binding force between ice and bitumen material, which meets the requirements of environmental protection while de-icing. Therefore, designing an intelligent long-term de-icing method has great economic value and social benefit.

---

* Corresponding author

2 THE PREPARATION OF SMART BIONIC DE - ICING BITUMINOUS MATERIAL

2.1 Synthesis of microcapsules

Autocrine microcapsules were prepared by in-situ polymerization. Polyvinyl alcohol (PVA)/ methoxylated hexylol melamine resin (HMMM) was used as composite shell material and glycerin with de-icing effect was used as core material to cover the microcapsule in the shell. The whole preparation process is divided into the following steps: (1) sodium hydroxide (2.0 wt%) solution is added to the styrene maleic anhydride copolymer (SMA) solution, the pH value of the solution is adjusted to 10, and the mixed solution reacts in the low temperature thermostatic bath at 50 °C for 1 h. Then glycerin was added into the mixed solution and stirred mechanically with a high-speed dispersing machine for 15 min. (2) The emulsion was reduced to room temperature, HMMM prepolymer and PVA solution were added at the same time, and the stirring speed was 400 r/min. (3) The temperature was slowly raised to 80 °C at 2 °C/min, and the PH of the system was adjusted by adding acetic acid solution (2.0 wt.%) drop by drop when the temperature was raised to 60 °C. After the temperature rose to 80 °C, the polymerization lasted for 2 h, and then slowly lowered the temperature to the ambient temperature (4) Finally, the microcapsules were filtered and washed with pure water and dried in a vacuum oven. A slow average heating rate and a constant rate of acetic acid are important during polymerization. At the same time, PVA and prepolymer have enough time to distribute evenly in the cyst wall. The microcapsule samples with PVA content of 1.0 wt.%, 2.0 wt.%, 3.0 wt.%, 4.0 wt.% and 5.0 wt.% in PVA/HMMM shell material were named PH-1, PH-2, PH-3, PH-4 and PH-5, respectively.

2.2 Preparation of microcapsules/bitumen composite samples

The 40/50 bitumen was blended with difference microcapsules (1-5 wt.%) using a propeller mixer for 30 min at 160 °C with a constant speed of 200 r/min.

3 TEST RESULTS AND ANALYSIS OF NEW INTELLIGENT DE-ICING BITUMEN

3.1 Physical structure of microcapsules

As shown in Figure 1(a-c), the microcapsules in this study were prepared by in-situ polymerization. The core material is dispersed into stable droplets by high speed stirring and SMA emulsification. The SMA is negatively charged in the hydrophilic end of the alkaline environment, and the molecular chain is fully extended. The hydrophilic end enters the oil core and the hydrophilic end is exposed to the solution. HMMM prepolymer molecules are adsorbed on the droplet by electrostatic interaction with SMA hydrophilic terminal [3]. Subsequent addition of PVA molecules also inserted the oil-wet end into the oil core material. Under appropriate polymerization conditions, HMMM prepolymer crosslinked to form a microcapsule skeleton, with a certain compressive resistance. PVA gel was embedded into HMMM polymer network to form polymer interpenetrating network with HMMM network. As shown in Figure 1 (d-f), PVA/HMMM composite microcapsule forms a ball with smooth surface and uniform size. Almost all microcapsules had particle sizes ranging from 10 to 50 μm. In general, the average diameter of microcapsules mainly depends on the stirring speed of the core material emulsion, and is not affected by the shell thickness and shell material. Therefore, the particle size of the composite microcapsule can be adjusted by stirring speed during the synthesis.

To verify the special structure of autocrine microcapsules, the dry microcapsules were eluted with ethanol using the solubility of PVA gel in ethanol. As shown in Figure 2, the SEM surface morphology of the microcapsules after elution. It can be seen from the figure that PVA gel in the microcapsule composite shell has been removed, and the surface of the microcapsule has a porous structure, which proves that PVA is embedded in the HMMM skeleton, forming an interpenetrating polymer network. HMMM crosslinked framework has certain

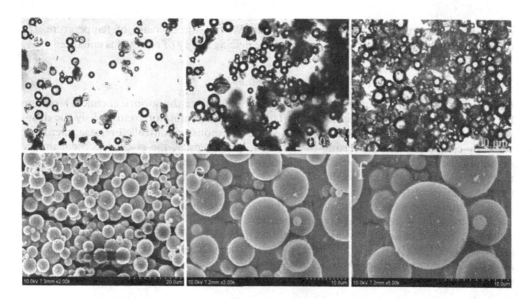

Figure 1. Optical morphology of microcapsules formation process: (a) core materials dispersed into droplets, (b) HMMM cross-lined, (c) PVA addition, SEM morphologies of microcapsules, (d-f) PVA/HMMM composite microcapsules of PH-1.

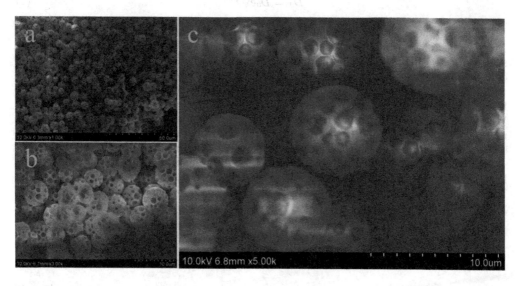

Figure 2. SEM morphologies of PVA/HMMM microcapsules (PH-1) after an alcohol elution process, (a, b) microcapsules still keeping global shape, and (c) porous structure appearing on shells of microcapsules.

strength. After elution, the microcapsules remained intact without damage or deformation. The special structure of microcapsules makes them have certain strength and keep the original shape in the process of processing and using. The hole on the microcapsule shell is the "outlet" of the core material. The PVA gel is the "door" of the microcapsule, and the temperature controls the opening and closing of the microcapsule. As the ambient temperature rises, the PVA gel expands, closing the microcapsule's "door" and sealing the core. As the temperature decreases, the volume of PVA gel decreases, the shell becomes thinner or the "door" opens, and the core material is gradually released. With the decrease of temperature, the release amount of deicing agent increases, and the deicing ability of bitumen is enhanced.

3.2 *Autocrine characteristic of microcapsules in bitumen*

FTIR-ATR can supply an approach to quantitatively analyze the surface chemical structure for a microcapsules/bitumen thin layer [4]. This method can monitor continuously the diffusion of core material in bitumen from temperature-responsive controlled-release microcapsules. By comparing and converting the infrared light intensity, the glycerin content on the surface of the thin bitumen layer can be calculated, so as to monitor the release of autocrine microcapsule core material. Usually, mean size and shell thickness of microcapsules greatly influence the property of shells. In order to eliminate the above effects, the microcapsules were prepared under the same conditions with the same core-shell weight ratio. In the test, an absorption band was specified as the measurement standard of D value (Diffusion coefficient). The absorbance is calculated by the logarithmic time scale of the integral peak area. Fick's law, represented by (1), is generally used to describe the value of D, where T is time, P is position, and C is concentration. D_T is described by D_0 and temperature T in Equation (2), where s_1 and s_2 are constants.

$$\frac{\partial c}{\partial t} = D \cdot \frac{\partial^2 c}{\partial p^2} \tag{1}$$

$$D_T = D_0 e^{\left(\frac{s_1}{T}\right) + s_2} \tag{2}$$

Figure 3(a) shows the diffusion coefficient value of glycerin in the pitch sample added with 1.0 wt.% microcapsules in the temperature range of - 20 ~ 40 °C for 30 days. Microcapsule samples (PH-1, PH-2, PH-3, PH-4, and PH-5) with the same mean particle size value and core-shell weight ratio were used. It can be seen that the diffusion coefficient increases linearly with the decrease of temperature. In addition, the diffusion coefficient of microcapsules with higher PVA content in bitumen was larger under the same temperature for different samples. When the ambient temperature reaches 10 °C, the diffusion coefficient value of bitumen mixed with PH-1, PH-2 and PH-3 tends to be the same. At the same temperature, the diffusion coefficient of PH-3/bitumen is close to 130% of PH-1/bitumen [5].

Figure 3(b) shows the diffusion coefficient values of bitumen samples with 2.0 wt.% PH-1, PH-2, PH-3, PH-4 and PH-5 microcapsules in the temperature range of -20-40 °C for 30 days. The diffusion coefficient increases linearly with the decrease of temperature. The more PVA

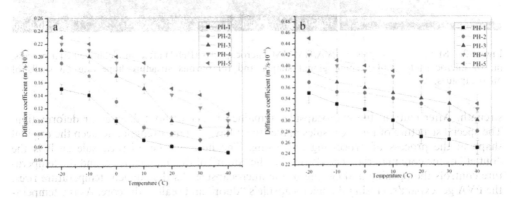

Figure 3. Diffusion coefficient values of in bitumen with PVA/HMMM microcapsules (PH-1, PH-2, PH-3, PH-4 and PH-5) under a temperature range of -20~40 °C.

"switch" in the shell material, the higher the release rate of microcapsules. The results show that with the increase of microcapsules in bitumen samples, the content of deicing agent on bitumen surface increases, the diffusion coefficient of microcapsules increases, and the deicing ability of bitumen improves. Previous studies have shown that the wall material of PVA/ HMMM microcapsule will form a chemical bond with bitumen. Therefore, adding appropriate amount of microcapsule will improve the mechanical properties of bitumen composite.

3.3 De-icing of bituminous material with autocrine microcapsules

In the cold winter, ice and bitumen are closely combined. With the increase of the bonding strength, the ice shear stress also increases. The interfacial adhesion between ice and bitumen is predictable, and the ice shear force is an important parameter to measure the adhesion performance of ice.

Figure 4 shows the detailed steps of deicing experiment. The ice shear force of bitumen samples mixed with different microcapsules (PH-1, PH-2, PH-3, PH-4, PH-5) was tested, and the maximum shear force was taken as its performance index. The weight of microcapsules in all bitumen samples was 1 %. As shown in the Figure 5, the ice shear force of pure bitumen is 590 kPa, and the ice is closely combined with the sample surface, which makes it difficult to remove the ice. In contrast, adding a small amount of PVA/HMMM microcapsules (1.0 wt.%) can significantly improve the de-icing performance of bitumen. Compared with pure bitumen, the maximum shear stress of PH-1/bitumen is only about half of that of pure bitumen under the same conditions. With the decrease of temperature, the maximum temperature of PH-1/ bitumen sample decreases. In other words, the binding force of ice and bitumen decreases with the decrease of temperature. In addition, with the decrease of temperature, the maximum temperature of PH-5/bitumen sample has no such downward trend. This suggests that PH-5 may be a threshold. When the value is higher than the threshold, the excessive addition of microcapsules has no significant effect on the performance of bitumen. A small amount of self secretory core material can reduce the adhesion between ice and bitumen. This is good news for using autocrine microcapsules to control the cost of de-icing pavement. In the future work,

Figure 4. Photographs of anti-icing capability evaluation through ice shearing force: aluminum plate coated with bitumen blending with microcapsules in a constant temperature box (-20 °C), (b, c) water pulled into the plastic hollow columns, ice form in the plastic hollow column under a low temperature, (d) ice shear force tested by the force sensor, (e) a vertical view of pure bitumen sample on ice column pushed over by the sensor probe, and (f) a vertical view of microcapsules/bitumen samples on ice column pushed over by the sensor probe.

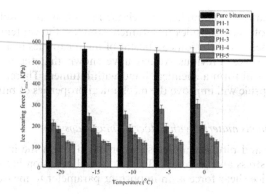

Figure 5. Ice shearing force values of microcapsules/bitumen samples (1.0 wt.% microcapsules) under a low temperature (-20, -15, -10, -5 and 0 °C), each sample maintaining at a low temperature for 7 days before measurement.

the optimal dosage of microcapsules will be considered and optimized from the aspects of energy consumption, road structure and de-icing effect.

4 CONCLUSIONS

In this work, an autocrine deicing microcapsule was designed. The microcapsule was prepared by in-situ polymerization with methyl methacrylate/polyvinyl alcohol as the shell. The autocrine properties of microcapsules were proved to be based on temperature response. The following conclusion can be concluded,

(1) HMMM/PVA microcapsule was successfully prepared by in-situ polymerization.
    All microcapsule samples are spherical.
    HMMM cross-linked microcapsule network has a certain strength. As the skeleton of microcapsule, PVA is embedded in the skeleton. After elution, PVA in the composite shell of microcapsule can be removed.
(2) The test results show that autocrine microcapsules are temperature-sensitive, and PVA acts as a "switch" to adjust the core material release amount according to the change of ambient temperature. Microcapsules with higher PVA content in the shell may have greater autocrine capacity in bitumen. At the same time, it is found that the core material has the ability to penetrate the shell even at the low temperature below zero.
(3) The results of ice shear test show that adding a small amount of PVA/HMMM microcapsules can significantly improve the de-icing performance of bitumen. The ice shear force of microcapsules/bitumen samples decreased with the decrease of temperature. The results show that the binding force of ice and bitumen decreases with the decrease of temperature, which meets the deicing requirements under low temperature conditions in winter.

REFERENCES

1 Su, J.F. et al. 2019. Smart bituminous material combining anti-icing and self-healing functions using electrothermal graphene microcapsules containing oily rejuvenator. *Construction and Building Materials* 224: 671–681.
2 Wang, X.Y. et al. 2020. Rheological behaviour of bitumen blending with self-healing microcapsule: Effects of physical and chemical interface structures. *Colloids and Surfaces A-Physicochemical and Engineering Aspects* 586: 124212.

3 Guo, Y.D. et al. 2019. Mechanical experiment evaluation of the microvascular self-healing capability of bitumen using hollow fibers containing oily rejuvenator. *Construction and Building Materials* 225: 1026–1035.

4 Guo, Y.D. et al. 2019. Microstructure and Properties of Self-Assembly Graphene Microcapsules: Effect of the pH Value. *Nanomaterials* 9: 587.

5 Su, J.F. et al. 2019. Experimental observation of the vascular self-healing hollow fibers containing rejuvenator states in bitumen. *Construction and Building Materials* 201:715–727.

Functional Pavements – Chen et al (eds)
© 2021 Taylor & Francis Group, London, ISBN 978-0-367-72610-2

Smart micro-containers for multi-functional asphalt pavement

Xin-Ming Xie, Li-Qing Wang & Jun-Feng Su*
School of Material Science and Engineering, Tiangong University, Tianjin, China

ABSTRACT: With the development of materials science and communication technology, the future road engineering is marching towards a new era. The intelligent material will give the road multi-functional performance bridging the gap between demand and reality. Smart micro-containers provide a new way to give road function including self-healing, anti-aging, anti-icing, storing energy, luminescence, electrothermal conversion, noise reduction and absorption of pollutant. In this lecture, microcapsule and microvascular were introduced for multi-functional asphalt pavement. Microcapsule and microvascular were fabricated containing core material through chemical reactions. These micro-containers can exist in situ in asphalt mixture and play a functional role. The preparation, performance characterization and function realization of the materials were introduced in detail.

1 SELF-HEALING ASPHALT USING MICROCAPSULES

Self-healing microcapsules containing rejuvenator now is an environmental-friendly and energy saving commercial product for long-life asphalt pavement successfully applied in China. The results of basic research and practical application are very excellent, which makes it as an ideal way to extend the service life of the asphalt pavement. The great success of the above basic research and practical application has aroused more researchers' interest. It has become a feasible approach to extend the life of asphalt by microcapsules containing rejuvenator. Even so, there are still two questions puzzle researches in this field. One of the questions is to determine if a broken microcapsule transform to be a empty hole after a self-healing process. In order to intuitively understand this problem, this study explored in detail the entire process of repairing bitumen microcracks with microcapsules, as shown in Figure 1. Among them, high methylether melamine formaldehyde (HMMM) polymer is used as wall material for self-healing microcapsule, which not only reduces the cost, but also ensures that the microcapsule has certain strength and toughness, and will not be damaged in the processing process. As shown in Figure 1(a), microcapsules containing regrowth are present in the bitumen binder. When a microcrack is triggered, several microcapsules on the growth path of microcrack are punctured by the tip-stress of this microcrack (Figure 1(b)). With the help of capillary action, the liquid rejuvenator flows into the microcracks rapidly (Figure 1(c)). The liquid rejuvenator then diffuses rapidly into the asphalt under the driving force of the concentration difference [1]. Softened bitumen can enhance the molecular movement ability and adhesion force of the microcrack interface, and then promote the movement and chain entanglement of the macromolecules, so that the microcrack can be healed (Figure 1(d)). When the self-healing process is completed, it is necessary to study and judge whether the capsule is hollow or solid. Naturally, some researchers and product users worry hat hollow microcapsules may become mechanical defects of asphalt materials. Litter holes in bitumen binder may damage the performance of bitumen [2].

---

* Corresponding author

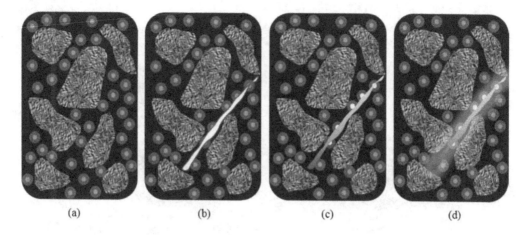

| (a) | (b) | (c) | (d) |

Figure 1. Illustration of self-healing process of asphalt using microcapsules in bitumen binder, (a) a microcrack triggered in bitumen binder, (b) several microcapsules on the path of microcrack growth punctured by the tip stress of the microcrack, the liquid rejuvenator flows into the microcracks, and (d) the microcrack healed under the diffusion of rejuvenator.

In this study, the self-healing microcapsules were uniformly dispersed in the bitumen. After the destruction of the microcapsules, the remediation agent diffused and mixed with the bitumen molecules, and the fragments of the shell formed chemical bonds with the bitumen. No holes were formed in the original position of the microcapsules. Due to the formation of new chemical bonds, the addition of shell fragments improves the mechanical properties of asphalt pavement.

2 ANTI-ICING ASPHALT USING AUTOCRINE MICROCAPSULES

Due to the low temperature in winter, the snow on the road is frozen due to snowfall. The limitation is obvious that the traditional passive de-icing methods can not prevent the formation of ice or can not remove ice with high efficiency and low cost (Figure 2(a,b)), it is necessary to develop a new active de-icing method. New de-icing method or material structure can be inspired in nature. American scientists have studied the structure of penguin wings and found that the feathers of penguins have water-proof effect [3]. There is a gland near the tail of this penguin, and this special body structure only has a very good waterproof effect, because it is precisely because this gland can let its body release a kind of grease, so when its gland releases the grease, this penguin will use its own mouth to lightly smear the grease on its feathers (Figure 2(c)). The inspiration above reminds us to apply this principle to de-icing pavement practice. The oil-like de-icing agent can remove snow and ice by reducing the binding force between ice and bituminous pavement. The de-icing agent can be stored in the container through packaging technology, and the bitumen can be mixed with the de-icing agent on the premise of not affecting the original properties of the bitumen. Under certain conditions, the container can be triggered to autocrine the de-icing agent to achieve the effect of de-icing, effectively extending the action time of the de-icing agent. From the perspective of bionics, this kind of microcapsule with controlled-releasing oil can be defined as autocrine microcapsules for de-icing pavement as illustrated in Figure 1(d). Microcapsules with autocrine function is a promising research direction of controlled-release microcapsules applied in pavement.

In this study, autocrine microcapsules with HMMM/polyvinyl alcohol (PVA) as wall material and glycerol as core material were prepared. Glycerin as a deicing agent can reduce the freezing point of water, weaken the bonding force between ice and asphalt pavement, and has

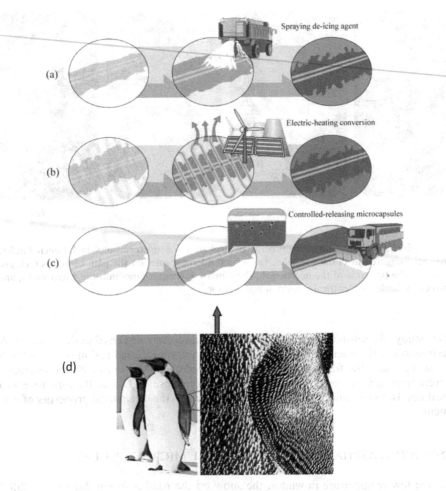

Figure 2. Illustration of de-icing methods of bituminous pavement, (a) spraying de-icing agent on pavement directly, (b) electric-heating conversion method based on the pre-buried metallic conductor, (c) controlled-releasing microcapsules containing de-icing agent base on temperature responsive character and (d) penguins in Antarctica with anti-icing wings.

excellent deicing effect. The composite wall material of the microcapsule takes HMMM polymer as the skeleton, which has a certain strength and plays a supporting role. PVA gel acts as the "door" of core material to control the release amount of core material. The PVA gel shrinks with the decrease of temperature, and the shell of the microcapsule becomes thin or has holes, which releases the deicing agent. When the ambient temperature rises, the volume of the PVA gel increases, the "door" of the microcapsule closes, and the core material is sealed. Autocrine microcapsule can adjust the release amount with temperature to achieve the effect of intelligent snow and ice removal.

3   ANTI-ICING ASPHALT USING GRAPHENE MICROCAPSULES

The problem of ice removal in winter and microcracks in asphalt pavement are difficult problems to be solved. The addition of snow removal agent to the road surface has too much environmental pollution, and the service life of super hydrophobic film is too short, so the use of hot melt de-icing method is an environmentally friendly and lasting method. However, heating

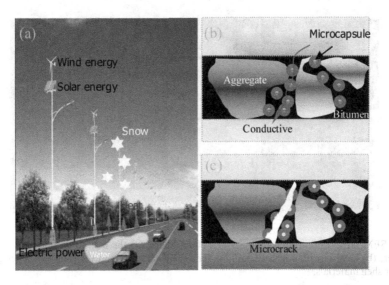

Figure 3. Schematic diagram of graphene/regenerant microcapsules-bitumen composite, (a) electricity provided by solar wind and power systems, (b) conductive pathways formed by graphene-containing microcapsules, and (c) bitumen regenerant flows from the microcapsules to repair microcracks.

asphalt pavement many times will accelerate the aging speed of asphalt and reduce its service life. graphene/remediation microcapsules asphalt pavement with self-melting and self-healing properties combines hot-melt technology with self-healing technology to prepare a conductive microcapsule coated with bitumen regenerating agent, and graphene provides the possibility to realize this combination idea (Figure 3(a)). Graphene is a two-dimensional (2D) atomic crystal with excellent properties such as high strength, high electrical conductivity and high thermal conductivity. These properties suggest that graphene can be used in combination with a variety of materials for a wide range of applications. In this study, HMMM polymer was used as the wall material. During the synthesis process, graphene was entangled with the polymer chain of HMMM prepolymer and adsorbed on the surface of the droplet by electrostatic attraction, and was cross-linked with the polymer chain [4] to form a composite shell, so that the microcapsule had certain thermal conductivity (Figure 3(b)). When a microcrack occurs inside the asphalt, the microcrack stress punctures the microcapsule, and the repair agent inside the microcapsule flows out and plays a role (Figure 3(c)). The graphene/regenerator microencapsulation-asphalt composite material converts wind and solar energy into heat energy to remove road snow and ice, and due to the addition of graphene/regenerator microcapsules, the composite material can self-healing and extend its service life.

4   SELF-HEALING ASPHALT USING MICROVASCULAR

Aging of asphalt pavement is an irreversible problem, how to make the aging asphalt through new intelligent materials or ways to restore the original state and excellent performance and can be repeated many times. In view of the disadvantages of microcapsules containing a small amount of repair agents that cannot be used for multiple repair of microcracks and the repair rate is slow, the hollow fiber containing oily repair agents for multiple repair of microcracks caused by asphalt pavement aging was developed and its working principle was studied. Firstly, a hollow fiber with oil remediation agent as core material and polyvinylidene fluoride (PVDF) as wall material was successfully prepared by dry-wet spinning technology. The hollow fiber wall material has the mechanical properties, good hydrophobicity, excellent thermal stability and permeability of slow release. An intelligent asphalt pavement material with

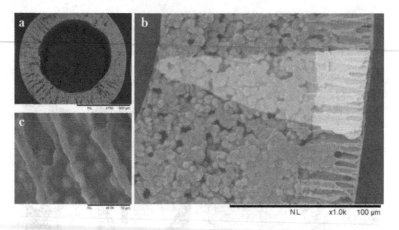

Figure 4. SEM cross-section morphologies of hollow fibers, (a) the overall outline of cross-section of hollow fiber, (b) local morphologies of hollow fiber shell material, and (c) enlarged finger pore of the hollow fiber shell materials.

multiple remediation is formed by doping hollow fiber containing oil remediation agent into asphalt mixture (Figure 4(a)). When asphalt aging occurs, the concentration difference between the hollow fiber wall and the hollow fiber wall is formed [5], which promotes the release of oil remediation agent through the hollow fiber wall to the asphalt aging area (Figure 4(b,c)). In addition, when the pavement aging appears microcracks, the microcracks extend forward to puncture the hollow fiber and release the repair agent into the microcracks, which are repaired by wetting the microcracks under the action of capillary tubes.

5 CONCLUSIONS

In order to solve the problems of asphalt pavement aging and icing, this study added micro-capsules, hollow fibers and other microvessels with deicing and anti-aging functions to asphalt binder. Its performance in pavement is verified by various tests. The following conclusions can be drawn,

(1) The microcapsules contained with the repair agent were evenly mixed into the bitumen. After the microcapsules were broken, the microcracks in the asphalt pavement were repaired. Fragments of the shell of the microcapsule form chemical bonds with the bitu-men molecules, and the original position of the microcapsule does not form a hole.
(2) The microcapsules with HMMM/PVA as composite shell material and glycerin as core material are evenly mixed into bitumen, and PVA gel is used as the "door" of microcap-sule, which can expand at high temperature to prevent core material from releasing, and to reduce at low temperature to accelerate core material release. According to the change of ambient temperature, intelligent deicing of pavement can be realized.
(3) The self-healing microcapsule with graphene/HMMM resin shell as the wall material has the self-healing function and certain thermal conductivity at the same time. It can convert wind and solar energy into heat energy and clear snow and ice on the road.
(4) PVDF hollow fiber material contained with repair agent is used for bitumen self-healing. The fiber has a porous microstructure, and after embedding in the bitumen, the repair agent can be permeated out under the drive of the difference in concentration between inside and outside. When the fiber is broken, the repair agent can also be rapidly diffused into the microcrack by capillary action to complete the self-healing process.

REFERENCES

Su, J.F. et al. 2019. Experimental observation of the vascular self-healing hollow fibers containing rejuvenator states. *Construction and Building Materials* 201:715–727.

Yang, P. et al. 2018. Design of self-healing microcapsules containing bituminous rejuvenator with nano-CaCO3/organic composite shell: mechanical properties, thermal stability and compactability. *Polymer Composites* 39: E1441–E1451.

Su, J.F. et al. 2019. Smart bituminous material combining anti-icing and self-healing functions using electrothermal graphene microcapsules containing oily rejuvenator. *Construction and Building Materials* 224:671–681.

Guo, Y.D. et al. 2019. Mechanical experiment evaluation of the microvascular self-healing capability of bitumen using hollow fibers containing oily rejuvenator. *Construction and Building Materials* 225:1026–1035.

Zhang, X.L. et al. 2018. Novel vascular self-nourishing and self-healing hollow fibers containing oily rejuvenator for aged bituminous materials. *Construction and Building Materials* 183:150–162.

Functional Pavements – Chen et al (eds)
© *2021 Taylor & Francis Group, London, ISBN 978-0-367-72610-2*

Effect of bio-oil on low-intermediate temperature properties of organosolv lignin-bitumen

Y. Zhang
Section of Pavement Engineering, Delft University of Technology, Delft, The Netherlands
School of Highway, Chang'an University, Xi'an, China

X. Liu, S. Ren, R. Jing, W. Gard, P. Apostolidis & S. Erkens
Section of Pavement Engineering, Delft University of Technology, Delft, The Netherlands

A. Skarpas
Section of Pavement Engineering, Delft University of Technology, Delft, The Netherlands
Department of Civil Infrastructure and Environmental Engineering, Khalifa University of Science and Technology, Abu Dhabi, United Arab Emirates

ABSTRACT: Lignin, one of the most abundant natural polymers, has been extensively studied as liquid or solid additive in bituminous binders. Despite the fact the organosolv lignin in bitumen improves the overall resistance against oxidative aging, lignin could lead to binders of high thermal cracking sensitivity. In this study, a bio-based oil is implemented in a lignin modified bitumen to ameliorate characteristics, such as fatigue and thermal cracking resistance. Pressure aging vessel conditioning was applied to new binders formulated by different proportions of bio-oil to simulate the long-term aging. A series of rheological tests were performed. Based on the linear amplitude sweep test results, fatigue damage of lignin-bitumen could be reduced by increasing the oil content. According to relaxation test results, the addition of oil significantly decreased the ratio of residual stress and relaxation time. This study has shown preliminary conclusions on the use of bio-oil to improve the low-intermediate temperature performance of lignin-bitumen binders.

1   INTRODUCTION

Bitumen is a complex petroleum-based material that is the most widely used binder for paving applications. However, considering the uncertainty in crude oil supply, alternative binders are encouraged to be used as a replacement of bituminous binders or bitumen modifiers. Especially, lignin, among others, has attracted considerable attention as a modifier (Xu et al., 2017, Su et al., 2018) or substitute (van Vliet et al., 2016) of bitumen. Lignin is one of the most abundant natural polymers on Earth, with the total amount of lignin present in the biosphere estimated to exceed 300 billion tons and with an annual increase of approximately 20 billion tons (Bruijnincx, 2016). Lignin can be found as well in co-products of timber production, or byproducts of paper and pulp industries. Thus, the utilization of lignin in binders specially designed for pavements may bring large economic benefits to sustainable development.

The addition of lignin can indeed improve the performance of bitumen (Batista et al., 2018) and play a pivotal role as partial replacement of bitumen (Arafat et al., 2019). Meanwhile, the results have proved that lignin in bitumen could lead to enhanced thermal cracking sensitivity (Zhang et al., 2019, Norgbey et al., 2020). In this study, bio-based oil is implemented in lignin modified bituminous binders, or called lignin-bitumen, to improve the low-intermediate temperature performance. Dynamic Shear Rheometer (DSR) was used to implement a series of tests to evaluate the improvement of bio-oil in lignin-bitumen. Linear Amplitude Sweep (LAS) tests were used to characterize the fatigue behavior of new binders. Relaxation tests were conducted to evaluate relaxation properties as well.

2 MATERIALS AND METHODS

2.1 Material preparation

The virgin binder which was used in this research was a 70/100 pen grade bitumen of 47.5°C softening point. The pure lignin was a kind of nutbrown powder provided by the Chemical Point UG (Germany). Lignin particles with purity over 87% are extracted by organosolv methods. Helium Pycnometer Test was used to accurately measure the density of lignin was 1.3774 g/cm^3. The surface measuring system (Dynamic Vapor Sorption) was used to calculate the specific surface area of lignin by Braunauer-Emmett-Teller (BET) method was 147.0593 m^2/g. The physical parameters of lignin were tested after aging as well. The apparent color of lignin particles became darker, the density was increased to 1.5029 g/cm^3 and the specific surface area was decreased to 65.0475 m^2/g (Zhang et al., 2020). Lignin particles became more compact with a smaller specific surface area after aging. The bio-oil used in this study was one of the most common rapeseed oil with saturated, monounsaturated, and polyunsaturated fatty acids.

The effect of different contents of lignin on bitumen had been studied. Here, the neat bitumen was the reference group, the lignin content by mass of bitumen was 30%, the contents of bio-oil were 2, 3, and 4% by mass of lignin-bitumen, the label was Bref, BL30, BL30+2, 3, and 4%, respectively. The lignin and bio-oil were added in the bitumen gradually. After premixing, the materials were mixed at a temperature of 163°C and a rate of 3000 rpm by the high shear mixer. Thirty minutes was the mixing time to ensure that the materials are fully mixed. The Pressure Aging Vessel (PAV) device was simulated to the long-term aging process according to the standard testing procedure (ASTM D 6521-19). It was performed after the short-term aging procedure. A bitumen film with 3.2-mm thickness was formed by pouring 50 ± 0.5 g materials into a standard 140-mm diameter PAV pan. Then the pans were placed in the PAV device at a temperature of 100°C under pressurized air at 2.10 MPa for 20 hours.

2.2 Experimental methods

2.2.1 Linear Amplitude Sweep
According to the standard testing procedure (AASHTO TP 101-14), a cyclic loading with linearly increasing strain amplitudes was used in the LAS test to assess the fatigue behavior of different binders. The 8-mm-diameter parallel plates with a 2-mm gap were used in LAS tests. The LAS test consisted of two steps; in the first step, the rheological properties of the sample were tested using a frequency sweep test at 20°C (applied load of 0.1% strain over a range of frequencies from 0.2–30 Hz). Afterward, the samples were tested by applying a strain sweep (10 Hz).

2.2.2 Relaxation test
The stress relaxation demonstrates the ability of a material to relieve stress under a constant strain. The relaxation tests were performed in a DSR by using a parallel-plate configuration of 8-mm diameter and a 2-mm gap under strain-controlled mode at 0°C. The tests were conducted as follows: firstly, the strain was increased from 0 to 1% shear strain in 0.1 s, and then the 1% shear strain was kept constant during a relaxation period of 100 s, while the change of shear stress was measured. Longer relaxation times imply that materials are more susceptible to stress accumulation. The relaxation time should be small enough to prevent high-stress accumulation in the asphalt pavement, caused by the continuous traffic load.

3 RESULTS AND DISCUSSION

3.1 *Linear Amplitude Sweep*

The LAS tests were used to measure the fatigue performance of lignin-bitumen. The number of cycles to failure (Nf) of different lignin modified bitumen at two typical applied shear strain levels, 2.5 and 5.0%, were illustrated in Table 1. Obviously, the dramatic decrease in fatigue life was associated with an increase in strain level.

Fatigue life decreased with the addition of lignin at both high and low strain levels (2.5 & 5.0%) by comparing the samples Bref and BL30. For fresh samples, the fatigue life of the BL30 was 24 and 23% of the Bref, under the applied strain level 2.5 and 5%, respectively. After the aging process, the fatigue life of lignin-bitumen became 61 and 45% of the neat binder. The lignin led to stiffer binders, and subsequently these binders have shown increased resistance to deformation at high temperature. However, it may be prone to brittle fracture, which affects the low-intermediate temperature performance, an observation consistent with the conclusions of previous studies. Even if the lignin and bitumen were mixed uniformly, vulnerable interfaces between these two materials still exist, making them more susceptible to damage under accumulated loads.

To reduce the hardening effect of lignin on bitumen and damage to fatigue performance, oil was added to the binder. Fatigue life gradually increased to 42, 55, and 68% of the virgin binder at 2.5 strain level and 50, 66, and 90% at 5.0 strain level with the increasing of oil. The improvement after aging was more obvious, but it was still lower than the neat bitumen. Oil could make the material softer, significantly improving and reducing the damage of lignin to fatigue performance.

3.2 *Relaxation test*

The ratio of residual and initial shear stresses of samples before and after aging were shown in Figure 1. This indicator represented the residual stress after the end of a relaxation test. The larger the ratio, the more residual stress, the worse the recoverability, and vice versa. The fresh specimens have shown better elasticity and recoverability than aged samples with a lower ratio, properties deteriorate with aging. The residual stress of the aged samples was more than the fresh after one load. Then under repeated loading, the real traffic vehicle loads, the more residual accumulated stress would not be released and recovered. The addition of lignin increased the stress ratio increased the ratio both in fresh and aged states by comparing Bref and BL30. It shows that lignin is detrimental to relaxation properties. As the oil was added proportionally, this damaging effect was eliminated. The ratio of lignin-bitumen with bio-oil was even smaller than the neat binder. In addition to the ratio, the speed and time of relaxation also needed to be analyzed. The relaxation time as the shear stress was reduced to 50% and 25% of the initial stress is shown in Figure 1.

Table 1. The fatigue life at the different applied strain of lignin–bitumen systems.

Samples		Bref	BL30	BL30+2%	BL30+3%	BL30+4%
N_f (2.5%)	fresh	1335	321	565	728	905
	PAV	1744	1057	1225	1300	1407
Ratio (versus Bref) [%]	fresh	-	24.0%	42.3%	54.5%	67.8%
	PAV	-	60.6%	70.2%	74.5%	80.7%
N_f (5.0%)	fresh	203	47	102	133	183
	PAV	94	43	61	71	92
Ratio (versus Bref) [%]	fresh	-	23.2%	50.3%	65.5%	90.2%
	PAV	-	45.7%	64.9%	75.5%	97.9%

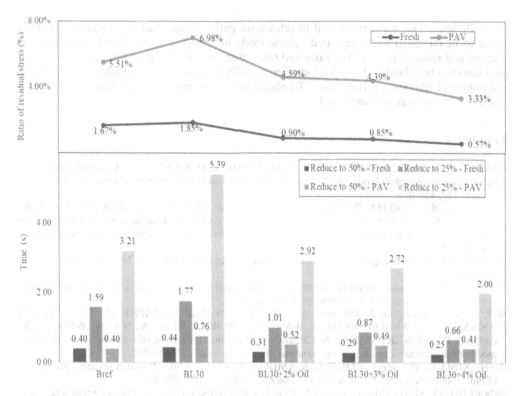

Figure 1. The ratio of stress and relaxation time of lignin–bitumen systems.

The index represented the time required for the stress to decrease to 50% and 25% of the initial stress (0.1s). The longer it took for the stress to decrease to a certain stress level, the worse the recoverability. Longer relaxation times revealed that samples are more susceptible to stress accumulation (Jing, 2019). The relaxation time of all samples increased with the aging, it was consistent with the real situation. By comparing Bref and BL30, the relaxation times of the stress reduction to 50% and 25% increased with the addition of lignin. Lignin prolonged the time that the stress decreases to a certain level, which had a negative effect on the relaxation properties. The results were consistent with the analysis of residual stress ratio. However, the addition of oil significantly declined the relaxation time and improved the situation. The lignin modified bitumen with oil even needed less time than the neat bitumen, it showed improved relaxation performance.

4 CONCLUSION

Organosolv lignin significantly improved the modulus and high temperature stability of bitumen, played a role in partly replacing bitumen, and had certain damage to the intermediate and low temperature performance at the same time. The research presented in this study had shown the preliminary conclusions on the use of a bio-oil in lignin-bitumen.

According to the result of the linear amplitude sweep tests, the addition of lignin and the aging process significantly declined the fatigue life at both the stress level of 2.5% and 5%. The hardening effect of lignin on bitumen and damage to fatigue performance could be reduced by increasing the oil content, the improvement was more obvious after aging. Moreover, the results of the relaxation test showed that the addition of lignin and the aging process increased the ratio of residual shear stress and relaxation time of the shear stress was reduced to a specific level. This was detrimental to relaxation performance, and the material was more prone to be damaged under repeated vehicle loads because the accumulated stress was not released and restored in time. The ratio and relaxation time of the lignin-bitumen with oil was even less than neat binder. The addition of oil signally decreased the ratio and relaxation time and improved relaxation performance. To obtain better low-intermediate temperature properties, 4% of bio-oil was recommended.

REFERENCES

ARAFAT, S., KUMAR, N., WASIUDDIN, N. M., OWHE, E. O. & LYNAM, J. G. (2019) Sustainable lignin to enhance asphalt binder oxidative aging properties and mix properties. *Journal of Cleaner Production*, 217, 456–468.

BATISTA, K. B., PADILHA, R. P. L., CASTRO, T. O., SILVA, C. F. S. C., ARA JO, M. F. A. S., LEITE, L. F. M., PASA, V. M. D. & LINS, V. F. C. (2018) High-temperature, low-temperature and weathering aging performance of lignin modified asphalt binders. *Industrial Crops and Products*, 111, 107–116.

BRUIJNINCX, P. (2016) Lignin Valorisation: The Importance of a Full Value Chain Approach. Utrecht University.

JING, R. (2019) Ageing of bituminous materials: Experimental and numerical characterization. *TU Delft Pavement Engineering*. Delft University of Technology.

NORGBEY, E., HUANG, J., HIRSCH, V., LIU, W. J., WANG, M., RIPKE, O., LI, Y., TAKYI ANNAN, G. E., EWUSI-MENSAH, D., WANG, X., TREIB, G., RINK, A., NWANKWEGU, A. S., OPOKU, P. A. & NKRUMAH, P. N. (2020) Unravelling the efficient use of waste lignin as a bitumen modifier for sustainable roads. *Construction and Building Materials*, 230, 116957.

SU, N., XIAO, F., WANG, J., CONG, L. & AMIRKHANIAN, S. (2018) Productions and applications of bio-asphalts–A review. *Construction and Building Materials*, 183, 578–591.

VAN VLIET, D., SLAGHEK, T., GIEZEN, C. & HAAKSMAN, I. (2016) Lignin as a green alternative for bitumen.

XU, G., WANG, H. & ZHU, H. (2017) Rheological properties and anti-aging performance of asphalt binder modified with wood lignin. *Construction and Building Materials*, 151, 801–808.

ZHANG, Y., LIU, X., APOSTOLIDIS, P., GARD, W., VAN DE VEN, M., ERKENS, S. & JING, R. (2019) Chemical and Rheological Evaluation of Aged Lignin-Modified Bitumen. *Materials*, 12, 4176.

ZHANG, Y., LIU, X., APOSTOLIDIS, P., JING, R., ERKENS, S., POERAN, N. & SKARPAS, A. (2020) Evaluation of Organosolv Lignin as an Oxidation Inhibitor in Bitumen. *Molecules*, 25.

Functional Pavements – Chen et al (eds)
© 2021 Taylor & Francis Group, London, ISBN 978-0-367-72610-2

Evaluation on aging characteristic of bio-asphalt containing high percentage of waste cooking oil residues with various aging methods

Yi Zhang
Shanghai urban construction design and research institute (group) co., Ltd., Shanghai, P.R. China

Tong Lu*, Daquan Sun, Mingjun Hu & Jianmin Ma
Key Laboratory of Road and Traffic Engineering of Ministry of Education, Tongji University, Shanghai, P.R. China

ABSTRACT: Bio-asphalt obtained from waste cooking oil residues (WCOR) is a potential environmentally friendly pavement material, but few studies have been done on high content WCOR based bio-asphalt (HWCORBA), especially on its aging characteristic. In this paper, HWCORBA containing 33.3 wt.% WCOR was prepared. Then three different aging methods such as rolling thin film oven (RTFO), pressure aging vessel (PAV) and weathering aging (WA) were applied on HWCORBA in the laboratory. The rheological, chemical and microscopic characteristics were evaluated after aging process using conventional asphalt tests (including penetration test, softening point test and ductility test), dynamic shear oscillatory test, fourier transform infrared spectroscopy (FTIR) and fluorescence microscope (FM). The results show that, HWCORBA with RTFO, PAV and WA shows higher softening point, complex modulus and better high temperature performance, but has lower penetration and ductility. Furthermore, the FTIR results prove the increment of $I_{C=O}$ and the reduction of $I_{B/S}$ during thermo-oxidative aging and weathering aging, but the change is not significant.

1 INTRODUCTION

Many studies on the partial substitution of bio-oil, derived from biomass wastes, for asphalt have been reported (Aflaki et al., 2014, Fini et al., 2012). Among them, the bio-asphalts containing waste cooking oil (WCO) or WCOR have received wide attention in recent years for its sustainability and abundance. Generally, use of WCO-based bio-oil in asphalt binders has proved to improve the cracking resistance at low temperatures, but has adverse effect on the high temperature performance (Dong et al., 2019, Sun et al., 2017b). Due to weak aging resistance and high temperature properties, the addition of WCO or WCOR is generally limited to less than 10%. Few studies about high content WCO-based bio-asphalt (HWCORBA) have been reported, especially the aging characteristics.

In this study, HWCORBA containing 33.3 wt.% WCOR was prepared, and various aging methods were employed to investigate the aging performance of HWCORBA, including rolling thin film oven (RTFO), pressure aging vessel (PAV) and weathering aging (WA). The basic properties, rheological, chemical characteristics were evaluated before and after aging by conventional asphalt tests, dynamic shear oscillatory test, and fourier transform infrared spectroscopy (FTIR).

---

* Corresponding author

2 MATERIALS AND METHODS

2.1 Materials

The specific materials selection, ratio and preparation process optimization of HWCORBA was described in our previous research (Sun et al., 2017a). The penetration (25 °C, 0.1 mm), softening point (°C), and ductility (5 °C, cm) of HWCORBA are 56.0, 62.2, and 14.7, respectively.

2.2 Aging methods

The short-term aging of HWCORBA during its mixture production and construction was simulated by RTFO aging test, according to ASTM D2872. The weathering aging was conducted for 72 h, which simulated 90 days average level outdoor in China. The test temperature, humidity and radiation intensity were set as 70°C, 50% and 1000 W/m^2, respectively. In this paper, the codes of unaged-BA, RTFO-BA, PAV-BA and WA-BA are used to represent these samples before and after RTFO/PAV/WA aging, respectively.

2.3 Asphalt property tests

The following conventional asphalt tests were carried out to evaluate the basic properties of RTFO-BA, PAV-BA and WA-BA: penetration (25 °C), softening point and ductility (5 °C) tests, according to ASTM D5, ASTM D36 and ASTM D113, respectively.

The dynamic shear oscillatory test was carried out using a TA dynamic shear rheometer (DSR) AR1500ex to obtain the complex modulus (G*), phase angle (δ) and rutting factor (G*/sin δ) of bio-asphalts. The test was performed at 10 rad/s utilizing a 25 mm parallel plate at 64 °C, 70 °C, 76 °C, 82 °C and 88 °C, according to AASHTO T315.

The chemical compositions test was conducted by a Bruker TENSOR FTIR spectrometer to record the infrared spectra values of bio-asphalts (Hofko et al., 2017). Carbonyl index $I_{C=O}$ was employed to reveal the thermal-oxidation and weathering aging of HWCORBA and LDPE (Hofko et al., 2018), and BS index $I_{B/S}$ was employed to obtain the degradation of SBS. These two indices were computed as follows:

$$\text{Carbonyl index}: I_{C=O} = A_{1740}/A_T \times 100\% \tag{1}$$

$$\text{BS index}: I_{B/S} = A_{965}/A_{700} \times 100\% \tag{2}$$

Where $A_{(XX)}$ represents the area of (XX) cm^{-1} peak, A_T represents the total peak area of the spectral bands between 600 cm^{-1} and 1800 cm^{-1}.

3 RESULTS AND DISCUSSION

3.1 Conventional asphalt tests

Figure 1 shows the influence of aging method on the basic properties of HWCORBA. The penetration values of RTFO-BA, PAV-BA and WA-BA are lower than that of unaged-BA, which indicates that HWCORBA becomes harder after thermo-oxidative aging and weathering aging. PAV-BA has the lowest penetration, followed by WA-BA, and finally RTFO-BA. The reason for this result is that PAV aging is more intense than RFTO and WA. Furthermore, there is little difference between penetration of RTFO-BA and WA-BA, which indicates that HWCORBA has excellent solar radiation resistance.

The softening point response of aged HWCORBA is quite similar, and the values of RTFO-BA, PAV-BA and WA-BA were 70.2 °C, 71.6 °C and 70.6 °C, respectively. Resin and hard asphalt particle in HWCORBA are processed by highly oxidized process, so it is reasonable that RTFO, PAV and WA have little effect on the softening point of HWCORBA.

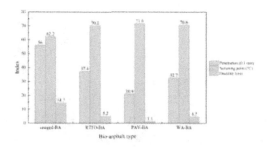

Figure 1.   Influence of aging method on the basic properties of HWCORBA.

RTFO-BA, WA-BA and especially PAV-BA show obvious extendibility deterioration response with low ductility values at 5°C. The ductility of HWCORBA decreases by 64.6% and 68.0% after RTFO and WA, and the ductility of PAV-BA decrease rapidly to 1.1 cm. This result indicates that all of short-term aging, long-term aging and solar radiation has a great adverse effect on the low temperature ductility of HWCORBA.

3.2   *Dynamic shear oscillatory test*

The complex modulus of HWCORBA with various aging methods are presented in Figure 2(a). With the increment of temperatures, the modulus decreases gradually, and the decline scale of the modulus decreases. PAV-BA has the largest complex modulus, followed by WA-BA and RTFO-BA at the same temperature. The difference of complex modulus decreases with the increase of temperature.

In Figure 2(b). PAV-BA has the smallest phase angle, followed by WA-BA, and finally RTFO-BA. This result indicates that the deepening of the aging degree reduces the phase angle, rendering HWCORBA more elastic. During aging, there are two aging behaviors: the hardening of bio-asphalt and the degradation of SBS and LDPE polymers. The hardening of bio-asphalt increases the modulus and elasticity of HWCORBA, while the degradation of polymers increases its viscosity. From phase angle results, the hardening of bio-asphalt

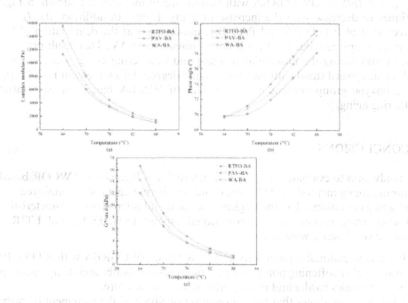

Figure 2.   (a) Complex modulus, (b) Phase angle and (c) Rutting factor of HWCORBA with various aging methods.

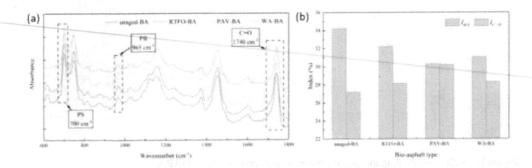

Figure 3.   FTIR spectra and aging indices of HWCORBA before and after various aging methods.

dominates, compared with the degradation of polymers, and thus phase angles decrease during RTFO, WA and PAV.

The rutting factor ($G^*/\sin \delta$) reflects the high temperature performance of asphalt binders, according to SHRP specification, and generally used to determine high temperature performance grade of asphalts. The rutting factor of HWCORBA with various aging methods are computed by complex modulus and phase angle, exhibited in Figure 2(c). It is noticed that PAV-BA has a largest rutting factor, followed by WA-BA, and finally RTFO-BA, indicating that the rutting factor response induced by WA is between RTFO and PAV. This result is consistent with conventional asphalt tests results.

3.3   FTIR

According to Figure 3(a), the HWCORBA has a more complex infrared spectrum than petroleum asphalt binders. The characteristic peak at 1740 cm^{-1} represents the carboxylic groups (C=O) in acids and esters. Carbonyl index $I_{C=O}$ reflects the oxidation of bio-asphalts as well as degradation of LDPE during various aging process. It should be noticed that SBS is composed of polybutadiene (PB) and polystyrene (PS). PB and PS possess characteristic peaks at 965 cm^{-1} and 700 cm^{-1}, respectively.

$I_{C=O}$ and IB/S of HWCORBA with various aging methods are shown in Figure 3(b). $I_{B/S}$ continues to decrease with the increase of aging degree. In addition, the $I_{B/S}$ of WA-BA is between that of RTFO-BA and PAV-BA, indicating that the degradation of SBS caused by WA can be captured, but its effect is not as obvious as PAV. This result proved little degradation of SBS during thermo-oxidative aging and weathering aging. The $I_{C=O}$ of HWCORBA displays an upward trend with increasing aging degree. This increment is still insignificant. So, the carboxylic group content on the surface of WA-BA barely increases after treated by weathering aging.

4   CONCLUSIONS

This study aims to evaluate the aging characteristics of high content WCOR based bio-asphalt via various aging methods. RTFO, PAV and weathering aging were employed to simulate different aging conditions. The rheological, chemical and microscopic characteristics were evaluated after aging process using conventional asphalt tests, DSR, and FTIR, respectively. Following conclusions were drawn:

(1) From conventional asphalt tests and DSR tests, HWCORBA with RTFO, PAV and WA shows higher softening point, complex modulus and better high temperature performance, but it becomes harder and more brittle at low temperature.
(2) FTIR analysis shows that the degradation of SBS and the increment of carboxylic group content can be obtained during weathering aging.

REFERENCES

AFLAKI, S., HAJIKARIMI, P., FINI, E. H. & ZADA, B. 2014. Comparing Effects of Biobinder with Other Asphalt Modifiers on Low-Temperature Characteristics of Asphalt. *Journal of Materials in Civil Engineering*, 26, 429–439.

DONG, R. K., ZHAO, M. Z. & TANG, N. P. 2019. Characterization of crumb tire rubber lightly pyrolyzed in waste cooking oil and the properties of its modified bitumen. *Construction and Building Materials*, 195, 10–18.

FINI, E. H., AL-QADI, I. L., YOU, Z. P., ZADA, B. & MILLS-BEALE, J. 2012. Partial replacement of asphalt binder with bio-binder: characterisation and modification. *International Journal of Pavement Engineering*, 13, 515–522.

HOFKO, B., ALAVI, M. Z., GROTHE, H., JONES, D. & HARVEY, J. 2017. Repeatability and sensitivity of FTIR ATR spectral analysis methods for bituminous binders. *Materials and Structures*, 50, 187.

HOFKO, B., POROT, L., FALCHETTO CANNONE, A., POULIKAKOS, L., HUBER, L., LU, X., MOLLENHAUER, K. & GROTHE, H. 2018. FTIR spectral analysis of bituminous binders: reproducibility and impact of ageing temperature. *Materials and Structures*, 51, 45.

SUN, D. Q., LU, T., XIAO, F. P., ZHU, X. Y. & SUN, G. Q. 2017a. Formulation and aging resistance of modified bio-asphalt containing high percentage of waste cooking oil residues. *Journal of Cleaner Production*, 161, 1203–1214.

SUN, D. Q., SUN, G. Q., DU, Y. C., ZHU, X. Y., LU, T., PANG, Q., SHI, S. Y. & DAI, Z. W. 2017b. Evaluation of optimized bio-asphalt containing high content waste cooking oil residues. *Fuel*, 202, 529–540.

Functional Pavements – Chen et al (eds)
© 2021 Taylor & Francis Group, London, ISBN 978-0-367-72610-2

Study on classical and rheological performance of high viscosity modified asphalt for porous asphalt mixture

Haiyan Zhang
Middle Road Construction (Beijing) Engineering Materials Technology Co. Ltd, Beijing, China

Hui Tang
School of Civil Engineering, Chongqing Jiaotong University, Chongqing, China

Bin Xu & Dongwei Cao
Zhong Lu Gao Ke (Beijing) Road Technology Co. Ltd, Beijing, China

Bo Yuan & Weitao Chang
Shijiazhuang Jiaojian Expressway Construction Management Co. Ltd, Shijiazhuang, China

ABSTRACT: The self-made SR particles and SBS (Styreneic Block Copolymers) particles were used as combined modifiers to prepare high viscosity modified asphalt (HVA modified asphalt) for porous asphalt pavement. The properties of HVA modified asphalt were characterized by conventional tests including penetration, Softening point, ductility, Brinell viscosity and aging, and compared with SK90 matrix asphalt and rubber asphalt and SBS modified asphalt. The results show that compared to the matrix asphalt the HVA modified asphalts with the high softening point and 5°C ductility, and low penetration; the dynamic viscosity of HVA modified asphalt reached 230000Pa•s and with moderate Brookfield viscosity which can effectively coat large void asphalt concrete and the asphalt concert do not need much high construction temperature. Dynamic shear rheological test and bending beam rheometer test showed that HVA modified asphalt has good cohesiveness and aging resistance, and exhibits good high temperature stability and low temperature flexibility. The HVA high viscosity modified asphalt prepared in this research can meet the requirements of asphalt materials for porous asphalt pavements.

Keywords: Porous asphalt mixture, High viscosity modified asphalt, Classical performance, Rheological performance

1 INTRODUCTION

With the rapid development of urbanization in China, urban waterlogging and flood disasters have become a serious challenge to urban construction and development [1]. The permeable pavement plays an important role in reducing and delaying the peak flow rate, increasing the duration of rain flow and reducing the water pollution [2,3]. The porous asphalt pavement, as a kind of permeable pavement, refers to the asphalt concrete layer which can form the drainage channels in the mixing material, so that the water seeping into the drainage layer can flow laterally to the pavement structure [4]. Another kind of permeable pavement refers to the rainwater into the internal structure layer, finally penetrated in subgrade soil, and the rainwater infiltrated in the soil, which can replenish groundwater, participate in the water cycle, reduce surface runoff and the risk of urban waterlogging, relieve urban heat island effect [5,6]. Meanwhile, porous asphalt pavement provides good skid resistance, noise reduction ability and reduces significantly splash and spray in wet conditions, as functional surface layer widely used all over the world.

The high viscosity modified asphalt is commonly used as the binder of porous asphalt mixture to increase the anti-raveling and rutting resistance property. However, because of the open gradation and rapid aging of the binder, raveling and raveling-induced pitting are still the most common type of destruction for porous asphalt pavement in long term service. Liu et al. studied the performance of the road surface of large interstitial porous pavement [7,8].A SBS/multi-walled carbon nanotube high viscosity modified asphalt was developed by Chen [9]. Shiguo Xu et al. developed a high viscosity asphalt modifier [10]. Visible, at home and abroad of drainage asphalt mixture design and road performance study more widely, the study of various composite high viscosity modified asphalt is also not uncommon, and for all through special type asphalt pavement asphalt materials research has shortcomings. The authors wish to develop a kind of modified asphalt with high viscosity which can be used for porous asphalt pavement with excellent performance.

In view of the current development status of drainage asphalt pavement materials, based on the existing research results at home and abroad, this paper has developed the high viscosity modified asphalt with excellent performance, and the conventional performance and rheology of the modified asphalt Performance was analyzed. The results show that the performance of this high-viscosity modified asphalt is better than that of conventional asphalt materials, and can meet the requirements of the use of drainage asphalt pavement.

2 EXPERIMENT

2.1 Materials

Rubber powder is purchased from Shandong hengfeng rubber co., LTD. SBS is purchased from Yue yang petrochemical, 791H linear. SR particles are commonly used as modifiers in laboratory. Matrix asphalt was Korean SK 90 and the base properties are shown in Table 1.

2.2 Preparation of modified asphalt

(1) The rubber asphalt was prepared by three steps. First, the matrix asphalt was heated to 180°C~185°C and melt in an iron container. Then, the rubber powders with 18% quality were added into matrix asphalt shearing 20min with a shearing speed of 3000 r/min. Last, slowly add stabilizer mixing and developing 1h at 175°C~180°C to form the stable system.

(2) The SBS modified asphalt was prepared by three steps. First, the matrix asphalt was heated to 180°C~185°C and melt in an iron container. Second, the oil of swelling oil and stir was added to a uniform state. Then, SBS with 6% quality were added into matrix asphalt shearing 20min with a shearing speed of 3000 r/min to obtain the no obvious grain state. Last, slowly add stabilizer mixing and development 3h to form stable system under 175°C.

(3) The HVA high viscosity modified asphalts were prepared by three steps. First, the matrix asphalt was heated to 180°C~185°C and melted in an iron container. Second, the oil of swelling oil and stir was added to a uniform state. Then, the asphalt modifier (5%SBS+3% SR) were added into matrix asphalt with a shearing speed of 3000 r/min to obtain a homogeneous phase. Last, slowly add stabilizer mixing and development 3h to form stable system at 175°C~180°C.

Table 1. Performance indicators of 90 matrix asphalt.

Penetration(25°C) /(0.1mm)	Softening point/(°C)	Ductility(10°C) /(cm)	Brookfield viscosity (135°C)/(Pa·s)	TFOT Quality change fraction/(%)	Ductility(10°C) /(cm)
80.6	46.3	>100	377.5	0.011	9.6

2.3 *Performance measurements*

The conventional performance test of asphalt mainly includes basic properties research (penetration, softening point and ductility), viscosity and aging characteristics research. The test methods are mainly carried out according to the test regulations of asphalt and asphalt mixture for highway engineering in China (JTG E20-2011) [11].

Asphalt, as viscoelastic material, displays different characteristics of application properties in paving industry. Thus, it is crucial to examine rheological properties of asphalt [10]. In this study, the frequency sweep and temperature sweep test of the matrix asphalt and modified asphalt were carried out. The test method of asphalt is carried out according to AASHTO standard (TP 70-12) [12].

3 RESULTS AND DISCUSSION

3.1 *Study on classical performance of high viscosity modified asphalt*

3.1.1 *Basic properties*
Basic properties were mainly composed of softening point, penetration, ductility; Basic properties of high viscosity modified asphalt were measured respectively according to the Standard Test Methods of Bitumen and Bituminous Mixtures for Highway Engineering (JTG E20-2011). It had also been compared with SK 90 matrix asphalt, rubber asphalt and SBS modified asphalt. The basic properties of the four asphalts are shown in Table 2.

It can be seen from Table 2 that the high and low temperature performance of the three modified asphalt is improved. The indicators of SBS modified asphalt and the high viscosity modified asphalt were similar, according to the analysis of preparation of process, due to the effect of SBS modified, two kinds of modified asphalt have decreased fluidity, become more sticky, showing better high temperature stability and low temperature flexibility, but the indicators of the high viscosity modified asphalt are better than that of SBS modified asphalt, it shows that SR particles to a certain extent improve the high-low temperature performance of asphalt. Powder particles absorbed light oil content in asphalt and swelling, cracking reaction under high temperature for a long time, the asphalt content is relative increase and make the rubber asphalts become "hard", showing the poor flexibility at low temperature.

3.1.2 *The study of viscosity*
The viscosity test of SK90 matrix asphalt, rubber asphalt, SBS modified asphalt and the high viscosity modified asphalt are shown in Table 3.

The dynamic viscosity is used to characterize the bond strength of asphalt and aggregates. The Brookfield viscosity is used to characterize the viscosity of bitumen, and the higher the viscosity, the higher the construction temperature. As can be seen from Table 3, the effect of asphalt modification is obvious, and the viscosity is significantly higher than that of matrix asphalt. Rubber powder particles in the asphalt can't melt completely, after swelling and cracking of the particle size still larger, and the compatibility of the matrix asphalt is poorer, with smaller dynamic viscosity, a poor binding of aggregate and flying off often occurred in the grain when used in the permeable pavement, and asphalts with less liquidity are relatively

Table 2. Basic properties of asphalts.

Asphalt type	Penetration(25°C) /(0.1mm)	Softening point /(°C)	Ductility(5°C) /(cm)
SK90 matrix asphalt	80.6	46.3	-
Rubber asphalt	45.6	71	12
SBS modified asphalt	53.5	92.0	41
The high viscosity modified asphalt	54.6	103.2	45

Table 3. Viscosity indicators of asphalts.

Asphalt type	Dynamic viscosity (60°C)/(Pa·s)	Brookfield viscosity (175°C)/(Pa·s)
SK90 matrix asphalt	213	0.083
Rubber asphalt	16700	3.418
SBS modified asphalt	123100	1.245
The high viscosity modified asphalt	232600	1.520

thick and need higher temperature in the construction. SBS blending ratio is higher in the SBS modified asphalts, the SBS asphalts with the substrate under high temperature develop full reaction for a long time and the cross-linking structure is formed finally, have a larger kinetic viscosity and moderate Brookfield viscosity, which is a kind of strong bonding of asphalt binder. High viscosity modified asphalt has the biggest dynamic viscosity, and Brookfield viscosity is between rubber asphalt and SBS modified asphalt. Full penetration type asphalt pavement structure for a long time in the wet state, the use of high viscosity asphalt binder can significantly improve the anti-stripping ability and anti-water damage performance of the mixture. And the high viscosity modified asphalts have higher dynamic viscosity and more bond strength with aggregate, suitable for full permeable asphalt pavement to ensure the durability of pavement. The asphalt has low viscosity and good fluidity, and has moderate mixing temperature and compaction temperature in the construction, which can reduce energy consumption and reduce exhaust emission.

3.1.3 Study on the aging characteristic

In this paper, according to the requirements of the film aging test (TFOT) in China's test code (JTG E20-2011), the HVA high viscosity asphalt samples were put into a 163°C film oven for aging test. The short-term aging of asphalt in practical application was simulated and compared with rubber asphalt and SBS modified asphalt. The permeability ductility and mass loss of asphalt samples were measured respectively. The test results are shown in Table 4.

The Table 4 shows that 25°C of rubber asphalt penetration and 5°C ductility after aging index attenuation degree is the weakest, showing excellent high temperature aging resistance of rubber particles. The performance indexes of SBS modified asphalts are attenuated and their aging resistance is poor. Aging causes the temperature of the bitumen to weaken, and the durability and the anti-cracking performance is reduced. Asphalt absorbs oxygen to produce free radicals in high temperature environment, part additive molecular chain rupture at the same time, the small molecule enters the asphalt organization to form a relatively stable large molecular group, its structure is not easy to damage at high temperature and high temperature performance was improved, but flexible sharply attenuation, low temperature crack resistance is poor. The performance indexes after aging of high viscosity modified asphalt was still better, the penetration ratio and ductility ratio can effectively respond the asphalt aging degree. The data in the table shows that the penetration and ductility indexes after aging of

Table 4. Results of modified asphalt aging test.

Asphalt type	Penetration (25°C)/(0.1mm) Before	After	Ratio of penetration/ (%)	Ductility (5°C)/(cm) Before	After	Ratio of ductility /(%)	Quality change /(%)
Rubber asphalt	45.6	44.1	96.7	12	9	74.2	0.2102
SBS asphalt	53.5	44.0	82.3	41	27	65.9	0.2358
HVA modified asphalt	55.7	51.0	91.6	45	31	68.9	0.1833

high viscosity is still very good, and the mass loss is small, it shows that it has excellent high temperature resistance to aging.

3.2 Study on rheological performance of high viscosity modified asphalt

3.2.1 Analysis of frequency scanning test

Frequency sweep test was used to investigate the relationship between the modulus of asphalt and the driving speed, which can describe the effect of wheel load on the asphalt pavement. High frequency represents high speed and low frequency represents low speed.

The frequency sweep experiment was tested by dynamic shear rheometer (DSR) which was made in Austria. The sample was fixed in parallel plate whose diameter is 8 mm and plate spacing is 2 mm. The total strain was 1%, the test temperature was fixed at 5°C, 20°C and 35°C with a loading frequency of 0.1-100 rad/s and each sample has 10 min of constant temperature. The Dynamic modulus (G*) and phase angle (δ) in different temperature were obtained.

By Figure 1 and Figure 2 analyze different asphalt in δ and G* under 5°C affected by frequency scanning. There is a good linear relationship between δ and scan frequency, and as the frequency increases the phase Angle becomes smaller. The decreasing rate of δ of matrix asphalt is obviously greater than that of modified asphalt, the δ changes of SBS modified asphalt, high viscosity modified asphalt are not very different, the δ of rubber asphalt is minimal. The G* increases with the loading frequency and the order of the G* of different asphalt are: SK90 > SBS modified asphalt > rubber asphalt > high viscosity modified asphalt. The results show that the modifiers in both kinds of modified asphalt improve the microstructure of asphalt, improve the viscoelasticity and deformation resistance of asphalt, and thus show better high and low temperature performance.

By Figure 3 and Figure 4 analyze different asphalt in δ and G* under 20°C affected by frequency scanning. The magnitude of the δ are: SK90 > SBS modified asphalt > high viscosity modified asphalt > rubber asphalt. There is a linear relationship between δ and scan frequency and the δ decreases with frequency. The δ of modified asphalt varies with frequency, and the δ increases with frequency first and then decreases. The order of the G* of different asphalt are: SK90 > rubber asphalt > SBS modified asphalt > high viscosity modified asphalt.

By Figure 5 and Figure 6, at 35°C tests temperature, the δ of the matrix asphalt decreasing along with the load frequency, the δ of rubber asphalt decreases first and then increases, the δ of SBS modified asphalt and high viscosity modified asphalt increasing with the load frequency. The δ and G* of different temperature change varied greatly by the comprehensive analyze to δ and G* of different asphalt changing with load frequency. The δ of modified asphalt increases with the experiment temperature, and the change of loading frequency is small, especially the high viscosity modified asphalt has the larger δ and the smaller G*. The results show that with the increase of the test temperature, the elastic components in the

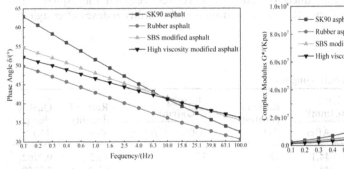
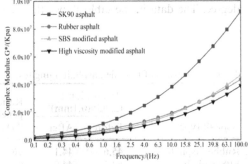

Figure 1. Phase angle δ change with frequency at 5°C.

Figure 2. Dynamic modulus G* change with frequency at 5°C.

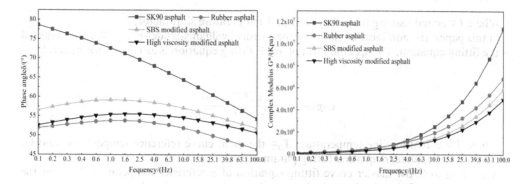

Figure 3.   Phase angle δ change with frequency at 20℃.

Figure 4.   Dynamic modulus G* change with frequency at 20℃.

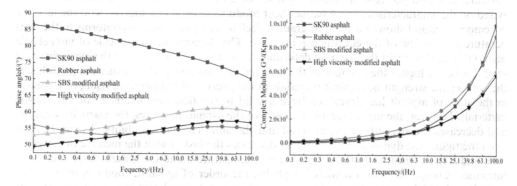

Figure 5.   Phase angle δ change with frequency at 35℃.

Figure 6.   Dynamic modulus G* change with frequency at 35℃.

asphalt material are increased; the asphalt deformation can be improved, showing good high-temperature anti-rutting performance.

According to the principle of equivalent time-temperature, at the same time with the aid of shift factor Log∅ conversion to translation, the frequency scanning test results under different temperature to horizontal displacement, can get pitch line characteristics of viscoelastic dynamic shear modulus master curve, the properties of viscoelastic materials can be in a wider time scale was calculated out. In this paper, Christensen-Anderson-Marasteanu (CAM) equation is the fitting equation of asphalt main curve.

$$|G^*| = \frac{\left|G_g^*\right|}{\left[1 + (f_c/f^*)^k\right]^{m/k}} \tag{1}$$

Where $|G^*|$= the glass state shear modulus of bitumen is 10^9Pa; f_c= the position fitting parameters of the main curve of loading frequency; k and m = the shape fitting parameters of the main curve; and f^*= conversion load frequency is calculated according to formula (2).

$$f^* = \phi_T \times f \tag{2}$$

Where f= actual loading frequency; and ϕ_T= temperature shift factor.

In this paper, the nonlinear equation of the classical williams-landel-Ferry (WLF) is selected as the fitting equation. The specific form of WLF fitting equation is as follows formula (3).

$$Log\phi_T = -\frac{C_1 \cdot (T - T_0)}{C_2 + (T - T_0)} \qquad (3)$$

Where T= actual loading temperature; T_0= the main curve reference temperature; and C_1 and C_2= WLF nonlinear equation fitting parameters.

According to CAM master curve fitting equation of a reference temperature 20°C for the curve fitting, after fitting the dynamic shear modulus master curve as shown in Figure 7.

It can be seen from Figure 7 that the dynamic shear modulus increases with the increase of scanning frequency in both matrix asphalt and modified asphalt. According to equivalent principle of time temperature can make its mechanics phenomenon observed under high temperature, can also get asphalt belongs to a viscoelastic material under the lower frequency. Based on this characteristic, the dynamic shear modulus of asphalt materials with the increase of temperature and showed a trend of gradual decline, this phenomenon conforms to the temperature change law of the equivalence principle. The dynamic modulus scale of viscoelastic material is used to resist the shearing deformation of materials. The higher the dynamic shear modulus is, the higher the strength of the material is. Conversely, the smaller the modulus is, the smaller the strength is. Asphalt materials in the process of temperature rising of a change in the state of asphalt has developed from a solid to the flow characteristics of viscoelastic materials, however, the unrecoverable viscosity of the asphalt increases, the elasticity weakens and decreases the strength of the material and the anti-deformation ability. As can be seen from the figure, the dynamic shear modulus decreases the fastest with the increase of temperature, and is more sensitive to the performance of temperature and the lowest temperature performance. Compared with 3 modified asphalts, the order of dynamic modulus of modified asphalt under high temperature and low temperature is not completely consistent, the order of dynamic shear modulus in high temperature is: high viscosity modified asphalt > rubber asphalt > SBS modified asphalt. The order of dynamic shear modulus in low temperature is: high viscosity modified asphalt > SBS modified asphalt > rubber asphalt. The high temperature performance of rubber asphalt is better than SBS modified asphalt, but the low temperature performance is less than SBS modified asphalt. Comparison of the high and low temperature performance of the three kinds of modified asphalt,the performance of high and low temperature of high viscosity modified asphalt is better, at the same time has good high temperature stability and low temperature crack resistance. Therefore, the wheel load has little effect on the modified asphalt, and shows good road performance.

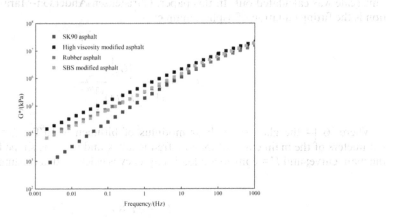

Figure 7. Master curve of different asphalt dynamic shear modulus.

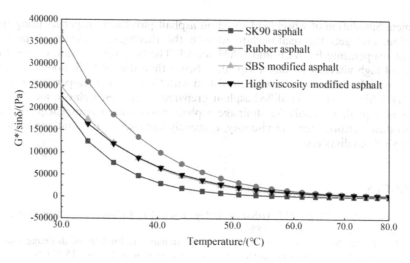

Figure 8. Change of rutting factor G*/sinδ of different asphalt with temperature.

3.2.2 *Analysis of temperature sweep test*

Asphalt is a temperature-sensitive material, and its viscoelasticity will change greatly with temperature. The temperature scanning test can be used to analyze the temperature sensitivity of asphalt. This paper uses the SmartPave101 Dynamic Shear Rheometer (DSR) manufactured by Anton Paar to scan the temperature. Set a fixed loading frequency of 10rad/s and select a stress level of 100Pa. For the selected base asphalt and three modified asphalts, through the temperature scanning range of 30°C~80°C, the asphalt sample is evenly smeared between 25mm parallel plates, parallel to each other. The plate spacing is 1mm for oscillation scanning. The test result parameters complex shear modulus G*, phase angle δ and rutting factor G*/sinδ are used to analyze the temperature sensitivity of different asphalts. The changes of rutting factor G*/sinδ with temperature is shown in Figure 8.

The research shows that the size of the rutting factor can describe the strength of the anti-rutting ability of the material, and the asphalt binder with strong anti-rutting ability has a larger rutting factor. From Figure 8, we can see that rutting factor of different bitumen is decreasing gradually with the increase of test temperature. Rutting factor of modified asphalt is significantly higher than that of matrix asphalt, indicating that rutting resistance of modified asphalt is outstanding. Due to the physical blending process of rubber powder and matrix asphalt, the asphalt system is not uniform, and the rheological test is greatly affected by the rubber powder particles. The test results cannot effectively respond to the road performance of materials. High viscosity modified asphalt has higher rutting factor, and with the scanning temperature rutting factor high viscosity modified asphalt decreased slowly, with stronger ability to resist deformation at high temperatures, affected by temperature change than other asphalt weaker, exhibit more excellent anti rutting performance of high temperature, consistent with the previous performance of conventional asphalt.

4 CONCLUSIONS

(1) In this study, the high viscosity asphalt with excellent performance for porous asphalt pavement was prepared. This material has high softening point and ductility at low temperature, showing the good high temperature stability and low temperature cracking resistance, and the dynamic viscosity can reach more than 200000Pa·s. The attenuating index of the index before and after aging is small, and the performance of heat-resistant pension is strong.

(2) The experimental results of different asphalt rheological mechanics show that the rheological properties of the modified bitumen are generally superior to that of the matrix

bitumen. Simulation of wheel load speed on asphalt pavement properties using frequency scanning, and gets the relationship between the rheological mechanical properties of asphalt temperature by WLF and CAM model. The high temperature resistant deformability of high viscosity modified asphalt is better than that of SBS modified asphalt and rubber asphalt, the test results were consistent with the results of temperature scanning.

(3) The HVA high viscosity modified asphalt prepared in this research can meet the requirements of asphalt materials for drainage asphalt pavements, has excellent high and low temperature performance and viscosity, especially suitable for the southern region of our country's flood disasters.

REFERENCES

1. Sang, Y.F. & Yang, M.Y. 2017. Urban waterlogs control in China: more effective strategies and actions are needed. *Natural Hazards* 85(2).
2. Zhou, J.H. & Yu, M.M. 2018. The research on maintenance technology of drainage asphalt pavement at home and abroad is summarized. *Modern transportation technology* 15(06): 6–11.
3. Drake, J.A.P. & Bradford, A. 2013. Review of environmental performance of permeable pavement systems: State of the knowledge. *Water Quality Research Journal of Canada* 48: 203–222.
4. Yang, Z.H. & Xu, B. 2019. Comparative Analysis of Performance of Porous Asphalt Pavement and SMA Pavement Based on Deck Pavement Structure. *IOP conference series. Earth and environmental science* 283, 12053.
5. He, D.J. & Wang, S.H. 2015. Research on application urban road pervious pavement based on the concept of sponge city. *Shanxi Architecture* 6:112–113.
6. Fan, B.W. & Ma, Q.W. 2018. Preparation and engineering application of high-viscosity composite modified asphalt. *Road construction machinery and construction Mechanization* 35(02): 81–84+88.
7. Liu, Y.D. 2014. *Research on Road Properties of Porous Drainage Asphalt Pavement*. Xi'an: Chang'an University.
8. Feng, X.J. & Xie, M.W. 2019. Preparation of TS high-viscosity modified asphalt and pavement performance of OGFC mixture, *Highway Traffic Technology* 36(01):8–15.
9. Chen, B.M. & Zhu, Y. 2019. Research progress of rubber powder modified asphalt for drainage pavement. *East China Highway* (06): 102–104.
10. Xu, S.G. & He, W.P. 2016. Research on the high-viscosity modified asphalt of porous asphalt mix-ture. *High Way* 61(03):166–170.
11. JTG E20–2011, S. *Standard Test Methods of Bitumen and Bituminous Mixtures for Highway Engineering*. China.
12. AASHTO Designation: TP 70–12, S. Standard Method of Test for Multiple Stress Creep Recovery (MSCR) Test of Asphalt Binder Using a Dynamic Shear Rheometer (DSR). America.

Functional Pavements – Chen et al (eds)
© 2021 Taylor & Francis Group, London, ISBN 978-0-367-72610-2

Application of infrared spectroscopy in prediction of asphalt aging time and fatigue life

Maoping Ran*, Yanmei Yang, Yuan Yan, Xinglin Zhou & Ruiqie Jiang
School of Automotive and Traffic Engineering, Wuhan University of Science and Technology, Wuhan, China

ABSTRACT: Based on Fourier transform attenuated total reflection infrared spectroscopy (ATR-FTIR) and principal component analysis (PCA), four kinds of asphalts (two kinds are matrix asphalts TPC70# and JL70#, two kinds are modified asphalts SBS-1 and SBS-2) were aged by rotating film oven test (RTFOT). The multiple stress repeated creep recovery test (MSCR) was carried out on AR1500ex dynamic shear rheometer (DSR). The PCA was carried out on the attenuated total reflection infrared spectrum of pretreated aged asphalt, the prediction model of aging asphalt recovery rate R was established based on functional group index, and the fatigue life of each aging asphalt based on the rate of change of dissipated energy (RDEC) was compared. The correlation between comprehensive index F and fatigue life of asphalt was analyzed too. The results show that the order of resistance to high temperature deformation of each aging asphalt is: SBS-2>SBS-1>TPC70#>JL70#; The prediction model of aging asphalt recovery rate R based on functional group index has a good reliability; Taking the fatigue life N_{RDEC} as the evaluation index, the fatigue life order of each aging asphalt is: SBS-1>SBS-2>JL70#>TPC70#. There is a positive correlation between fatigue life of aged asphalt and comprehensive index F of asphalt, and the correlation degree R^2 is 0.8534.

Keywords: infrared spectrum, aging time course, fatigue life, asphalt recovery rate, principal component analysis

1 INTRODUCTION

Asphalt aging is an important reason for shortening the life of asphalt pavement. Research on asphalt aging time and fatigue life is of great significance to improve the durability of asphalt pavement. However, asphalt performance is the most difficult factor to monitor in the construction process of asphalt pavement, and its technical indicators need to be tested according to the specifications. Traditional testing not only consumes a lot of time, but also may cause different degrees of damage to the road surface. Infrared spectroscopy technology is a non-destructive rapid detection technology. The test process does not require sample pretreatment, the test speed is fast, and there is no pollution, and the collection of spectral data can analyze multiple index data at the same time, thus greatly reducing the test time and analysis Cost [1]. Some scholars [2–6] when using infrared spectroscopy to study aging asphalt, select characteristic peaks such as carbonyl and sulfoxide groups to study the characterization of asphalt aging behavior, the effect of modifier on asphalt, and the optimal amount of modifier, only considering Characteristic absorption peaks that have undergone significant changes, using characteristic peak height (ratio) or peak area (ratio) and characterization parameters to establish

---

* Corresponding author

45

a simple linear regression relationship, the infrared spectrum analysis of asphalt aging time and rheological performance is not deep enough.

With the development of chemometrics, the combination of spectral technology and chemometrics has been widely used in various fields. Some scholars have applied this method to study the anti-aging performance of asphalt. Literature [7,8,9] used principal components analysis (PCA), linear discriminant analysis (LDA), partial least squares (PLS) and other methods to analyze the infrared spectrum data of asphalt, the results show that based on partial least squares (PLS) The establishment of infrared spectroscopy quantitative prediction model can achieve rapid detection of asphalt properties such as wax content, softening point, penetration, etc. However, the above tests have higher requirements on the source and type of asphalt, and have certain limitations on the parameters of different types of asphaltene content, softening point, logarithm of penetration, composite shear modulus and phase angle at different temperatures.

More and more studies have shown that the three traditional indicators of asphalt (penetration, ductility and softening point) cannot accurately characterize the anti-aging properties of asphalt. Currently widely used evaluation indexes of asphalt aging include quality change, residual penetration ratio, residual ductility ratio, residual viscosity ratio, viscosity increase, softening point increase, etc. Literature [10,11,12] has carried out relevant research on these indexes, and used each aging index to analyze the aging trend of asphalt, so as to evaluate the anti-aging performance of asphalt, and achieved certain results. In addition, the literature [13,14] studies believe that the colloidal instability index I_c can well reflect the peptizing ability of soft asphalt. The more unstable, the worse its anti-aging ability.

In this study, based on Fourier transform attenuated total reflection infrared spectroscopy (ATR-FTIR) and principal component analysis (PCA), the infrared spectrum data of four kinds of aged asphalt were qualitatively and quantitatively analyzed, the comprehensive index f was calculated, the prediction model of asphalt aging time history was established and verified, and the prediction model of asphalt recovery rate R was established by functional index. Analyze the relationship between the recovery rate R of each aging asphalt and the aging time course, compare the fatigue life of each aging asphalt based on the rate of change of energy dissipation, and analyze the correlation between asphalt fatigue life and aging time course in order to achieve the aging time and fatigue life of asphalt Rapid prediction.

2 TESTS

2.1 *Rotating Film Oven Test (RTFOT)*

In this test, two base asphalts of TPK 70# and JL 70# and SBS (I-D) modified asphalts of two different base asphalts were selected, which were denoted as SBS-1 and SBS-2, respectively. According to the current "Testing Regulations for Asphalt and Asphalt Mixtures of Highway Engineering" (JTGE20-2011), the technical indicators of the above four types of asphalts are tested, as shown in Table 1, and their technical indicators are in compliance with the specifications.

Five different aging time courses were selected to carry out the rotary film oven test (RTFOT) on the above four asphalt samples. The test device is shown in Figure 1.

The aging time of five RTFOT are as follows: a. the aging time is 0 min; B. the aging time is 85min; C. the aging time is 120min; D. the aging time is 240min; E. the aging time is 360min; A total of 20 asphalt samples were obtained, as shown in Table 2.

2.2 *Infrared spectrum acquisition*

Use thermo scientific nicolettis50 integrated Fourier transform infrared spectrometer, and select ATR (Attenuated Total Reflection) accessory to collect infrared spectrograms of aged asphalt samples, as shown in Figure 2. The Fourier infrared spectrometer needs to be preheated for at least 30 minutes in advance. Before each measurement, background scan is

Table 1. Basic physical performance index of asphalt.

Test project		Unit	TPC70#	JL70#	SBS-1	SBS-2	Test method
Penetration (25°C, 100g, 5s)		0.1mm	62.1	61.3	55.2	54.6	T0604
Ductility (5°C, cm/min)		cm	19.5	20.8	34.6	35.2	T0605
Softening point		°C	49.4	50.6	77.8	78.2	T0606
Brookfield viscosity(135°C)		Pa*s	0.492	0.486	1.38	0.486	T0625
Rotating film heating	Mass loss	%	0.08	0.07	0.07	0.08	T0610
Test (RTFOT) residue	Ductility (5°C, cm/min)	cm	13.7	14.3	17.2	17.6	T0605
163°C,85min	Penetration atio	%	84.4	82.8	79.2	78.9	T0604

Figure 1. Rotary film oven test (RTFOT).

Table 2. Aging asphalt sample.

Asphalt brand		Aging simulation method	Aging tempera-ture/°C	Aging time/ min	Aging asphalt samples
Matrix asphalt	JL70#	RTFOT	163	0	JL-A
				85	JL-B
				120	JL-C
				240	JL-D
				360	JL-E
	TPC70#	RTFOT	163	0	TPC-A
				85	TPC-B
				120	TPC-C
				240	TPC-D
				360	TPC-E
Modified asphalt	SBS-1	RTFOT	163	0	SBS1-A
				85	SBS1-B
				120	SBS1-C
				240	SBS1-D
				360	SBS1-E
	SBS-2	RTFOT	163	0	SBS2-A
				85	SBS2-B
				120	SBS2-C
				240	SBS2-D
				360	SBS2-E

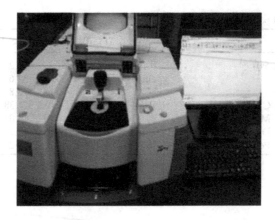

Figure 2.   Infrared spectrum acquisition.

performed. The acquisition parameters are set to a resolution of 4 cm^{-1}, the number of scans is 32, and the test range is 500 to 4000 cm^{-1}. The aging asphalt in flowing state at the end of the aging simulation experiment can be directly sampled to avoid reheating. Smear melted asphalt samples on a dry and clean SiO2 glass sheet of 20mm*20mm*1mm, so that the glass sheet is completely covered and the asphalt surface on the glass sheet is as smooth as possible. Each sample is duplicated for 3 times and repeatedly loaded for 3 times during collection. Each sample repeatedly records 9 spectra, and a total of 180 spectrograms (20*9) are obtained, thus ensuring the repeatability of asphalt samples and reducing the instability of measurement environment.

2.3   *Multiple Stress Creep Recovery Test (MSCR test)*

In order to evaluate the ability of the aging asphalt to resist permanent deformation, the aging residue of the asphalt rotary oven was subjected to a multi-stress creep recovery test, or MSCR test. The AR1500ex dynamic shear rheometer (DSR) was used for MSCR. referring to ASTMD7405, a 25mm rotor was selected, and the gap between the rotor and the lower parallel plate of DSR fixture was set to 1mm. The multi-stress creep recovery test is carried out at five temperatures: 64°C, 70°C, 76°C, 82°C and 88°C. The dynamic shear rheometer will apply creep stress of 0.1kPa and 3.2kPa to asphalt samples successively, and each stress level takes loading 1s and unloading 9s as one cycle, with each loading 10 cycles[15]. Repeated tests of three asphalt samples are carried out at different temperatures to ensure the repeatability of the tests, and the final test result is the average of the three repeated test results.

The final evaluation indexes of MSCR test are unrecovered creep compliance J_{nr} and recovery rate R. The smaller the J_{nr} value and the larger the R value, the better the high temperature performance of asphalt, that is, the better the deformation resistance and recovery ability. The calculation formulas of the two are shown below.

$$J_{nr}(\sigma, N) = \frac{\varepsilon_{10}}{\sigma} \qquad (1)$$

$$R(\sigma, N) = \frac{\varepsilon_1 - \varepsilon_{10}}{\varepsilon_1} \times 100\% \qquad (2)$$

$$J_{nr} = \sum\nolimits_{N=1}^{10} J_{nr}(\sigma, N)/10 \qquad (3)$$

$$R = \sum\nolimits_{N=1}^{10} R(\sigma, N)/10 \qquad (4)$$

Where: σ is the applied stress level, which is 0.1kPa and 3.2kPa; N is the loading period; $J_{nr}(\sigma,N)$ is the unrecovered creep compliance of the nth cycle. R is the recovery rate of the nth cycle; ε_{10} is the unrecovered strain at the end of the recovery phase of the nth cycle (i.e., the 10s) relative to that before loading; ε_1 is the strain at the end of loading for the nth cycle (i.e., the first 1s) relative to that before loading; J_{nr} is the average unrecovered creep compliance; R is the average recovery rate.

3 RESULTS AND ANALYSIS

3.1 *Infrared spectrum analysis of aged asphalt*

3.1.1 *Qualitative analysis*
In order to eliminate the interference of noise and physical factors on spectral data and make spectral information reflect the truest spectral characteristics of samples to the greatest extent, the original spectral data should be preprocessed first. According to the interference of spectrum, 180(20*9) spectrograms obtained in section 2.2 above are preprocessed by SNV smoothing (the number of smoothing points is 5) and baseline correction. Finally, the average absorption intensity of 9 preprocessed infrared spectrograms of each asphalt sample is taken as the final spectrogram data of the asphalt sample. Import the spectral data of 20 asphalt samples into Origin software, and draw the attenuated total reflection Fourier infrared spectrogram of all asphalt, as shown in Figure 3.

It can be seen from Figure 3 that the infrared spectra of 20 samples of the selected four types of asphalt are not only similar in shape, but also have the same characteristic absorption peaks and positions of functional groups. Different samples have different properties, which leads to great differences in peak heights of some characteristic absorption peaks, that is, absorbance, which can distinguish asphalt types to a certain extent.

In order to analyze the change of infrared spectrum characteristics of asphalt samples with different aging degrees, the infrared spectrum comparison diagrams of four types of asphalt with different aging degrees are drawn, as shown in Figure 4.

According to Figure 4, JL70# and TPK70# have the same peak positions and changes, and there are 13 obvious characteristic absorption peaks.The SBS-1 and SBS-2 asphalt have the

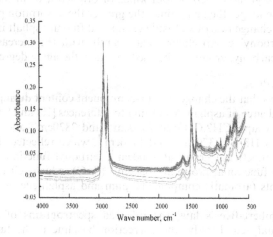

Figure 3. Attenuated total reflection Fourier infrared spectroscopy of 20 kinds of asphalt samples.

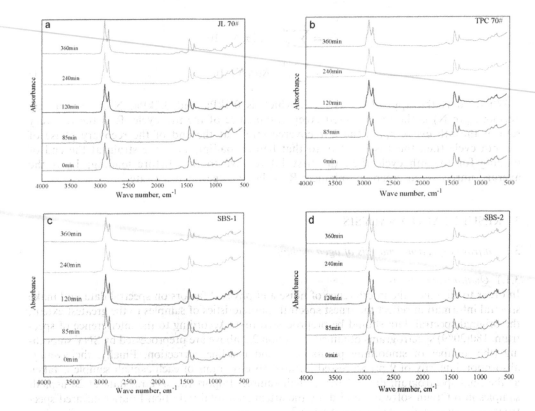

Figure 4.   The infrared spectrum diagrams of four types of asphalt with different aging time.

same peak position and change. Compared with the base asphalt, in addition to the 13 absorption peaks in total, there are two more obvious characteristic absorption peaks at 699cm^{-1} and 966cm^{-1}. Therefore, the base asphalt and SBS modified asphalt can be distinguished by the presence or absence of these two characteristic absorption peaks;There is no absorption peak at 1700cm^{-1} for the four types of original asphalt, but after a short-term aging of 85 minutes, due to the oxidation of carbon and oxygen, a carbonyl (C=O) absorption peak appears at this place. As the aging time increases, the carbonyl group The absorption peak is more and more obvious, but there are differences in the speed of the carbonyl absorption peak of different asphalts;The sulfoxide group (S=O) of four kinds of original asphalt has a weak absorption peak at 1030cm^{-1}. The longer the aging time, the greater the absorption peak intensity of sulfoxide group, and the change rate of sulfoxide group of different asphalt is different. Both sulfoxide group and carbonyl group change significantly with the increase of aging time, so sulfoxide group and carbonyl group can be used to reflect the aging degree of asphalt.

3.1.2 Quantitative analysis
Research [16,17] shows that the change of four-component content during the aging of asphalt can reflect the aging degree of asphalt.According to references [18] and [19], this study selected aliphatic functional groups (CH2,CH3) at 2920cm^{-1} and 2820cm^{-1} and asymmetric aliphatic functional groups (C-CH3) at 1376cm^{-1} and 1456cm^{-1}, which reflected the change of asphalt saturation, aromatic functional group at 1600cm^{-1}, carbonyl functional group at 1700cm^{-1} (C=O) and sulfoxide functional group at 1030cm^{-1} (S=O), which reflect the change of aromatic ring components (aromatic components, gum and asphaltene) as the main objects of quantitative analysis.

According to Lambert-Beer's law, the infrared spectrograms of 20 asphalt samples are quantitatively analyzed. Firstly, the correction baseline is the tangent of the lowest point on both sides of the characteristic absorption peak, and the area enclosed by the baseline and the spectral curve is the peak area of the absorption peak [20]. The peak areas of 15 obvious characteristic absorption peaks of each asphalt were calculated by OMNIC software. In order to explore the best quantitative analysis method, the selected benchmark for the study is [2,3]: A1, the peak area of the fingerprint area in the range of 650 ~ 1400cm^{-1}; A2, sum of peak areas of group stretching vibration region in the range of 1400 ~ 4000 cm^{-1}; A3, full spectrum in the range of 650 ~ 4000cm^{-1}; A4, sum of peak areas in the range of 2920 and 2850cm^{-1}. Comparing different analysis methods, the peak area ratio of each characteristic absorption peak, i.e. functional group index, is defined as follows:

$$\begin{cases} I_{C=O} = A_{1700}/\sum A_i \\ I_{S=O} = A_{1030}/\sum A_i \\ I_B = (A_{2920} + A_{2850})/\sum A_i \\ I_{B,a} = (A_{1376} + A_{1456})/\sum A_i \\ I_{Ar} = A_{1600}/\sum A_i \end{cases} \tag{5}$$

Where $I_{C=O}$ and $I_{S=O}$ are carbonyl and sulfoxide functional group index respectively; I_B and $I_{B,a}$ are aliphatic functional group and asymmetric aliphatic functional group index; I_{Ar} is aromatic functional group index; A_{2920}, A_{2850}, A_{1700}, A_{1600}, A_{1456}, A_{1376} and A_{1030} are the corresponding peak areas at wave numbers of 2920, 2850, 1700, 1600, 1456, 1376 and 1030 cm^{-1} respectively.$\sum A_i$ is the sum of peak areas under the i th benchmark (i=1,2,3,4), and the specific values are shown in Table 3.

From the above analysis, it can be seen that the peak area of functional groups has nothing to do with the selection of reference standards, and selecting different reference standards will make the change rule of functional group index different or even opposite with the increase of aging time. Asphalt reacts with oxygen to produce carbonyl and sulfoxide groups, which makes the functional index of carbonyl and sulfoxide groups increase with the aging time[21]. According to formula (2), the functional group indices of asphalt samples under different benchmarks are calculated, and the change trend of carbonyl and sulfoxide functional group indices is shown in Figure 5.

It can be seen from Figure 5 that the carbonyl index of the four types of asphalt shows an increasing trend under the selected four reference standards, while the sulfoxide group of TPC 70# asphalt shows a negative growth within the aging time range of 120~240 min under the benchmarks A2, A3 and A4, which does not conform to the aging mechanism of asphalt. Therefore, This work selects A1 (650 ~ 1400 cm^{-1}) as the reference standard, so the changes of functional groups in asphalt aging process can be accurately and reasonably characterized. The functional group indexes of 20 kinds of asphalt under A1 standard are shown in Table 4.

Table 3.   The values of spectral peak area under each benchmark.

Peak range of reference spectrum, cm^{-1}	The sum of peak area
650 ~ 1400	$\sum A_1 = A_{699}+A_{721}+A_{744}+A_{807}+A_{863}+A_{965}+A_{1030}+A_{1159}+A_{1306}$ $+A_{1376}$
1400 ~ 4000	$\sum A_2 = A_{1456}+A_{1599}+A_{1700}+A_{2850}+A_{2920}$
650 ~ 4000	$\sum A_3 = A_{699}+A_{721}+A_{744}+A_{807}+A_{863}+A_{965}+A_{1030}+A_{1159}+A_{1306}+$ $A_{1376}+A_{1456}+A_{1599}+A_{1700}+A_{2850}+A_{2920}$
2920, 2850	$\sum A_4 = A_{2850}+A_{2920}$

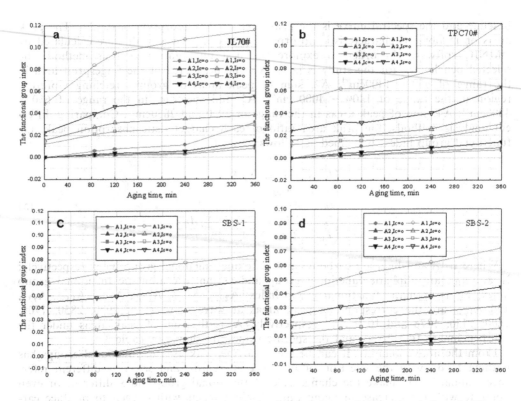

Figure 5. The change trend of carbonyl and sulfoxide functional group indices of asphalts with different aging time.

Table 4. Functional group index of aged asphalt.

Asphalt samples	$I_{C=O}$	$I_{S=O}$	$I_{B,a}$	I_B	I_{Ar}
JL-A	0	0.048	1.248	2.141	0.091
JL-B	0.005	0.083	1.210	2.114	0.111
JL-C	0.007	0.094	1.242	2.048	0.084
JL-D	0.011	0.107	1.210	2.127	0.085
JL-E	0.031	0.115	1.167	2.105	0.108
TPC-A	0	0.049	1.390	2.037	0.131
TPC-B	0.007	0.061	1.404	1.919	0.138
TPC-C	0.010	0.062	1.415	1.976	0.142
TPC-D	0.017	0.077	1.403	1.938	0.118
TPC-E	0.026	0.119	1.342	1.901	0.132
SBS1-A	0	0.061	0.849	1.368	0.103
SBS1-B	0.002	0.067	0.848	1.418	0.100
SBS1-C	0.003	0.070	0.858	1.429	0.092
SBS1-D	0.014	0.076	0.841	1.378	0.091
SBS1-E	0.029	0.082	0.815	1.317	0.089
SBS2-A	0	0.038	0.885	1.604	0.089
SBS2-B	0.005	0.050	0.875	1.638	0.101
SBS3-C	0.007	0.054	0.889	1.700	0.098
SBS2-D	0.011	0.061	0.865	1.640	0.085
SBS2-E	0.015	0.072	0.853	1.627	0.078

3.2 *MSCR test results and analysis of aged asphalt*

According to Formulas (1)~(4), the unrecovered creep compliance J_{nr} and recovery rate R of 20 aged asphalt samples under different temperature and stress conditions are calculated. Comparing and analyzing the influence of temperature and stress on asphalt in Figure 6, it is found that $J_{nr0.1}$ of JL 70# and TPC 70# are slightly smaller than $J_{nr3.2}$ and $R_{0.1}$ is slightly larger than $R_{3.2}$ under any temperature, which shows that the sensitivity of matrix asphalt to applied stress is weak. However, after 70°C, J_{nr} and R of the two modified asphalts differ greatly under different stress levels, and change obviously with temperature, which indicates that the modified asphalt is sensitive to temperature and stress. Therefore, this work, selected 70°C and 3.2kPa as the test conditions to study the deformation resistance of asphalt at high temperature.

It can be seen from Figure 7 that the Jnr value of all asphalt decreases and the r value increases with the aging time at 70°C and 3.2kPa, indicating that the high temperature deformation resistance of asphalt gradually increases with the aging time increasing; Among them, the J_{nr} value of JL 70# is larger than that of TPC 70# when they are not aged, and there is little difference between them. With the increase of aging time, the J_{nr} value of TPC 70# increases obviously after 120min, but the R value of JL 70# has no obvious change, which indicates that TPC 70# has better anti-aging ability than JL 70#. The R value of SBS-1 decreases rapidly, while the R value of SBS-2 has a fluctuation region and does not change much, which indicates that SBS-2 has better resistance to deformation at high temperature than SBS-1. Within the aging time range of 0~360min, the J_{nr} of modified asphalt is always much smaller than that of base asphalt, and its recovery rate R is much larger than that of base asphalt, that is, after aging, its resistance to deformation at high temperature gradually increases, and the resistance to permanent deformation of modified asphalt is still significantly better than that of base asphalt, which is consistent with the analysis results of unaged asphalt samples. Moreover, the order of resistance to deformation at high temperature of four types of asphalt is SBS-2>SBS-1>TPC 70# >JL 70#.

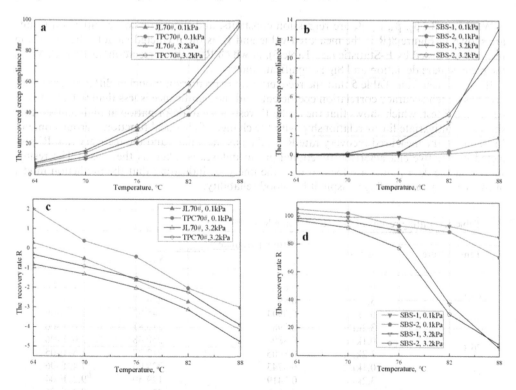

Figure 6. The unrecovered creep compliance J_{nr} of matrix asphalts and modified asphalts (a)(b), and the recovery rate R of them (c)(d) under different temperature and stress conditions.

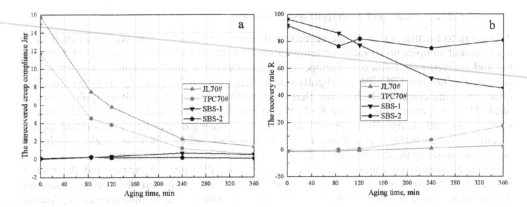

Figure 7. The unrecovered creep compliance J_{nr} (a) and the recovery rate R (b) of asphalts with the aging time of 70°C and stress condition of 3.2kPa.

3.3 Correlation analysis between functional group index and MSCR test results

The quantitative analysis standard proposed in section 3.1.2 is used to calculate the five characteristic functional group indexes of 20 kinds of aged asphalt spectra. Based on Levenberg-Marquardt method and general global optimization algorithm, SPSS23 is used to carry out multivariate statistical regression analysis on the asphalt recovery rate R of MSCR test under different temperature and controlled stress and the functional group change index of infrared spectrum test. R can be predicted by the following mathematical model,

$$R = \alpha_1 + \beta_1 I_{C=0} + \beta_2 I_{S=0} + \beta_3 I_{B,\alpha} + \beta_4 I_B + \beta_5 I_{Ar} \qquad (6)$$

Where α_1, β_1, β_2, β_3, β_4, β_5 are regression constants, and the regression comparison results are shown in Figure 8(R is the measured value and R' is the predicted value). The regression results were tested by F-Statistic (see Table 5), in which R^2 is the correlation coefficient, MS is the mean square deviation and Sig is significant index.

It can be seen from Table 5 that the recovery rate R prediction model at different temperatures has a high accuracy correlation coefficient, and the significance is less than 0.05, that is, it passes the F test, which shows that the asphalt's resistance to deformation at high temperature shows a multivariate linear relationship with the change of chemical functional group content. In Figure 8, the measured recovery rate R is taken as the x-axis, and the recovery rate R' predicted by the five characteristic functional group indexes is taken as the y-axis. The results show that the data points are located near the line y=x, indicating that the prediction model of the recovery rate R of aged asphalt has good reliability.

Table 5. Fitting results of the recovery rate R model.

Temperature	Stress	Fitting parameters		
		R^2	MS	Sig
64°C	0.1kPa	0.8954	264.27	2.12E-06
	3.2kPa	0.8959	234.77	1.62E-06
70°C	0.1kPa	0.8994	245.93	1.62E-06
	3.2kPa	0.8988	215.81	1.69E-06
76°C	0.1kPa	0.8898	250.48	3.01E-06
	3.2kPa	0.8605	212.27	1.51E-05
82°C	0.1kPa	0.8743	238.96	7.43E-06
	3.2kPa	0.7419	129.50	9.27E-04
88°C	0.1kPa	0.8607	224.36	8.43E-06
	3.2kPa	0.6538	108.31	10.32E-04

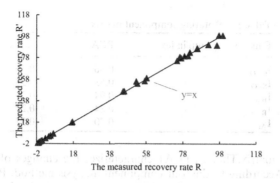

Figure 8.   The prediction result of aging asphalt recovery rate.

3.4   *Establishment and verification of prediction model of asphalt aging time*

3.4.1 *Model establishment*

The PCA method is a kind of chemo-metrics method, which aims at data dimensionality reduction, and can solve the problem of spectral matrix collinearity and the limitation of the number of variables to a certain extent [22]. The combination of infrared spectroscopy and PCA can extract more effective information in infrared spectroscopy measurement data, thereby establishing a stable and highly reliable prediction model.

According to the quantitative analysis results of mid-infrared spectroscopy in section 3.1.2, the PCA of the five main characteristic absorption peak functional group index values of 20 asphalt samples was performed using SPSS23 software. The results are shown in Tables 6~8. Table 6 shows that the correlation between the original variables is relatively high. The correlation between the aliphatic functional group index and the asymmetric aliphatic functional group index is as high as 0.946, and the correlation between the carbonyl index and the sulfoxide functional group index also reaches 0.842. Therefore, it is necessary to establish a principal component analysis model.

It can be seen from Table 7 that the eigenvalues of two components are greater than 1, and the cumulative contribution is as high as 81.43%. Therefore, selecting the first two components as the main components can represent most of the information of the original five data indicators. Table 8 shows that $I_{B,a}$, I_B and I_{Ar} have a large load on PCA1 by analyzing the

Table 6.   Correlation matrix of five characteristic functional group indices.

	$I_{C=O}$	$I_{S=O}$	$I_{B,a}$	I_B	I_{Ar}
$I_{C=O}$	1	0.84	0.18	0.14	0.16
$I_{S=O}$	0.84	1	0.36	0.40	0.13
$I_{B,a}$	0.18	0.36	1	0.95	0.82
I_B	0.14	0.40	0.95	1	0.48
I_{Ar}	0.16	0.13	0.82	0.48	1

Table 7.   Total variance explanation.

Component	Initial eigenvalue		
	Total	Variance percentage, %	Cumulative contribution rate, %
1	2.47	49.35	49.35
2	1.60	32.08	81.43
3	0.67	13.32	94.74
4	0.21	4.12	98.87
5	0.06	1.14	100

Table 8.　Rotating component matrix.

Functional group index	PCA1	PCA2
$I_{C=O}$	0.36	0.86
$I_{S=O}$	0.53	0.78
$I_{B,a}$	0.94	-0.30
I_B	0.84	-0.22
I_{Ar}	0.70	-0.37

rotation component matrix.That is, PCA1 characterizes the changes of saturated and aromatic components. According to the four-component analysis method, PCA1 can be defined as a "component change factor". IC=O and IS=O have a greater load on PAC2. IC=O and IS=O are the functional indexes of carbonyl and sulfoxide groups respectively, so PCA2 reflects the changes of functional indexes of carbonyl and sulfoxide groups, and the changes of carbonyl and sulfoxide groups are caused by the absorption of oxygen and oxidation of asphalt. Therefore, PCA2 can be defined as "oxidation factor".

The results of principal component analysis of 20 kinds of asphalt with different aging degrees show that when asphalt is aging, PCA1 and PCA2 can represent most information of functional group change in infrared spectrum, that is, they both can represent the change of characteristic absorption peak of attenuated total reflection infrared spectrum of aging asphalt, which can reflect the aging degree of asphalt to a certain extent.

$$PCA1 = 0.15ZX1 + 0.21ZX2 + 0.38ZX3 + 0.34ZX4 + 0.28ZX5 \tag{7}$$

$$PCA2 = 0.53ZX1 + 0.49ZX2 - 0.18ZX3 - 0.13ZX4 - 0.23ZX5 \tag{8}$$

$$F = \frac{49.35}{81.43} \times PCA1 + \frac{32.08}{81.43} \times PCA2 \tag{9}$$

Where ZX1 ~ ZX5 are the standardized data of $I_{C=O}$, $I_{S=O}$, $I_{B,a}$, I_B and I_{Ar} functional group indices of each asphalt sample.

According to Formula (9), calculate and analyze the comprehensive index F value of each asphalt sample. It is found that the comprehensive index F value increases with the deepening of aging degree, therefore, the aging of asphalt can be characterized by comprehensive index F. Establish the prediction model of F and asphalt aging time as follows:

$$F = 0.0048t - 0.2518 \tag{10}$$

$$F = 0.0028t - 0.9293 \tag{11}$$

Figure 9 Shows that F value of asphalt has a linear relationship with aging time in the range of 0~360 min, and the correlation degree is high. The linear fitting correlation of the two matrix asphalts is 0.9508, and the modified asphalts is slightly lower, mainly because the addition of modifiers makes the aging process more complicated and has more positive influence factors than the matrix asphalt.

3.4.2 *Model validation*
The existing SK90# matrix asphalt and its SBS modified asphalt with known aging time are predicted by the above prediction model, and the steps are as follows:

Infrared spectra of six groups of asphalt samples were collected by attenuated total reflection infrared spectrometer, and the measured infrared spectral data were preprocessed according to section 3.1.Selecting A1 as basis reference, the spectral data after pretreatment was quantitatively analyzed, and the functional index of five characteristic functional groups was obtained and standardized. The comprehensive index F is calculated according to Formula (7) ~(9). By substituting the comprehensive index F value into the prediction model, the predicted

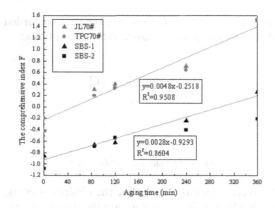

Figure 9. The linear relationship between F and aging time.

aging time histories of six asphalt samples were obtained. It can be seen from Figure 10 that all data points are near y=x, and the variability of data at 120min, 240min and 360min is 6.6%, 4.0% and 6.2%, respectively, which shows that the prediction model is reliable.

3.5 *Fatigue life analysis of aging asphalt based on infrared spectroscopy*

3.5.1 *Establishment of relationship model between fatigue life and aging time of asphalt*

Aging of asphalt is the main cause of its fatigue performance decay, and fatigue life is usually used to evaluate the fatigue cracking resistance of asphalt [23]. Fatigue life refers to the loading period required for fatigue failure of asphalt in time scanning test. At present, there are many methods to define the loading period, and the fatigue life under different definition methods is different. Therefore, choosing the appropriate definition method of fatigue life is the key to accurately analyze the fatigue cracking resistance of aging asphalt. A large number of studies show that the fatigue life determined by the change rate curve of dissipative energy can accurately characterize the anti-fatigue performance of asphalt [24]. The change rate of dissipated energy is a fatigue life evaluation index based on the law of energy change. It evaluates the development process of fatigue damage according to the speed of dissipated energy change of asphalt in the process of loading. The change rate curve of asphalt dissipation energy is roughly divided into three stages. The rate-of-dissipation energy change rate curve drops sharply from the initial stage to the gentle fluctuation stage, and finally reaches the stage of rapid increase.The loading period corresponding to the sudden increase inflection point of the dissipation

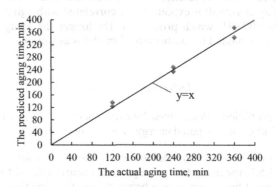

Figure 10. The validation of asphalt aging time prediction model.

57

energy change rate curve is the asphalt fatigue life, which is recorded as N_{RDEC}, and the calculation formula of dissipation energy change rate is as follows:

$$RDEC_a = \frac{|\omega_a - \omega_b|}{\omega_a(b-a)} \tag{12}$$

$$\omega_a = \pi\varepsilon\sigma\sin\delta = \pi\varepsilon^2 G^*\sin\delta \tag{13}$$

Where $RDEC_a$ is the average rate of change for the dissipated energy of the a-th loading cycle relative to the b-th loading cycle; ω_a and ω_b are the dissipated energy corresponding to the a-th and b-th loading cycles respectively, G^* is complex shear modulus, ε, δ are strain and stress respectively.

In order to get the inflection point of the dissipative energy change rate curve, the dissipative energy change rate needs to be further processed, and the calculation formula is as follows:

$$RDEC'' = \frac{RDEC_b - RDEC_a}{RDEC_a(b-a)} \tag{14}$$

With N_{RDEC} as the evaluation index, the fatigue life of four types of asphalt under the same aging time is compared (see Figure 11). It can be seen that the fatigue life of four kinds of asphalt is ranked as SBS1>JL70#>TPC 70#>SBS2 when the aging time is 0 and 85min. The modified asphalt SBS1 has the largest anti-fatigue performance, mainly because the asphalt has been aged during transportation and the hardness of the asphalt has increased, so its anti-fatigue performance is greater than that of matrix asphalt. But after aging for 120min, the order is TPC70#>SBS1>JL70#>SBS2, and aging for 240min and 360min is TPC70#>JL70#>SBS2>SBS1. The chaning orders show that with the increase of aging time, the fatigue resistance of matrix asphalt is greater than that of modified asphalt, among which the fatigue resistance of TPC70# asphalt increases rapidly, exceeding JL70#, while that of SBS1 increases slowly, and finally is lower than that of SBS2.The anti-fatigue performance of asphalt reflects the anti-aging ability of asphalt to a certain extent. Therefore, the anti-aging ability of four types of asphalt is ranked as SBS1>SBS2>JL70#>TPC70#.

According to the analysis of the fatigue life change law of asphalt with different aging time, it can be seen in Figure 11 that the fatigue life of four types of asphalt increases with the aging time in the range of 0-360min, that is, the fatigue resistance of asphalt increases with the deepening of aging. Figure 12 shows the relationship between fatigue life and aging time. It reflects that the fatigue life of aged asphalt is exponentially correlated with aging time, and the correlation degree R^2 reaches 0.8631, which proves that the longer the aging time of asphalt, the greater the fatigue life of asphalt.The mathematical model is as follows:

$$LnN_{RDEC} = \alpha T + \beta \tag{15}$$

Where α and β are coefficients determined by experiments; N_{RDEC} is the fatigue life determined by the rate of change of dissipated energy; T is the aging time.

3.5.2 Correlation analysis between comprehensive index f and fatigue life
According to section 3.5.1, the aging time of asphalt is linearly related to the logarithm of its fatigue life. There is a good linear correlation between asphalt aging time and asphalt comprehensive index F(see section 3.4.1). Therefore, there must be a certain relationship between

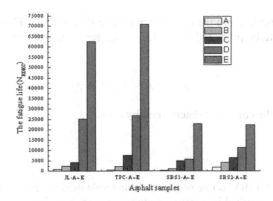

Figure 11.   The fatigue life(N_{RDEC}) of all asphalt samples.

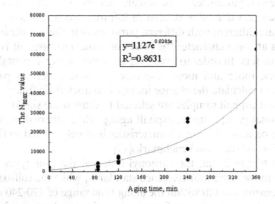

Figure 12.   The relationship between fatigue life N_{RDEC} and aging time.

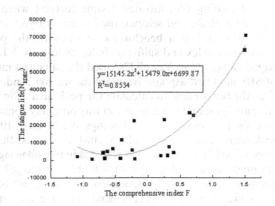

Figure 13.   The fitting curve between N_{RDEC} and F.

fatigue life of aging asphalt and its infrared spectrum principal component analysis result. The fitting result between N_{RDEC} and F is shown in Figure 13. It can be seen that there is a positive correlation between them, that is, with the increase of comprehensive index F, the

N_{RDEC} value also increases, and the correlation degree R^2 is 0.8534. Therefore, the fatigue life NRDEC of asphalt can be predicted by the following mathematical model,

$$N_{RDEC} = \gamma_1 F^2 + \gamma_2 F + \gamma_3 \tag{16}$$

Where $\gamma_1, \gamma_2, \gamma_3$ are the constants determined by tests.

4 DISCUSSIONS

Infrared spectroscopy (FTIR), as one of the main methods used to characterize the aging properties of asphalt, has been applied by more and more scholars to evaluate the structure and properties of asphalt. However, the complexity of asphalt structure and the instability and variability of performance of each component after being affected by the environment lead to the diversity and disunity of asphalt research parameters. The researchers choose different evaluation methods, so the comparison of absolute value of peak area in infrared spectrum data is not referential. The relative value changes are different with different parameters in the calculation process. The absolute value of peak area at some characteristic peaks of some aged asphalt is too small due to the great influence of parameters. In order to avoid the influence of experimental errors, human operation and other factors, more and more researchers choose a certain part of peak value as a reference benchmark to calculate the change index of characteristic functional groups. Different benchmarks and different asphalt samples are selected to draw some similar conclusions about the change index of functional groups during asphalt aging, while others draw opposite conclusions. Therefore, the research on asphalt aging characteristics is closely related to the artificially selected different asphalt samples and different benchmarks[17].

It can be seen from Figure 5 that the carbonyl index of the four types of asphalt shows an increasing trend under the selected four reference standards, while the sulfoxide group of Taipuke 70# asphalt shows a negative growth within the aging time range of 120-240 min under the benchmarks A2, A3 and A4, which does not conform to the aging mechanism of asphalt. Therefore, in this study, the peak value at A1 (650 ~1400cm⁻¹) is selected as the calculation benchmark, in order to accurately and reasonably characterize the changes of functional groups during asphalt aging.

Qiu Longliang [25] used coating film infrared testing method when studying the aging mechanism of SBS modified asphalt, and selected methylene antisymmetric stretching vibration absorption peak at 2924 cm⁻¹ as a benchmark to calculate the peak area ratio, and obtained the change of carbonyl index and sulfoxide index as follows: NEA (natural exposure aging) > RTFOT> unaged. When Zhao Yongli [26] used the same test method to analyze the aging characteristics of SBS modified asphalt, he chose the methyl bending vibration absorption peak at 1460 cm⁻¹ as the benchmark to calculate the peak area ratio and carry out infrared spectrum quantitative analysis, and the conclusion was probably consistent with Qiu.

Pang Ling[27]calculated the indexes of functional groups at 1700cm-1, 1030cm-1 and 968cm-1 based on the sum of peak areas between 2000 and 600 cm-1, and found that the indexes of carbonyl and sulfoxide groups of AH-70 matrix asphalt changed in the following order: NEA(natural exposure aging)>PAV (asphalt accelerated aging test) > RTFOT > unaged. Zhang Feng[28] and others used infrared spectroscopy to study the changes of SK-90 asphalt before and after aging with SBS and SBS/sulfur modified asphalt, and concluded that the change rule of carbonyl index was the same as that of the above scholars, but the change rule of sulfoxide index was not obvious. Compared with the original one, the short-term thermal oxidation aging index increased obviously, but most of them showed a decreasing trend in simulating PAV aging for 5 years. Therefore, the research on asphalt aging characteristics is a complex and important subject, which requires more scholars to develop new technologies and means to implement the research.

Finally, this work still exists some possibilities for in-depth study. For example, in the process of establishing and verifying the prediction model, the types and quantities of selected asphalt samples need to be further enriched; The study realized the rapid prediction of asphalt aging time and fatigue life by infrared spectrum, and further considered the integration of infrared spectrum pretreatment process, quantitative analysis, principal component analysis and prediction model, and designed a whole set of prediction system. As long as the infrared spectrum is input, it is possible to get the prediction results directly by integrated software.

5 CONCLUSIONS

This work established a prediction model of asphalt aging time history and fatigue life based on infrared spectroscopy technology and principal component analysis method. Firstly, the qualitative and quantitative analysis of the attenuated total reflection infrared spectrum after pretreatment was carried out. Then the high-temperature deformation resistance of various aged asphalts was analyzed based on the infrared spectrum quantitative analysis, and the relationship between the functional group index and the recovery rate R was discussed. Taking the main component analysis results as original datas, the asphalt aging time prediction model was established and verified. The fatigue life of various aged asphalt samples was analyzed by adopting NRDEC as the fatigue life evaluation index. Finally, this work explored the correlationship between the comprehensive index F and asphalt fatigue life. The main research conclusions are as follows:

After the asphalt is aged, its recovery rate R increases, and the high temperature resistance to deformation is enhanced. The high temperature resistance to deformation of four types of asphalts are ranked as follows: SBS-2>SBS-1>Taipuke 70#>Jinling70#. The recovery rate R and the functional group index showed a multivariate linear correlation and a high degree of correlation. As can be seen from Figure 8, the data points of the recovery rate R predicted by the five characteristic functional group indexes are all located near the straight line of y=x, indicating the aging based on the functional group index. The prediction model of asphalt recovery rate R has good reliability, and can achieve rapid prediction of aging asphalt high temperature resistance to deformation to a certain extent.

Principal component analysis method was used to determine two principal component factors, namely oxidation factor and component change factor, both of which can be used to characterize the aging degree of asphalt. The asphalt aging time prediction model based on the principal component comprehensive evaluation index F has a certain degree of reliability, indicating that the asphalt infrared spectroscopy can quickly predict its aging time.

Taking N_{RDEC} as the fatigue life evaluation index, the fatigue life rankings of the 4 types of asphalt are: SBS1>SBS2>Jinling70#>Taipuke70#, and N_{RDEC} is exponentially correlated with the aging time, the correlation reaches 0.8631. The fatigue life of aging asphalt is positively correlated with the comprehensive index F of asphalt, and the correlation degree R^2 is 0.8534, that is, the comprehensive index F of asphalt increases, and its fatigue life also increases. Therefore, it is feasible to use infrared spectroscopy to predict the aging time and fatigue life of asphalt, and it can provide a rapid and non-destructive prediction method for the practical application of asphalt.

ACKNOWLEDGMENTS

The authors would like to acknowledge the financial support from the National Natural Science Foundation of China (NSFC; 51827812, 51778509, 51578430), the Hubei Provincial Natural Science Foundation of China (2018CFB293) and the overseas study program for young teachers of Hubei Province (201659194).

REFERENCES

1 Chu, X.L.; Xu, Y.P.; Lu, W.Z. Research and application progress of chemometrics for near infrared spectroscopy [J]. Analytical Chemistry, 2008(05): 702–709.

2　Zhang, Z.Y.; Shen, J.N.; Shi P.C.; Zhu, H. Micro-mechanism of asphalt aging based on nano-mechanics and functional groups [J]. Highway Traffic Science and Technology, 2017, 34(05): 19–27.

3　Li, P.; Nian, T.F.; Wei, D.B.; Lin, M. FTIR quantitative analysis method and new exploration of rheological parameters of aged asphalt [J]. Journal of Huazhong University of Science and Technology (Natural Science Edition), 2018, 46(02): 34–39.

4　Wang, Y.B.; Luo, A.L.; Lu, W.Z.; Yuan, H.F. Rapid determination of wax content in asphalt by near infrared analysis [J]. Acta Petrolei Sinica (Petroleum Engineering), 2001(03): 68–72.

5　Tang, J.Q. Research on Rapid Identification and Analysis of Asphalt Materials [D]. Beijing: beijing university of chemical technology, 2015.

6　Wang, Y.J.; Chen, Q.T.; Zhao, W. Study on Rapid Detection of Asphalt Properties by Attenuated Total Reflection Infrared Spectroscopy [J]. Highway Traffic Technology (Applied Technology Edition), 2016,12(11): 74–77.

7　Ye Y.F.; Zhen, Y.; Zhang, X.R.; Wu H.N. Study on rapid detection method of asphalt penetration [J]. petroleum refining and chemical industry, 2014(6): 13–16.

8　Weigel, S.; Stephan, D. The prediction of bitumen properties based on FTIR and multivariate analysis methods[J]. Fuel, 2017, 208: 655–661.

9　Hao P.L. Evaluation and Analysis of Aging Detection of Asphalt Mixture for Road Use [J]. Northern Transportation, 2015(7): 71–73.

10　Zeng, M.; Pan, H.; Zhao, Y.; et al. Evaluation of asphalt binder containing castor oil-based bioasphalt using conventional tests[J]. Construction and Building Materials, 2016, 126: 537–543.

11　Siddiqui, M.; Ali, M. Studies On The aging behavior of the Arabian asphalts[J]. Fuel, 1999, 78(9): 1005–1015.

12　Lin, L.; Li, X.L.; Zheng, G.Y.; et al. A new discussion on the relationship between aging performance of petroleum asphalt and its chemical composition [J]. Material Guide, 2013, 27 (16): 123–126.

13　Ma, Z. Performance evaluation of sulfur modified asphalt based on DSR test [J]. traffic science and engineering, 2017, 33(1): 23–26.

14　Huo, Y.; Zhu, J.; Li, J.; et al. An active La/TiO2 photocatalyst prepared by ultrasonication-assisted sol-gel method followed by treatment under supercritical conditions[J]. Journal of Molecular Catalysis A:Chemical, 2007, 278(1): 237–243.

15　Bahia, H.U.; Hanson, D.I.; Zeng, M.; et al. Characterization of Modified Asphalt Binders in Superpave Mix Design [R]. NCHRP Report 459,Transportation Research Board, National Research Council, 2001.

16　Lamontagne, J.; Dumas, P.; Mouillet, V.; et al. Comparison by fourier transform infrared (FTIR) spectroscopy of different ageing techniques: application to road bitumens[J]. Fuel, 2001, 80: 483–488.

17　Zhang, B.L. Study on Asphalt Structure Characterization Based on Infrared Spectroscopy [D]. Wuhan: Wuhan University of Technology, 2014.

18　Zhang, D.; Zhang, H.; Shi, C. Investigation of aging performance of SBS modified asphalt with various aging methods[J]. Construction and Building Materials, 2017, 145: 445–451.

19　Yut, I.; Zofka, A. Correlation between rheology and chemical composition of aged polymer-modified asphalts[J]. Construction and Building Materials, 2014, 62: 109–117.

20　Chen, Y.D.; Tu, J.; Zhang, B.; et al. Study on the influencing factors of SBS modified asphalt by infrared spectroscopy [J]. Petroleum Asphalt, 2014, 28 (1): 67–72.

21　Liu, B.; Shen, J.N.; Shi, P.C. Nano-scale microscopic characteristics and functional group properties of aged asphalt [J]. Highway Traffic Science and Technology, 2016, 33 (2): 6–13.

22　Lu, W.Z. Modern Near Infrared Spectroscopy [M].The second edition. Beijing: Sinopec Press, 2007.

23　Luo, R.; Xu, Y.; Liu, H.Q.; et al. Rheological mechanical properties of DCLR modified asphalt [J]. china journal of highway and transport, 2018, 31(6): 165–171.

24　Zhou, X.L.; Yang, Y.M.; Guan, J.X.; Yan, Y. Prediction of asphalt aging time history based on infrared spectrum[J]. Sino-foreign Highway, 2020, 40(01): 218–223.

25　Qiu, L.L. Study on Aging and Regeneration Mechanism of SBS Modified Asphalt [D]. Master's Degree Thesis, Dalian: Dalian University of Technology, 2012.

26　Zhao, Y.L.; Gu, F.; Huang X.M. Analysis of aging characteristics of SBS modified asphalt based on FTIR [J]. Journal of Building Materials, 2011,14 (5): 620–623.

27　Pang, L. Study on UV Aging Characteristics of Asphalt [D]. Doctoral Dissertation, Wuhan: Wuhan University of Technology, 2008.

28　Zhang, F.; Yu, G.Y.; Han, J. The Effects of thermal oxidative ageing on dynaminc viscosity,TG/DTG,DTA and FTIR of SBS-and SBS sulfur-modified asphalts [J]. Construction and Building Materials, 2011, 25(1): 129–137.

Functional Pavements – Chen et al (eds)
© *2021 Taylor & Francis Group, London, ISBN 978-0-367-72610-2*

Research on mechanical characteristics of asphalt pavement based on gradient aging

Huimin Chen, Xudong Shi, Guilin Lu, Xingyu Yi & Jun Yang
Southeast University, Nanjin, China

ABSTRACT: The asphalt will gradually deteriorate due to the influence of the external environment, and the distribution of asphalt aging in the pavement is uneven and has a gradient. Aging will cause the performance and mechanical characteristics of asphalt mixtures to change. Therefore, based on the premise of gradient aging, this paper researches the stress characteristics of common semi-rigid base asphalt pavements. The constitutive model parameters of asphalt mixture with different aging duration are obtained by dynamic modulus test and applied to Abaqus. It is found that the deepening of the aging degree will also increase the tensile stress at the bottom of the surface to some extent. As the load moves, the stress and strain at the bottom of the asphalt surface layer will have the transformation phase of " compression-tension-compression ", and the horizontal stress at the center of the two-wheel load will also change from compression to tension with the increase of the depth.

1 BACKGROUND

Asphalt pavement is affected by oxygen, ultraviolet rays and other factors during use, which will cause aging. The distribution of aging on the road surface is uneven and gradient. The intuitive performance is that the modulus gradually decreases as the depth increases, and finally reaches a stable value at a certain depth. For asphalt mixtures, aging is inevitable. The existence of aging makes the asphalt pavement more prone to cracking and other diseases, thereby reducing its service life. At present, scholars at home and abroad have conducted a lot of research on the aging of asphalt and asphalt mixture, including the overall performance before and after aging, such as changes in fatigue properties or water damage resistance, the influence of different aging modes, and the influence of porosity on the degree of aging [1–4]. In view of the gradient aging of asphalt mixture, a phenomenon that actually exists during the use of pavement, researchers at home and abroad have also carried out research. In theory, Mirza and Witczak first proposed the "global aging model" [5], which was then adopted by AASHTO and added to the MEPDG design method. Through a large number of core drilling samples, the model establishes the relationship between the asphalt viscosity and the dynamic modulus of the asphalt mixture, and proposes the variation of the viscosity at different stages with time. Combined with the viscosity-depth model, the mixture viscosity of the pavement system at different times and depths can be obtained. The dynamic modulus can be predicted based on the volume characteristics of the mixture. Qijian Zhang applied the model to the analysis of pavement structure and found that considering aging and not considering aging have a great influence on the normal stress in the X direction of the pavement [6]. In terms of tests, Lu Jun used centrifugal extraction to extract pavement asphalt with different service life in layers, and conducted index tests on the extracted asphalt. It was found that the overall asphalt aging degree gradually decreased from top to bottom. However, due to the greater shear stress at the pavement depth of 4cm, the degree of aging is also more serious [7]. When Luo studied the actual pavement core samples, he also found that under long-term aging conditions, uneven aging mainly appeared within 4cm below the road surface [8].

However, the current research on gradient aging of asphalt pavement focuses on characterizing the aging gradient, and characterizing aging by selecting a certain variable such as the change of modulus. This type of method has the following two problems: 1. The selected variable does not have representativeness: The selected aging variables are often the results obtained at a specific temperature and frequency, and cannot accurately reflect the material properties of the material and the force characteristics under load; 2. The obtained aging gradient is limited in the actual application process and cannot provide substantial guidance for the force analysis and design of the pavement structure. Therefore, this article will conduct material performance research on the asphalt mixture before and after aging under the assumed gradient aging situation, and then import the material parameters into the finite element, and analyze the force characteristics of the asphalt pavement under the gradient aging state in the finite element.

2 VISCOELASTIC PARAMETERS OF ASPHALT MIXTURE

The dynamic modulus test is used to obtain the material parameters of the asphalt mixture. In order to simulate different degrees of aging, the dynamic modulus specimens were subjected to short-term aging (135°C, 4h), and then subjected to different long-term aging (85°C, 2d, 5d, 8d). Short-term aging means that the asphalt mixture is mixed and aged in an oven at a constant temperature of 135°C for 4 hours, and then the specimen is formed; long-term aging means that the asphalt mixture is formed and aged in an oven at a constant temperature of 85°C for 2d, 5d, 8d. The specific parameters of Prony series of relaxation modulus under different aging time are shown in Table 1.

3 ASPHALT PAVEMENT STRUCTURE MODELING

The common semi-rigid base asphalt pavement in China is modeled by finite element method to research its mechanical characteristics. The materials and thickness of each layer of the pavement structure are shown in Table 2 from top to bottom. The material parameters of the asphalt surface layer are represented by Prony series as Table 1.

In the analysis of the pavement force, the double-circle vertical uniform load is adopted in the Chinese design code, according to which the mechanical response of different positions is calculated and the maximum value is taken as the design index. However, some studies [9,10] pointed out that the shape of the Tire-road Contact is more similar to a rectangle when the actual vehicle load is applied, the realization of a circular moving load in finite element is also more complicated than a rectangular one. Considering the actual conditions of the load, according to the equivalence principle of total stress, the load is simplified to two rectangular loads with a length of 19.2 cm and a width of 18.6 cm. The interval of load area is 12.8cm and the load size is 0.7MPa.

Table 1. Prony series parameter value.

N	ρ_i/s	4h E_i/MPa	2d E_i/MPa	5d E_i/MPa	8d E_i/MPa
0	∞	8	33	26	59
1	10-6	510	2741	3177	3397
2	10-5	1342	3381	3730	3768
3	10-4	3600	4969	5281	5282
4	10-3	5142	5226	5351	5450
5	0.01	3632	3899	3977	4267
6	0.1	1802	1929	2061	2310
7	1	649	715	832	919

Table 2. Structure and materials characteristics of asphalt pavement.

Material	Thickness/cm	Material constitutive	Modulus/MPa	Poisson ratio
asphalt surface	18	Linear viscoelasticity	——	0.3
cement stabilized macadam	20	Elastic	1000	0.2
lime-ash soil	30	Elastic	400	0.3
soil	>100	Elastic	45	0.4

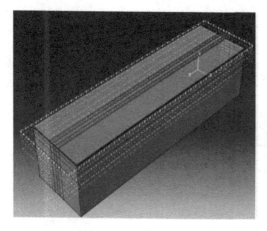

Figure 1. Schematic diagram.

Figure 2. Horizontal tensile stress cloud diagram of cross section.

The finite element model of the pavement is established by Abaqus, as shown in Figure 1. The dimensions of the three-dimensional model are 2.5m wide, 2m deep, and 9.216m along the driving direction. The load is set to be symmetrically distributed along the driving direction. The connection of each layer is set to be completely continuous. The boundary conditions are set as follows: the lower boundary is three-direction fixed, the z-direction displacement is set on both sides along the driving direction, the x-direction displacement is set on both sides perpendicular to the driving direction, and the load action surface is not restricted. During the division, the load action area and the surface area are meshed and encrypted, and the element type is selected as an 8-node solid stress element (C3D8R).

4 BASIC RESPONSE OF PAVEMENT STRUCTURE

When the aging gradient is considered, Figure 2 shows the horizontal tensile stress response cloud diagram along the driving direction when the traffic load acts on the pavement structure at a speed of 90km/h at noon under normal temperature field. The selected section is the cross section of the pavement when the load reaches the midpoint in the Z direction.

It can be seen from the Figure 2 that the maximum tensile stress is about 0.26 MPa, which appears at the load center of the asphalt surface layer bottom. At this time, the bottom of the base layer is also under a tensile stress with a value of 0.07 MPa, therefore, for horizontal tensile stress, the most unfavorable force position appears at the bottom of the asphalt surface layer.

The curve of stress and strain at the bottom of the asphalt surface layer relative to the load locations is plotted in Figure 3. At the time shown in Figure 2, the curve of the stress and strain at the center of the two-wheel load with depth is plotted in Figure 4 .

As shown in Figure 3, when the load is approaching, compressive stress or compressive strain will appear at the bottom of the asphalt surface layer. When the load continues to approach,

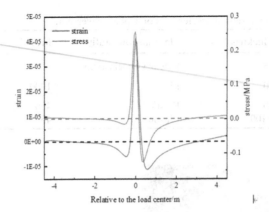

Figure 3. The variation curve of the horizontal stress and strain at the bottom of asphalt pavement with the loading position(left).

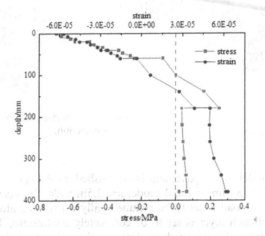

Figure 4. The variation curve of the stress and strain in the center of double wheel load with the direction of depth(right).

both stress and strain become tensile; when it is just above the calculation point, the tensile stress and tensile strain at the calculation point will reach the peak value; as the load moves away, the stress and strain decrease rapidly, the tensile stress/tensile strain is transformed into compressive stress/compressive strain under the action of viscoelastic recovery. Under the action of load and viscoelastic effect, the peak value of compressive strain/compressive stress during load driving away will be slightly larger than the peak value when approaching. After the load drives away, a certain residual stress appeared at the bottom of the asphalt surface layer. During the driving period, the stress/strain changes with the relative position of the load center are not symmetrical. Due to the tension-compression transition process during the load movement, this alternating behavior will lead to fatigue failure under long-term action more likely.

It can be seen from Figure 4 that the stress and strain changing with the depth will also show the phenomenon of tension-compression transition, and there is a sudden change of stress/strain at the interface. When the stress and strain are compressed on the surface of the pavement, as the depth increases, they gradually convert from compressive stress/compressive strain to tensile stress/tensile strain. The tensile stress reaches a peak value of 0.26MPa at the bottom of the asphalt surface layer. When the depth continues to increase, the tensile stress at the interface between the surface layer and the base layer changes from 0.26 MPa to 0.03

66

MPa. Even though the tensile stress value will increase slightly as the depth continues to increase, it cannot reach the tensile stress level at the bottom of the asphalt surface layer; at the interface of the base layer and the bottom base layer, the stress decreases suddenly again. Therefore, the horizontal tensile stress will reach a peak at the bottom of the asphalt surface layer. In the process of tension-compression transition, stress and strain are not synchronized: at the moment shown in Figure 4, the position where the strain is converted is closer to the road surface, which also reflects the viscoelastic properties of the asphalt mixture.

5 CONCLUSION

1. Combined with the existing research and the actual situation, it can be said that the distribution of aging on the pavement is uneven and gradient. Therefore, it is more reasonable to study the performance of asphalt mixture before and after aging under the assumption of gradient aging.
2. The aggravation of the asphalt pavement aging increases the tensile stress at the bottom of the surface layer to a certain extent.
3. In the process of moving loads, the stress and strain at the bottom of the asphalt surface layer will show a "compression-tension-compression" transition as the load approaches and drives away. The horizontal stress at the center of the two-wheel load will also change into tensile as the depth increases.

REFERENCES

[1] CHEN Huaxin, JIANG Yi, LI Shuo, et al. Low-temperature properties of aged asphalt mixtures [J]. Journal of Chang 'an university: natural science, 2010, 30 (1): 1–5. (in Chinese)
[2] MA Li Xing. Research on the Viscoelastic Properties of Asphalt and Asphalt Mixture in the process of ageing [D]. Hu Bei: Wuhan University of Technology, 2012. (in Chinese)
[3] ZHANG Jinxi, JIANG Fan, WANG Chao, et al. Dynamic Modulus Evaluation of Indoor and Outdoor Aging Asphalt Mixture. Journal of Building Materials,2017,20(6): 937–942. (in Chinese)
[4] XU Zhiyuan, LIAN Kefang, XU Wei. Analysis of the Aging Experiment Methods of Asphalt and Asphalt Mixture [J]. Highway Engineering, 2011,36(2): 176–177,182. (in Chinese)
[5] Mirza, M. W., M. W. Witczak. Development of a Global Aging System for Short and Long Term Aging of Asphalt Cements [J]. Journal of the Association of Asphalt Paving Technologists, 1995, 64: 393–430.
[6] ZHANG Qijian. Analysis of temperature and aging Gradient Characteristics of Asphalt Pavements [D]. Zhejiang: Zhejiang University, 2011. (in Chinese)
[7] LU Jun. Research on Asphalt Pavement Aging Behavior and Recycling Technology [D]. Shanxi: Chang 'an university, 2008. (in Chinese)
[8] Luo X, Gu F, Lytton R L . Prediction of Field Aging Gradient in Asphalt Pavements [J]. Transportation Research Record: Journal of the Transportation Research Board, 2015, 2507: 19–28.
[9] HUANG Qinghong. Research on mechanical response of asphalt pavement under non-uniform load [D]. Jiangsu: Nanjing University of Aeronautics and Astronautics, 2008. (in Chinese)
[10] HU Xiaodi, SUN Lijun. Measuring Tire Ground Pressure Distribution of Heavy Vehicle [J]. Journal of Tongji University(Natural Science), 2005, 33(11): 1443–1448. (in Chinese)

Functional Pavements – Chen et al (eds)
© 2021 Taylor & Francis Group, London, ISBN 978-0-367-72610-2

Morphology characteristics of SBS modifier in the asphalt mixture

Xiaoshan Zhang, Yue Xiao & Mujaheed Yunusa
State Key Laboratory of Silicate Materials for Architectures, Wuhan University of Technology, Wuhan, China

ABSTRACT: Fluorescence microscopy technology is used to study the microscopic morphology of SBS modifier in asphalt mixture, and the effect of different SBS content, in the binder, on the microscopic morphology of SBS is presented. Cross section of asphalt mixture was specially designed for the fluorescence microscopy study purpose. Results show that fluorescence microscopy image analysis can be used to determine the SBS morphological characteristics in asphalt mixture, combined with cross section preparation. SBS content mainly affects its volume fraction in asphalt mortar. With the increase of SBS content, SBS average particle area, area weighted average axis ratio and its coefficient of variation increase.

1 INTRODUCTION

Research on the internal morphology and structure of asphalt mixture is of great significance for improving the service performance of asphalt pavement. There are many literatures on the internal characteristics of asphalt mixture, and a series of research results have been obtained. Wang Feng et al. (Wang et al., 2020) used digital image processing technique to quantify aggregate skeleton structure including orientation, distribution and contact properties of aggregate. In collaboration with other laboratories, N.A. Hassan (Hassan et al., 2015) used X-ray computed tomography and optical microscopy to observe the internal structure of the mixture, especially the internal voids and the distribution of recycled asphalt. Vibeke Wegan (Wegan and Bredahl Nielsen, 2000) improved the preparation process of mixture slices, making it possible to observe the real structure of the polymer modified binder directly in the asphalt mixture. Julien Navaro (Navaro et al., 2012) proposed a microscopic observation technique for observing the interface between fresh binder and aged binder in recycled asphalt pavement, then quantified the influence of production parameters, material temperature and mixing time on the surface morphology of binder clusters. In these literatures, image processing and analysis techniques were developed and used to study the air void, aggregate distribution and fresh-aged binder interface. But the modifier morphology in asphalt mixture is seldom studied.

Fluorescence microscopy technology, which is also image based analysis technic, undoubtedly played a vital role during the research of polymer modified asphalt. This study used fluorescence microscopy to characterize the SBS modifier in asphalt mixture with cross section specimens, with the aim of providing a good technic to enhance the quality of SBS modified binder and its corresponding mixture by means of improving the SBS morphologies in the mixture.

2 MATERIALS AND RESEARCH METHODS

Cross section of asphalt mixture with flat surface was prepared by epoxy filling, cutting and polishing. Epoxy was filled into the voids under vacuum condition to avoid micro-cracks during the cutting stage. After the asphalt mixture slices were prepared, fluorescence microscope is used to analyze the dispersion form of the SBS modifier in the mortar transition zone

1. Asphalt mixture;   ▪ Cross section of   • Fluorescence image of
2. Epoxy filling;   asphalt mixture   mortar in asphalt mixture
3. Cutting; 4. Polishing;   with flat surface

Figure 1.   Preparation of cross section for fluorescence image analysis.

between aggregates in asphalt mixture. The microscopic morphology characteristics of SBS modifier in asphalt mixture under different SBS content, used in prepare SBS modified asphalt binder, were studied. At the end, the morphology characterization method of SBS modifier in the asphalt mixture was concluded. The experimental flowchart is shown in Figure 1.

3   CHARACTERIZATION OF SBS MODIFIER IN MIXTURE

MATLAB was used for the performance of RGB image analysis and processing with microscopic fluorescence images of asphalt mixture. Figure 2 presents the SBS morphology characterization steps. The distribution of SBS modifier in mortar area of asphalt mixture is shown in Figure 2(a). This area contains a very small amount of coarse aggregate (dark black-green part), mineral powder, fine aggregate, asphalt, and SBS modifier (bright green-yellow area). Firstly, in order to reduce the influence of background caused by uneven light, the imopen function in MATLAB was used to get the background of analyzed image, and then the imsubtract function was used to remove this back ground, which is presented as Figure 2(c). Secondly, RGB color image analysis and processing was conducted to define the pixel threshold of SBS phase, mineral powder and fine aggregate. As Figure 2(d) indicates, the black and white map of SBS distribution is obtained by self-programming, and the right one shows the mineral powder and fine aggregate's distribution.

Six parameters of area ratio, box dimension, particle area, coefficient of variation in area, the area weighted average axis ratio and the coefficient of variation in axis ratio are used to characterize the SBS microscopic morphologies. And these six indicators are calculated by self-programming in MATLAB. Their representative physical meanings are presented in Table 1. Among them, the more uniform the particle distribution is, the closer the box dimension D of the region is to 2.

4   RESULTS AND DISCUSSION

4.1   *Fluorescence images*

SBS morphologies in AC-13 asphalt mixtures, with SBS content of 2wt%, 3wt%, 4wt%, 5wt% and 7wt% in modified binder, were described in this study. Figure 3 shows the fluorescence microscopy images of such SBS modifier in the asphalt mixture under blue excitation light. Magnification of 200 times was presented. It illustrated that the cross section of asphalt

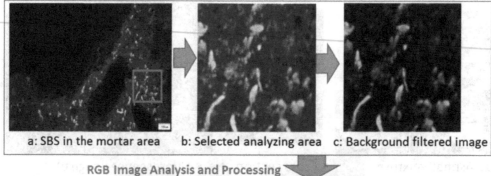

RGB Image Analysis and Processing

a: SBS in the mortar area   b: Selected analyzing area   c: Background filtered image

d: SBS modifier   e: Filler

Figure 2.   Explanation of SBS morphology characterization steps.

Table 1.   Physical meaning of the used microscopic morphological parameters.

Microscopic morphological parameters	Physical meaning
Area ratio	Volume fraction of SBS modifier in modified asphalt
Box dimension (D)	Distribution uniformity of SBS modifier
Area weighted average axis ratio	SBS particle shape
Coefficient of variation in axis ratio	SBS particle shape distribution
Average particle area	SBS particle size
Coefficient of variation in average particle area	SBS particle size distribution

mixture with flat surface can be successfully used in fluorescence microscopy analysis. The distribution of SBS modifier in the mortar area can be clearly detected under blue excitation light. Higher SBS content in asphalt binder will present with more bright green-yellow area in the fluorescence microscopy images, representing higher area ration of SBS modifier in the asphalt mortar from mixture.

4.2   *SBS morphologies*

Morphology parameters were than plotted from Figure 3 for further analysis. Figure 4 presents how the six parameters of SBS modifier vary with modifier content. It can be seen from Figure 4 (a) that with the increase of SBS content, SBS area ratio in asphalt mortar first rises and then decreases, reaching the highest value of 7.859% at 5wt% content, indicating that asphalt mixture preparation process of different SBS content has different effects on area ratio. Figure 4 (b) shows that with the increase of the SBS content, the box dimension goes up a lot first. The reason is that the number of polymer particles in the 3wt% SBS sample significantly increased compared with the 2wt%, which improves the dispersion uniformity of SBS in asphalt mortar. When SBS content increased to

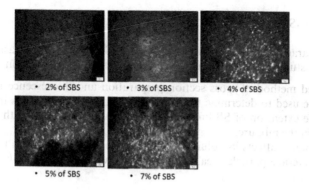

Figure 3. Fluorescence image of SBS modifier in mixture with different SBS ratio (magnification: 200).

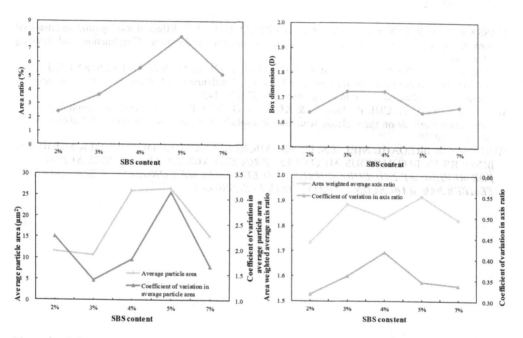

Figure 4. Influence of modifier content on the morphology characteristics.

5wt%, the SBS particle size in asphalt mortar is larger and the distribution is denser compared with the lower content samples. However, due to the dispersion of SBS in asphalt mortar becomes worse, which is related to the determination method of asphalt mortar range, the box dimension decreases significantly.

Figure 4 (c) presents that when the content increases from 3wt% to 7wt%, the particle area first rises sharply and then decreases, so it has no linear relationship with SBS content in the asphalt mortar. The trend of the area variation coefficient is similar to particle area curve. When the mixing content increases from 5wt% to 7wt%, the size difference of SBS particles in asphalt mortar becomes smaller and the dispersion is better. In addition, Figure 4 (d) shows that area weighted average axis ratio in asphalt mortar fluctuates in the range of 1.732 to 1.916, and the coefficient of variation remains at a low level, indicating that the content has little effect on the shape of polymer particles in the asphalt mortar.

5 CONCLUSIONS

A distribution characteristics analyzing method of SBS modifier in the asphalt mixture was introduced in this study. With the discussed result, the following conclusions can be drown:

1) The introduced method of cross section preparation and fluorescence microscopy image analysis can be used to determine the SBS morphology characteristics in asphalt mixture. So that the life extension of SBS asphalt mixture can be forwards with better distributed SBS modifier in the mixture.
2) SBS content mainly affects its volume fraction in asphalt mortar. With the increase of SBS content, SBS average particle area, area weighted average axis ratio and its coefficient of variation increase.

REFERENCES

HASSAN, N. A., KHAN, R., RAABERG, J. & PRESTI, D. L. 2015. Effect of mixing time on reclaimed asphalt mixtures: An investigation by means of imaging techniques. *Construction and Building Materials*.

NAVARO, J., BRUNEAU, D., DROUADAINE, I., COLIN, J., DONY, A. & COURNET, J. 2012. Observation and evaluation of the degree of blending of reclaimed asphalt concretes using microscopy image analysis. *Construction & Building Materials*, 37, 135–143.

WANG, F., XIAO, Y., CUI, P., MA, T. & KUANG, D. 2020. Effect of aggregate morphologies and compaction methods on the skeleton structures in asphalt mixtures. *Construction and Building Materials*, 263, 120220.

WEGAN, V. & BREDAHL NIELSEN, C. 2000. MICROSTRUCTURE OF POLYMER MODIFIED BINDERS IN BITUMINOUS MIXTURES. *PROCEEDINGS OF THE PAPERS SUBMITTED FOR REVIEW AT 2ND EURASPHALT AND EUROBITUME CONGRESS, HELD 20-22 SEPTEMBER 2000, BARCELONA, SPAIN. BOOK 1 - SESSION 1.*

Functional Pavements – Chen et al (eds)
© 2021 Taylor & Francis Group, London, ISBN 978-0-367-72610-2

Effect of ozone aging on asphalt properties

M. Guo, X. Liu & Y. Jiao*

The Key Laboratory of Urban Security and Disaster Engineering of Ministry of Education, Beijing University of Technology, Beijing, China

ABSTRACT: In order to better simulate the actual aging process of asphalt pavement in the laboratory, this study conducted an exploratory study on the ozone aging of asphalt based on the thermal oxygen aging of asphalt and the ozone aging of rubber. Compared with aging in vacuum, dynamic shear rheometer (DSR) was used to study the effect of asphalt on the rheological properties of ozone aging, and the Fourier transform infrared spectroscopy (FTIR) was used to study the microscopic mechanism of the asphalt after ozone aging. The results show that the complex modulus of asphalt increases slightly and the phase Angle decreases slightly after ozone aging. FTIR experiment showed that the position and shape of the absorption peak of the ozone aged asphalt were similar to that of the virgin asphalt, and the carbonyl index and sulfoxide index increased.

1 INTRODUCTION

Asphalt pavement is a kind of pavement type with excellent road performance and widely used. However, during the production and service period of asphalt mixture, a series of physical and chemical reactions will occur due to the comprehensive effect of natural conditions and traffic load, which will change the internal molecular structure and chemical components of asphalt, resulting in the deterioration of pavement performance and the induced pavement diseases (Araujo et al. 2013). Aging is an important factor affecting the performance of asphalt pavement (Xu et al. 2015). Scholars at home and abroad have explored the aging behavior and mechanism of modified asphalt and virgin asphalt. The main reason of asphalt aging is oxidation (Rebelo et al. 2014). Other studies have shown that asphalt aging is mainly caused by high temperature and oxidation reaction (Zhang et al. 2019, Hou et al. 2020). By using infrared spectroscopy to analyze asphalt, the change of oxygen-containing functional groups such as carbonyl group and sulfoxide group can be judged by the position and intensity of absorption peak, so as to explore the aging mechanism.

In recent years, the problem of environmental pollution has become increasingly serious. Due to the increasing automobile exhaust emissions and illegal emission of ozone, the concentration of ozone in the environment is much higher than before, and the ozone concentration in the atmosphere is increasing year by year. However, ozone oxidation is very strong and more active than oxygen, which is an important factor leading to rubber aging in the atmosphere (Radhakrishnan et al. 2006). Unsaturated rubber, such as natural rubber, contains more unsaturated double bonds, but these $C = C$ bonds are very unstable, which can easily react with ozone to form molal ozonation and peroxide, and then regenerate into ozonation (Razumovsky et al. 1986). The bad aging resistance of asphalt is mainly due to the fact that there are more active groups and easily oxidized double bonds in asphalt molecules. In view of the fact that the research on thermal oxidative aging of asphalt has become mature at home and abroad, the research on ozone aging of asphalt is relatively lacking. Therefore, on the

---

* Corresponding author

basis of thermal oxygen aging of asphalt and ozone aging of rubber, this study has carried out exploratory research on the ozone aging of asphalt, and the research results provide certain reference for the laboratory simulation of the actual aging process of asphalt pavement.

2 METHODOLOGY

2.1 Sample preparation

The virgin asphalt was heated to 135°C in the oven to melt completely, and then (50±0.5) g asphalt was poured into each of the five sample trays. Two of them were put into ozone testing machine for 24h and 48h respectively, and the other two were put into vacuum drying oven for 24h and 48h respectively.

2.2 Aging procedure

The OZ-0500AH ozone testing machine produced by Taiwan Gotech Testing Machines is selected. The temperature is 60°C, the humidity is lower than 65%, and the ozone concentration is 100 pphm. In order to detect the influence of ozone on asphalt aging more intuitively, a constant temperature vacuum drying oven filled with nitrogen was selected for comparison, and the experimental temperature and humidity were the same as above.

2.3 Test method

In order to study the influence of ozone on viscoelastic properties of Asphalt during aging, DSR were carried out on five asphalt samples with frequency scanning mode of DSR. The aluminum parallel plate with a diameter of 25 mm was selected. The distance between the upper and lower parallel plates was set at 1 mm, the angular frequency range was set at 100-0.1 rad/s, and the test temperature was 35°C, 45°C and 55°C.

In order to quantitatively characterize the change of chemical structure during asphalt aging, FTIR was selected. The change rule of the content of functional group can be quantitatively analyzed by the area of characteristic peak of functional group in infrared spectrum.

3 RESULTS AND DISCUSSION

3.1 DSR experiment results

DSR is an instrument for measuring the high temperature stability of asphalt binder in Superpave system. The complex shear modulus G* and phase angle were measured by DSR at a certain speed to characterize its viscoelastic properties. G* is an index to characterize the total deformation resistance of asphalt binder. The higher the value, the stronger the resistance to deformation at high temperature. The phase angle reflects the ratio of viscosity and elasticity of asphalt binder. The higher the value, the more likely it is to produce high temperature permanent deformation.

Based on 45°C, the modulus frequency curve and phase angle frequency curve at each temperature are translated according to the principle of time temperature equivalence (Cardone et al. 2017), and the corresponding main curves are obtained.

It can be seen from Figure 1 (left) that the frequency is positively correlated with the complex modulus. Compared with the virgin asphalt, the complex modulus of asphalt has no obvious change after 24h and 48h under vacuum condition. The results show that 60°C has no effect on the deformation resistance of asphalt under vacuum condition; the complex modulus increases slightly after ozone aging, which indicates that ozone aging can improve the stiffness of asphalt and enhance the anti-deformation ability of asphalt. Among them, the complex modulus increased more obviously after 48h than 24h. It can be seen from Figure 1 (right) that the frequency is negative correlated with the phase angle. Compared with the virgin

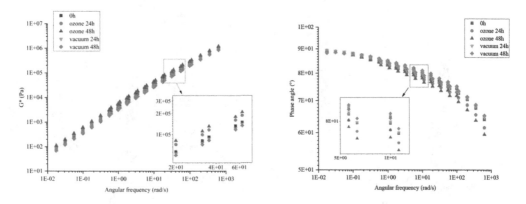

Figure 1.   Modulus master curve (left) and phase angle master curve (right) of aging 0h/24h/48H.

asphalt, the change of phase angle is not obvious after 24h and 48h under vacuum condition. It shows that 60°C has no effect on the deformation resistance of asphalt under vacuum condition. After ozone aging, the phase angle decreases slightly compared with the virgin asphalt, which indicates that ozone aging can increase the elastic composition of asphalt, improve the elastic recovery performance and anti deformation ability.

3.2   FTIR experiment results

The FTIR test is mainly used to analyze the changes of molecular structure and functional groups of asphalt. In this study, the infrared absorption spectra of asphalt samples under vacuum and ozone conditions were tested, as shown in Figure 2.

It can be seen from Figure 2 that compared with the virgin asphalt, the positions and shapes of absorption peaks after 24h and 48h are similar under vacuum or ozone conditions, indicating that the chemical composition of asphalt after ozone aging has not changed significantly; there are no different absorption peaks, indicating that no new functional groups have been formed.

In order to further explore the influence of aging conditions on the functional groups of asphalt, the principle of infrared spectrum quantitative analysis of asphalt in this part is based on Lambert Beer law, using baseline method, and using the analysis function of OMNIC software for data analysis. The carbonyl index and sulfoxide index of asphalt before and after aging are shown in Figure 3.

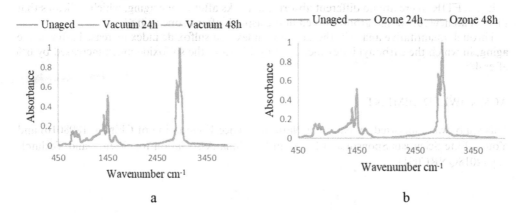

Figure 2.   Infrared spectra of aging 0h/24h/48H and a: Under vacuum condition; b: Under ozone condition.

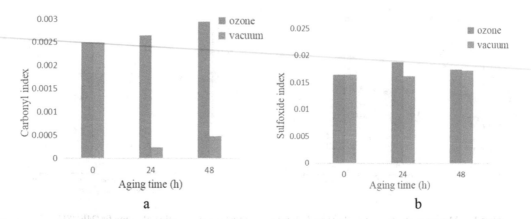

Figure 3. Functional group index of asphalt after different aging time and a: Carbonyl index; b: Sulfoxide index.

It can be seen from Figure 3a that compared with the virgin asphalt, the carbonyl index has a downward trend after vacuum aging; among them, the carbonyl index decreases by 90% after 24h and 80% after 48h. After ozone aging, carbonyl index increased by 6% at 24h and 18% after 48h.

It can be seen from Figure 3b that compared with the virgin asphalt, the sulfoxide index decreases by 2% after 24h of vacuum aging and increases by 4% after 48h. After ozone aging, the sulfoxide index increased by 14% at 24h and 6% after 48h.

The above results show that compared with the virgin asphalt, the carbonyl index and sulfoxide index increase after ozone aging, indicating that the aging degree is deepened.

4 CONCLUSION

Based on the testing and analysis presented herein, the conclusions of the study are summarized as follows:

DSR test shows that the complex modulus has no change in vacuum compared with the virgin asphalt; after ozone aging, the complex modulus increases slightly. The results show that ozone aging can improve the stiffness of asphalt and enhance the anti-deformation ability of asphalt.

The infrared spectrum test results show that the position and shape of the absorption peak of the asphalt after ozone aging are similar to that of the virgin asphalt, indicating that the chemical composition of the asphalt after ozone aging has not changed significantly.

In the FTIR, there are no different absorption peaks after ozone aging, which indicates that there is no new functional group formed in the asphalt after ozone aging.

Through quantitative analysis, the carbonyl index and sulfoxide index increased after ozone aging, in which the carbonyl index increased by 18% and the sulfoxide index increased by 6% after 48h.

ACKNOWLEDGEMENT

This study was supported by National Natural Science Foundation of China (51808016) and Young Elite Scientists Sponsorship Program by China Association for Science and Technology (2018QNRC001).

REFERENCES

Araujo, M. & Lins, V. & Pasa V. & Leite, L. 2013. Weathering aging of modified asphalt binders. *Fuel Processing Technology*, 115:19–25.

Graziani, A. & Canestrari, F. & Cardone, F. & Ferrotti, G. 2017. Time–temperature superposition principle for interlayer shear strength of bituminous pavements. *Road Materials and Pavement Design*. 18: 12–25.

Hou, X. & Liang, B. & Xiao, F. & Wang, J. & Wang, T. 2020. Characterizing asphalt aging behaviors and rheological properties based on spectrophotometry. *Construction and Building Materials*, 256: 119401.

Radhakrishnan, C. & Alex, R. & Unnikrishnan, G. 2006. Thermal, ozone and gamma ageing of styrene butadiene rubber and poly(ethylene-co-vinyl acetate) blends. *Polymer Degradation and Stability*, 91(4): 902–910.

Razumovsky, S.D. & Podmasteriyev, V.V. & Zaikov, G. 1986. Kinetics of the growth of cracks on polyisoprene vulcanizates in ozone. *Polymer Degradation and Stability*, 16(4): 317–324.

Rebelo, L.M. & De Sousa, J.S. & Abreu, A.S. & Baroni, M.P.M.A. & Alencar, A.E.V. & Soares, S.A. & Mendes Filho, J. & Soares, J.B. 2014. Aging of asphaltic binders investigated with atomic force microscopy. *Fuel*, 117(part A): 15–25.

Xu, O. & Cong, L. & Xiao, F. & Amirkhanian, S.N. 2015. Rheology investigation of combined binders from various polymers with GTR under a short term aging process. *Construction and Building Materials*, 93: 1012–1021.

Yang, Z. & Zhang, X.N. & Yu, J.M. & Xu, W. 2019. Effects of Aging on the Multiscale Properties of SBS-Modified Asphalt[J]. *Arabian Journal for Science and Engineering*, 44(5): 4349–4358.

Zhang, P. & Guo, Q. & Tao, J. & Ma, D. & Wang, Y. 2019. Aging mechanism of a diatomite-modified asphalt binder using Fourier-Transform infrared (FTIR) spectroscopy analysis. *Materials*, 12(6): 988.

Functional Pavements – Chen et al (eds)
© 2021 Taylor & Francis Group, London, ISBN 978-0-367-72610-2

Evaluation of the effects of different modifiers on the rheological properties of bitumen

Bangwei Wu, Xing Wu, Peng Xiao, Aihong Kang & Zhengguang Wu
College of Civil Science and Engineering, Yangzhou University, Yangzhou, China

ABSTRACT: More and more different polymer modifiers are being used in China nowadays. In order to evaluate the effects of them on the rheological properties of bitumen to better guide the application of these modifiers, this paper chose three kinds of widely used polymer modifiers, namely high viscosity modifier (HVM), anti-rutting agent (ARA), and high modulus modifier (HMM) to modify the pure bitumen. The rheological properties of pure and modified bitumen were tested with dynamic shear rheometer (DSR), and the rutting factor and phase angle of different bitumen samples were compared and analyzed. Fourier Transform Infrared Spectrometer (FTIR) test was also adopted to analysis the chemical functional groups of bitumen, HVM, ARA, HMM. The results show that these modifiers increase the elasticity and the anti-rutting ability of the pure bitumen, and different modifiers have different effects on the rheological performance of the bitumen. The pure bitumen and the modifiers share some similar chemical functional groups.

Keywords: bitumen modifiers, rheological properties, rutting factor, phase angle, FTIR

1 INTRODUCTION

In China, bitumen pavement diseases such as rutting, raveling, and cracking [1,2] are quite common because of the overloading traffic [3]. Therefore, researchers are trying to use different modifiers, such as high viscosity modifier (HVM), anti-rutting agent (ARA), and high modulus modifier (HMM), to enhance the pavement performances [4–6]. Although the properties of bituminous mixtures with these additives have been studying a lot, the research on the performances of these corresponding modified bitumen is very limited. Thus, this paper conducted research about the properties of the bitumen modified by these additives to better explain the strengthening mechanism of these additives.

There are several researches show that the rheological performance of the bitumen can be used to better evaluate the properties of the bitumen materials than the normal evaluation indexes such as softening point and penetration etc. [7–9] Fan et al. [7] found that the normal evaluation indexes of bitumen such as softening point, penetration are not accurate enough because the results will be affected by many factors, including the testing time, the heating speed and so on. Morea et al. [8] also found that the performance of bituminous mixtures depends on the rheological properties of bitumen mortar.

Therefore, this paper conducted experiments about the rheological properties of the bitumen modified by HVM, ARA and HMM using the dynamic shear rheology test [10]. Different kinds of modified bitumen were compared with each other and with the bitumen. FTIR test [11] was used to study the chemical functional group of the different bitumen samples. This paper has a certain significance on the guidance of the application of these modifiers.

2 MATERIALS AND TEST METHODS

2.1 *Raw materials*

Bitumen used in this paper is provided by Jiangsu Tiannuo Road Materials Technology Co., Ltd. The properties of the bitumen are shown in Table 1 and the performance indexes meet the requirements of Chinese specification. HVM is made by China Road High Tech (Beijing) Highway Technology Co., Ltd., ARA and HMM used in this paper is provided by Faon Transportation Technology (Shanghai) Co., Ltd.

2.2 *Methods*

2.2.1 *Sample preparation*

In this paper, the dosage of polymer modifiers is 8% of the weight of bitumen. Existing research [12] show that the best ratio of the HVM and the bitumen is 8:92 which is very close to 8%. The bitumen and the modifiers are blended at 175°C and then are stirred at a speed of 1200r/min for 60min. The test samples have four combinations, bitumen, bitumen + HVM, bitumen + ARA and bitumen + HMM, and they are marked as A, B, C and D respectively.

2.2.2 *Test methods*

The rheological properties of different bitumen samples were tested using the dynamic shear rheometer produced by Malvern Instruments Co., Ltd., Malvern, UK. The test parameters were in accordance with the requirements of ASTM D7175. The diameter of the parallel plate is 25mm, and the gap value is 1000μm. The angular velocity is 10 rad/s, and the strain is set as 12%. The test temperatures are 52°C, 58°C, 64°C, 70°C and 76°C. Phase angle is used to describe the elasticity of the samples and rutting factor is used to reveal the anti-rutting ability of the samples. FTIR Spectrometer made by Perkinelmer Instruments Co., Ltd. is used to test the chemical functional groups of different modifiers and bitumen samples.

3 RESULTS AND DISCUSSION

3.1 *Rheological properties*

The test results of rheological properties are shown in Figure 1 and Figure 2. the two figures present the phase angle and rutting factor of bitumen samples respectively. It can be seen from Figure 1 that the phase angle of bitumen decreases after adding the modifiers, which means that the elasticity of the bitumen increases when combining with the modifiers. The elasticity of sample B (bitumen +HVM) increase the most, and this means that HVM could increase the elasticity recovery ability of bitumen the most. When the temperature is between 52°C to 70°C, the phase angle of sample B keeps decreasing, and it reaches the lowest point at 70°C. The phase angle of sample C (bitumen +ARA) is the biggest at 58°C, and when the temperature is from 58°C to 76°C, the phase angle decrease (elasticity increase). The changing value of D (bitumen +HMM) is similar to that of A (bitumen), yet, the phase angle of D is lower.

The reasons for these phenomena can be explained as follows. Firstly, the HVM combined with the bitumen (it has been melted before in the sample preparation process and it is

Table 1. Properties of bitumen.

Index	Penetration (25°C, 100g, 5s), 0.1mm	Softening point ($T_{R\ \&\ B}$), °C	Ductility (15°C, 5cm/min), cm	Dynamic viscosity (60°C), Pa·s
Value	65	53	102	182
Requirements	60~80	≮46	≮15	≮180

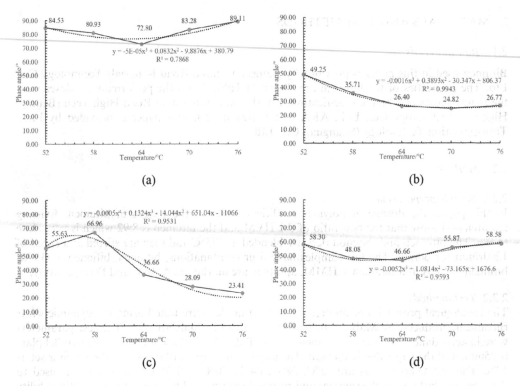

Figure 1.   Phase angle: (a) bitumen; (b) bitumen +HVM; (c) bitumen +ARA; (d) bitumen +HMM.

hardened after the sample preparation) will be further melted when the temperature rises, thus it can combine tighter with the bitumen. Secondly, the maximum value of the phase angle of sample C at 58°C means that the ARA (melted before and get hardened after the sample preparation) do not melt when the temperature is below 58°C, and it will not exert its modifying effect on the bitumen. However, when the temperature is from 58°C to 76°C, the ARA will be further melted which will increase the elasticity of bitumen. Thirdly, the same changing pattern of sample D as that of the sample A means that the HVOA and bitumen exhibit a similar property. Yet, HMM also increase the elasticity of bitumen. The above results show that different modifiers have different effects on the elasticity and the temperature sensitivity of bitumen.

Figure 2 shows that the modifiers increase the anti-rutting ability of the bitumen. The rutting factor of sample B is very stable when the temperature changes, which means that the HVM has a relative stable property. As for sample C, the rutting factor at 52°C is the highest. This is mainly because that ARA do not melt well when the temperature is below 58°C, the highest rutting factor at 52°C is mainly caused by the ARA itself. When the temperature reaches 58°C, the ARA is softer (a little bit melted), and the bitumen begin to show its effect on the rutting factor. When the temperature is between 58°C to 76°C, the ARA begin to be softer and will make the modified bitumen transfers the stress more effectively, which will make the rutting factor bigger. This phenomenon means that the ARA is more suitable in areas where the temperature is high. The rutting factor of sample D is the biggest when the temperature is at 58°C, and when the temperature is between 58°C to 76°C, the rutting factor of sample D keeps decreasing. Generally, the modifiers and the bitumen will have a synergic effect on the modified bitumen. The complexity of different modifiers and the bitumen make the rheological performances of these test samples different.

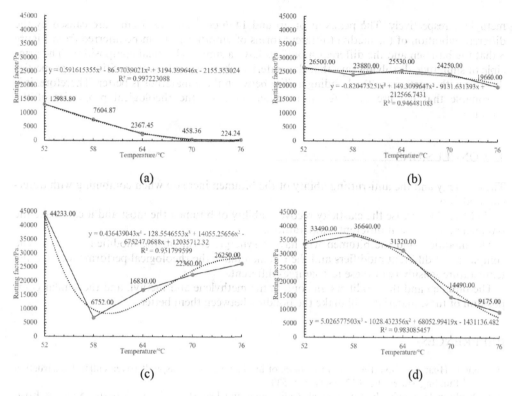

(a)

(b)

(c)

(d)

Figure 2.   Rutting factor: (a) bitumen; (b) bitumen +HVM; (c) bitumen +ARA; (d) bitumen +HMM.

3.2   *FTIR test*

The results of FTIR test is shown in Figure 3. Different absorption peaks in Figure 3 are caused by different chemical functional groups. The absorption peaks at 2920 cm^{-1}, 2850 cm^{-1} are caused by the antisymmetric stretching vibration, symmetric stretching vibration of

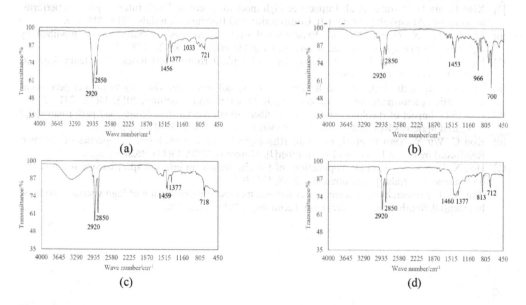

(a)

(b)

(c)

(d)

Figure 3.   FTIR spectrum: (a) bitumen; (b) HVM; (c) ARA; (d) HMM.

methylene respectively. The peaks at or around 1456 cm^{-1} and 1377 cm^{-1} are caused by the different vibration of the methyl (different forms of vibration). It can be inferred from Figure 3 that the bitumen and the different modifiers have a similar chemical composition. The principle of similitude dissolving says that the materials sharing similar compositions or properties dissolve with together, so the bonding ability between these materials is better. Therefore, it is reasonable that the use of different modifier increases the rheological properties of the bitumen.

4 CONCLUSIONS

The elasticity and the anti-rutting ability of the bitumen increase when combining with different modifiers.

HVM could increase the elasticity recovery ability of bitumen the most and it can make the bitumen more stable at different temperatures.

The modifiers and the bitumen will have a synergic effect on the modified bitumen. The complexity of different modifiers and the bitumen make the rheological performances and the temperature sensitivity of these test samples different.

The bitumen and the modifiers all contain the methylene and methyl, and the similar composition of these materials will make the bonding between them better.

REFERENCES

[1] Xu T, Huang X. Investigation into causes of in-place rutting in asphalt pavement[J]. Construction and Building Materials, 2012, 28(1): 525–530
[2] Bendtsen H, Kohler E, Lu Q, et al. Californian and Danish Study on Acoustic Aging of Road Pavements[C]// Compendium of Papers, TRB 89th Annual Meeting, Washington, DC. 2010.
[3] Rys D, Judycki J, Jaskula P. Analysis of effect of overloaded vehicles on fatigue life of flexible pavements based on weigh in motion (WIM) data[J]. International Journal of Pavement Engineering, 2016, 17(8): 716–726.
[4] Luo Y, Zhang K, Li P, et al. Performance evaluation of stone mastic asphalt mixture with different high viscosity modified asphalt based on laboratory tests[J]. Construction and Building Materials, 2019, 225: 214–222.
[5] Xiao F, Ma D, Wang J, et al. Impacts of high modulus agent and anti-rutting agent on performances of airfield asphalt pavement[J]. Construction and Building Materials, 2019, 204: 1–9.
[6] Ma T, Ding X, Zhang D, et al. Experimental study of recycled asphalt concrete modified by high-modulus agent[J]. Construction and Building Materials, 2016, 128: 128–135.
[7] Fan, L., Lin, J. T., & Li, Y. Z. Recognition of asphalt routine test index. Petroleum Asphalt, 2019, 33(01),60–66. (In Chinese)
[8] Morea F, Agnusdei J O, Zerbino R. The use of asphalt low shear viscosity to predict permanent deformation performance of asphalt concrete[J]. Materials and structures, 2011, 44(7): 1241–1248.
[9] Guo D D. Impact of rheological properties of fiber asphalt mortar on mixture road performance[C]// Advanced Materials Research. Trans Tech Publications Ltd, 2013, 734: 2287–2291.
[10] Kou C, Wu X, Xiao P, et al. Physical, Rheological, and Morphological Properties of Bitumen Reinforced by Basalt Fiber and Lignin Fiber[J]. Materials, 2020, 13(11): 2520.
[11] Hou X, Lv S, Chen Z, et al. Applications of Fourier transform infrared spectroscopy technologies on bitumen materials[J]. Measurement, 2018, 121: 304–316.
[12] Zhu, M. The research on performance evaluation method and application of high viscosity modified bitumen[D]. South China University of Technology, 2015, Shanghai.

Functional Pavements – Chen et al (eds)
© *2021 Taylor & Francis Group, London, ISBN 978-0-367-72610-2*

Optimization of low-temperature vacuum evaporation test for emulsified asphalt

J. Shi & C. Li
Research Institute of PetroChina Fuel Oil Company Limited, Beijing, China

H. Xu, X. Yang & H. Li
School of Transportation Science and Engineering, Harbin Institute of Technology, Harbin, Heilongjiang Province, China

ABSTRACT: Emulsified asphalt is a common material for pavement engineering. Obtaining emulsified asphalt evaporation residues is the precondition for evaluating the field performance of emulsified asphalt which is essential for proper use. In this study, to obtain unaged evaporation residues, a test method at low temperature and low pressure is developed and optimized. The influence of five factors (vacuum degree, container material, stirring method, sample quality, and test temperature) on test time and the test error is analyzed to improve the procedure. All the factors have a remarkable influence on test time except stirring method, while only stirring method, container and temperature affect the error of results obviously. The criterion for judging the end of the test is also confirmed.

1 INTRODUCTION

Emulsified asphalt has been widely used in pavement construction or maintenance. In construction, the evaporation residues containing asphalt and emulsifier adhesive aggregates or layers as the binder. Thus, the properties of residues relate to the performance of emulsified asphalt directly, and it's essential to ensure high correlation of residue properties between laboratory and construction.

At present, the methods for obtaining evaporation residue of emulsified asphalt are mainly divided into four categories: direct heating evaporation, distillation, oven heating evaporation, and low-temperature evaporation (Niu et al. 2016). Direct heating evaporation is the simplest method using an electric furnace or gas stove to heat samples, but operators have to judge if the test has been finished by themselves, resulting the risk of overheating (JTG E20-2011). Oven heating evaporation specifies the temperature and test time, but the risk of aging still exist due to the high temperature at 163°C (ASTM D6934 2008). To obtain unaged evaporation residue, low-temperature evaporation is developed, evaporate the sample at a temperature range from 25°C to 60°C (BS 2000-493 2002, AASHTO PP72 2011). Due to the decrease of temperature, test time is extended and only few residues can be obtained from each test. Moreover, distillation is also a regular method, divided into two methods namely distillation method at 240°C and low-temperature vacuum evaporation at 135°C, also having the risk of aging (ASTM D6997 2004).

High temperature is the major source of the aging during the test. To reduce the aging further, a test method at lower temperature is required. In this study, a method is developed to measure the solid content and obtain the evaporation residues of emulsified asphalt at low temperature by vacuum drying oven. Tests are carried out with different sets of conditions including vacuum degree, container material, stirring method, sample mass and test temperature. The test time and test error are measured as indicators to investigate the influence of these factors to optimize experimental conditions. In the end, test time of samples with different solid content is quantified to confirm the criterion for judging the end of the test.

2 MATERIALS AND METHODS

2.1 Sample preparation

Matrix asphalt of penetration 70 and four different types of emulsifier namely two slow-setting emulsifiers (LH-206 and MQ3), a medium-setting emulsifier (LH-101Z) and a slow-setting emulsifier (803Y) were used for the preparation of emulsified asphalt. The manufacturers, types and recommended usage of emulsifiers are listed in Table 1, and the performances of emulsified asphalt are listed in Table 2, actual solid content is measured before each test.

2.2 Low-temperature vacuum evaporation test

Reduce the temperature is effective to avoid aging during the test, but the test time would be extended. As a solution, sample is heated in a vacuum environment to accelerate the evaporation in this study. Firstly, a primary test method had been developed which would be optimized in this paper. In the primary method, 100g emulsified asphalt sample is weighed into 1000ml metal beakers containing a glass stirring rod. In the heating process, sample may boiling suddenly due to the increasing temperature, so a preheating procedure is needed to heat sample to the test temperature before raising vacuum. After preheating, vacuum is adjusted every 30min to keep a mild boiling of sample. The mass loss is weighed after each adjusting. When the mass loss is less than 0.2g per 30min, the test should be finished, then the solid content and evaporation residues are obtained.

In this study, tests were carried out in conditions with differences in vacuum degree, container material, stirring method, sample quality, and test temperature to investigate the influence of these factors on test time and test error. A vacuum drying oven produced by Shanghai Shuli Instrument Limited Company was used to heat samples in vacuum environment, has a temperature range for room temperature (or 10°C) to 200°C and vacuum range for ≤133Pa. To compare the test error between different conditions, the solid contents obtained by ASTM D6934 before each test were regarded as the standard value in this paper.

Table 1. The properties and recipes of emulsifiers.

Emulsifier	Manufacturer	Ingredient	Usage/%	Solid content/%
LH-206	ArrMaz Custom Chemicals	Cationic-acidamides	2	60
MQ3	MeadWestvaco	Cationic-acidamides	2	60
LH-101Z	ArrMaz Custom Chemicals	Cationic-quaternary Ammonium salts	0.8	60
803Y	Tianjin Kangzewei	Anion	2	50

Table 2. The performances of emulsified asphalt.

Emulsified asphalt	Demulsification speed	Sieve test(1.18mm)/%	Storage stability(1d)/%
LH-206	Slow setting	0.1	1
MQ3	Slow setting	0	0.7
LH-101Z	Medium setting	0.05	0.3
803Y	Slow setting	0	0.5

3 RESULTS AND DISCUSSION

3.1 *Vacuum control*

Vacuum has a significant impact on the boiling point of water. In general, high vacuum promotes the evaporation of emulsified asphalt. During the test, vacuum decreases as the water evaporating to air in the hermetic oven. Therefore, vacuum should be adjusted regularly. To obtain the best interval, two tests with adjusting intervals of 30min and 10min respectively were carried out. Both 803Y and LH-206 emulsified asphalt were tested.

The residue content variation curves obtained by two methods are illustrated in Figure 1, a similar trend can be observed. It indicates that the evaporation process of the emulsified asphalt was divided into two obvious stages. For LH-206 emulsified asphalt with 10min interval, the first stage from beginning to 3.5h and the second stage from 3.5h to ending are distinguished by the inflection point at 3.5h, similarly curves of LH-206 with 30min interval, 803Y with 10min and 30min interval had the inflection points at 4.0h, 3.5h, 4.5h respectively. Both stages were approximately linearly changing, indicating that the evaporation rate remained at a constant value. The curve declined faster in the first stage, meaning a stronger evaporation conducted in the first stage. Viewing at the residue content between two stages, it can be observed that the main part of the total evaporation conducted in the first stage, only a little water evaporated in the second stage. The reason is that the water in emulsified asphalt is composed by the free water between micelles and the bound water adsorbed on the surface of micelles. The free water has weak interaction between molecules while the bound water has strong interaction. In the evaporation process, firstly, the free water evaporated from emulsion rapidly in the first stage due to the weak interaction. When free water had been evaporated, bound water began to evaporate at a lower speed.

10min interval always showed curves under the 30min ones. The difference got enlarge as the tests went on. For 803Y emulsified asphalt, the difference between two curves increased from 2% to 5.3% as the test time went from 1.5h to 3.5h. It suggested that adjusting vacuum frequently will keep high vacuum in the oven and reduce the test time. Thus, 10min interval should be selected.

In the second stage, the disparity of evaporation rate between the two methods gradually disappeared, same to the decrease rate of vacuum. That means it is not necessary to adjust vacuum frequently in the second stage. Thus, a variable-frequency vacuum regulation was ruled: adjust vacuum every 10min from the beginning till the vacuum decreases less than 2 kPa per 10 min, then adjust vacuum every 30min.

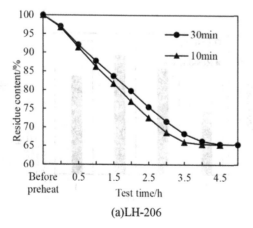

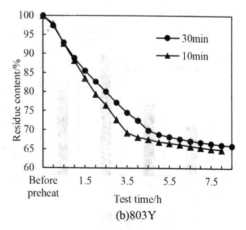

(a)LH-206   (b)803Y

Figure 1. The residue content curves of LH-206 and 803Y with different adjusting interval.

3.2 Sample containing

The container, sample mass, stirring method have influence on testing time and the validity of results. Container is the medium for heat transporting to sample, affect the rate of heat transportation. The sample mass influence how much heat is needed, and stirring makes different parts of sample evaporate evenly. To obtain the optimum container, sample mass and stirring method, 803Y and LH-206 emulsified asphalt were tested under conditions illustrated in Table 3. The conditions were combined to four groups, including group 2 as the control group and group 1, 3, 4 as experimental groups. The vacuum adjusting method described in 3.1 was used in this section.

The residue content and testing time of four groups were illustrated in Figure 2. To analyze the influence of stirring, group 1 can be contrasted with group 2 with difference in whether to stir. Group 1 yielded less error with less test time. For LH-206 emulsified asphalt, the test time and test error were 5h and 1.0%, better than group 2 for 1h and 0.6%, respectively. For 803Y emulsified asphalt, group 1 was better than group 2 for 0.5h and 0.6%. Therefore, stirring regularly helped to reduce test time and test error.

The influence of the container was analyzed by group 2 and group 3 which used 1000ml metal beakers and 500ml glass beakers, respectively. Comparing the two groups, group 2 always took less test time and yielded less errors than group 3. For LH-206 emulsified asphalt, comparing to group 3, the test time and test error of group 2 reduced for 2h and 1%. For 803Y emulsified asphalt, the reducing range was 2h and 0.2%. It showed that using 1000 metal beaker to contain the sample reduced test time and test error. 1000ml metal beaker could be the better container.

In general, more sample leads to overtime of test. The sample mass is 100g for group 2 and 200g for group 4. As we have speculated, comparing group 2 and group 4 of LH-206

Table 3. The groups of test conditions.

Temperature/°C	Group	Emulsifier	Whether to stir	Container
90°C	1	LH206	yes	1000ml metal beaker
	2		no	1000ml metal beaker
	3		no	500ml glass beaker
	4		no	1000ml metal beaker
	1	803Y	yes	1000ml metal beaker
	2		no	1000ml metal beaker
	3		no	500ml glass beaker
	4		no	1000ml metal beaker

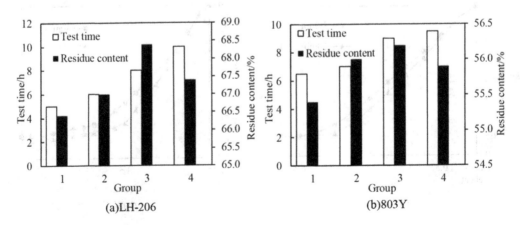

(a)LH-206　　　　(b)803Y

Figure 2. The residue content and testing time of four condition groups.

86

emulsified asphalt, the test time extended for 4h and test error increased by 0.1% due to the sample mass changed from 100g to 200g. For 803Y emulsified asphalt, the increase in test time was 2.5h, but there was a small decrease in test error by 0.1%. It indicated that sample mass affect test time, but have little influence on test error.

From the above, container have an obvious impact on both test error and test time, while stirring and sample mass mainly affect test error and test time, respectively. To conduct the test in best conditions, the stirring procedure, 1000ml metal beaker and sample mass for 100g should be selected.

3.3 Test temperature

The temperature has an important influence on the test process and the performance of the evaporation residue. High test temperature increases the evaporation rate and reduce the error of results, but brings out the risk of thermo-oxidative aging; on the other hand, testing at low temperature avoids aging, but extends test time and leaves more water in the residues. To determine the optimum test temperature, 803Y, LH-206, MQ3, LH-101Z emulsified asphalt were tested under the conditions obtained in 3.2 at four temperatures (80°C, 90°C, 100°C and 110°C). The test time and test error of the four emulsified asphalts at different temperatures were illustrated in Figure 3 respectively. Figure 4 showed the variation of residue content during the test.

As shown in Figure 3(a), the test time of the four emulsified asphalts decreased as the temperature increased. While the test temperature rose from 80°C to 110°C, the test time of four emulsified asphalts reduced 1h, 3h, 1.5h and 1h, respectively. According to Figure 4, at high temperature, the time to reach the bound point of two stages increased for 1.5h, 2.5h, 1h, 2h. It indicated that high temperature reduced the time of stage1, but may had no effect, or even opposite effect to stage 2. That means the free water evaporated out early at high temperature, and the evaporation of bound water was not affected by temperature. High temperature reduces the test time by promoting the evaporation of free water.

According to Figure 3(b), the test error increased as the temperature rose. When the test temperature rose from 80°C to 110°C, the test error of the four emulsified asphalts decreased by 0.4%, 0.6%, 0.1% and 0.5%. Raising temperature helped to reduce the test error.

Figure 5 showed the covariance of test error and test time with temperature. Both test time and test error had a negative covariance with temperature, means the negative correlation between temperature and test time or test error. LH-206 emulsified asphalt had the minimum covariance, indicating that LH-206 emulsified asphalt was most affected by temperature.

Overall, raising the test temperature helped to improve test efficiency and validity. However, considering the thermo-oxidative aging at high temperature, 90°C is selected as the test temperature to obtain the residues having same properties in construction.

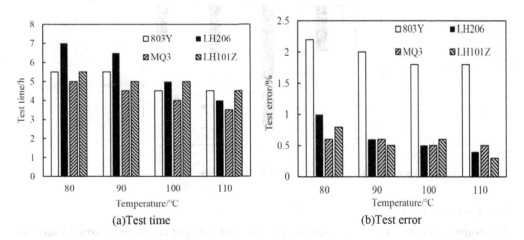

Figure 3.   The test time and error of four types of emulsified asphalt at different temperatures.

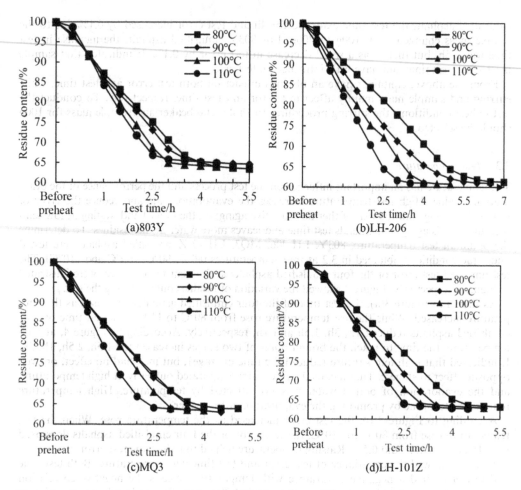

Figure 4. Residue content curves of four emulsified asphalts at different temperatures.

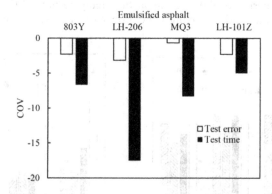

Figure 5. Covariance of test error and test time with temperature.

3.4 *Criterion of test ending*

The evaporation of water is a continuous process, and the light components could volatilize from residues after the water has evaporated. Therefore, it is necessary to define the criteria of whether evaporation is complete. LH-206 and 803Y emulsified asphalt samples with different solid content were prepared and tested under the determined test conditions, then the test time were counted. The results were shown in Table 4:

Table 4. The results of samples with different solid content.

Emulsifier	Actual solid content/%	Test time/h	Residue content/%
803Y	62.7	6.0	64.7
	55.2	6.5	56.7
MQ3	62.4	4.5	63.0
	54.5	6.0	55.1

It could be seen from Table 4 that the test time of 62.7% solid content 803Y emulsified asphalt increased for 0.5h compared with 55.2% solid content, and test time of 62.4% solid content MQ3 emulsified asphalt increased for 1.5h compared to 54.5% solid content. It was obvious that samples with high solid content needs longer test time, and different types of emulsified asphalts with similar solid content may have different test times. Therefore, it is not appropriate to use the same test time uniformly. By testing, when drying the completely evaporated residues in the vacuum drying oven, the mass loss was about 0.2g every 30min. So, the criterion for judging the end of the test was determined: when the mass loss of sample is less than 0.2g per 30min, the test is finished.

4 CONCLUSIONS

In this study, an evaporation test in vacuum environment was performed on four kind of emulsified asphalt in different conditions to analyze the influence of vacuum, stirring, container, sample mass and temperature, some important points can be summarized as follows.

The evaporation process is divided into two stages, and the stage 1 have a faster evaporation rate. Evaporation get faster at high vacuum. The container and sample mass have a remarkable impact on test time while the container and stirring method mainly effects test error. High temperature reduces the test time and test error. The test time of various types and content emulsified asphalt is quite different and causes difficulty to specify the test duration.

In conclusion, the optimal conditions are:

Contain 100g sample in a 1000ml metal beaker, put the container in vacuum drying oven at 90° C and adjust the vacuum at 10min(beginning) or 30min(ending) intervals. Samples should be stirred every 30min.When the mass loss is less than 0.2g per 30min, the evaporation test is finished.

REFERENCES

AASHTO. 2011. Recovering Residue from Emulsified Asphalt Using Low-Temperature Evaporative Techniques. AASHTO PP72. Washington, DC: AASHTO.
ASTM. 2004. Standard Test Method for Distillation of Emulsified Asphalt. ASTM D6997.
ASTM. 2008. Standard Test Method for Residue by Evaporation of Emulsified Asphalt. ASTM D6934.
ASTM. 2009. Standard Test Methods and Practices for Emulsified Asphalts. ASTM D244.
ASTM. 2009. Standard Test Method for Determination of Residue of Emulsified Asphalt by Low Temperature Vacuum Distillation. ASTM D7403.
BSI. 2002. Petroleum Products-Bitumen and bituminous binders-Recovery of binder from bitumen emulsions by evaporation. BS 2000-493.
JTG E20-2011 Standard Test Methods for Evaporation Residue Content of Emulsified Asphalt. *JTG E20-2011 T0651-1993, Standard Test Methods of Bitumen and Bituminous Mixtures for Highway Engineering*, Beijing: Ministry of Transport of the People's Republic of China.
Niu, X.W. 2016. Research on Obtaining Methods for Asphalt Emulsion Residues. *Modern Transportation Technology* 13(3):1–4.

Functional Pavements – Chen et al (eds)
© *2021 Taylor & Francis Group, London, ISBN 978-0-367-72610-2*

Morphology and size analysis of polymer phase in PMA based on 3D reconstruction

X. Zhu, G. Hao & Y. Wang
The Hong Kong Polytechnic University, Hung Hom, Hong Kong

ABSTRACT: The morphology and size of polymer phase affect the storage stability and engineering performance of polymer modified asphalt (PMA). Polymer phase in PMA is commonly investigated by using fluorescence microscopy (FM). However, some studies suggest that the two-dimensional (2D) images obtained from FM cannot reveal the real internal structure of polymer phase in PMA, because 2D images are formed by the projection of the three-dimensional (3D) polymer phase on a 2D plane. Using confocal fluorescence microscopy and image processing techniques, this study reconstructed the polymer phase in PMA of various polymer concentration. The research results indicate that the polymer phase likely increases to about 2.5 times by volume after being mixed with asphalt binder, and connected polymer network is not formed. Compared with the reconstructed 3D images, the 2D images would overestimate the swell effect and misinterpret the morphology of polymer phase in PMA.

1 INTRODUCTION

Polymer modified asphalt (PMA) has been wildly used in pavement construction over the last decades for its superior performance such as rutting, fatigue, cracking resistance (Behnood & Modiri Gharehveran 2019, Zhu et al. 2014). Various studies have been conducted on PMA, including but not limited to: modification mechanisms, anti-aging property, microstructure, storage stability, etc (Ye et al. 2015, Hao et al. 2017). Among these research topics, the morphology and size of the polymer phase in PMA have attracted a great amount of attention. Indeed, polymer size and morphology play an essential role in determining the interaction and compatibility between asphalt binder and polymer as well as the mechanical performance of PMA.

Fluorescence microscopy (FM) is a widely used tool to identify the polymer phase in PMA. When a PMA sample is illuminated through a blue light for excitation in FM, the polymer phase will emit fluorescent light while the asphalt phase remains dark (Fu et al. 2007). Based on the excitation pattern, FM can be categorized into five types. Widefield microscopy is the most widely used one in the analysis of PMA morphology. However, two-dimensional (2D) images of PMA obtained from widefield microscopy are formed by light from both the in-focus plane and those out-of-focus planes. As a result, the image does not reflect specific information from the in-focus plane of the sample, but the projection of the polymer phase within the range of excited depth. Moreover, it is reported that light from the out-of-focus planes creates noises and blurs (Bankhead 2014), which significantly degrade image quality. Therefore, widefield microscopy cannot accurately characterize the morphology of the polymer phase in PMA and may also overestimate its proportion. In this regard, reconstructing the polymer phase in PMA in the three-dimensional (3D) space may be a more effective way to investigate the internal structure of PMA. The confocal fluorescence microscopy, i.e. laser scanning confocal microscope (LSCM) in this study, would be a good alternative to achieve this goal.

Compared with widefield microscopy, only light from the in-focus plane is detected by LSCM (Bankhead 2014). In addition, LSCM is capable of recording the X-Y plane images of the sample at different Z-positions. All these X-Y plane images contain the morphological information of the sample at a certain depth, which can be used for 3D reconstruction of the polymer phase.

This study reconstructed the polymer phase in styrene-butadiene-styrene (SBS) modified asphalt with various SBS contents by using LSCM and image processing techniques. The objectives are to investigate the morphological difference of PMA with different polymer concentrations, and to determine the swell effect of polymer particles after being mixed with asphalt binder.

2 MATERIALS AND METHODS

2.1 Materials and sample preparation

Base asphalt binder 60/70 and SBS polymer were selected for the preparation of PMA samples, whose density is 1.03 and 0.94, respectively. The concentrations of SBS polymer are 1.5%, 3%, 4.5% and 6% by the mass of the base asphalt binder, and the corresponding volume fractions are 1.62%, 3.18%, 4.70% and 6.17% for each type of PMA sample, respectively. Additionally, the freeze fracture method was adopted to prepare the LSCM sample (Oliver et al. 2012). Five replicates were produced for all PMA samples.

2.2 Image processing and 3D reconstruction of polymer phase in PMA

In this study, the Leica TCS SP8 MP Multiphoton/Confocal Microscope was used to record the images of PMA samples at room temperature with the spectral reception ranging from 535 nm to 625 nm. The resolution of the microscope is 212 nm, and frame averaging (with four scans) was employed to reduce the noise of the images.

The X-Y plane images of the polymer phase in PMA at different Z-positions (1.79 μm thickness) were firstly processed by the Imaris software (version 9.3). Subsequently, the processed images were used for reconstructing the 3D morphology of the polymer phase and for further image analysis. In addition, the 2D projections of PMA samples at the X-Y plane were captured by the LAS_X_Small_3.7.1 software to calculate the area fraction of polymer phase in PMA.

3 RESULTS AND DISCUSSION

3.1 3D reconstruction of polymer phase in PMA

The 3D reconstruction of polymer phase in PMA, as shown in Figure 1, was performed with the processed images by Imaris software. It can be observed that the polymer particles in the PMA sample with 1.5% SBS content are mainly spherical and disperse homogeneously in the matrix phase. The morphology of 3% and 4.5% SBS PMA sample is similar: Larger polymer particles with some holes on the surfaces become more prevalent, and the number of small particles becomes relatively less. For 6% SBS PMA, the size of large polymer particles increases considerably while the number decreases, and more evident porous structures can be found. As the SBS content rises, the polymer phase cannot be fully swollen and sheared, leading to the agglomeration of particles. However, network of polymer phase cannot be detected in any of the samples. The interaction force between polymer particles may be not strong enough to form the chemical bonds.

3.2 Quantitative analysis of 3D polymer phase in PMA

Figure 2 presents the average quantity distribution and volume fraction of polymer particles in PMA with different concentrations. In general, variations in particle size can be relatively

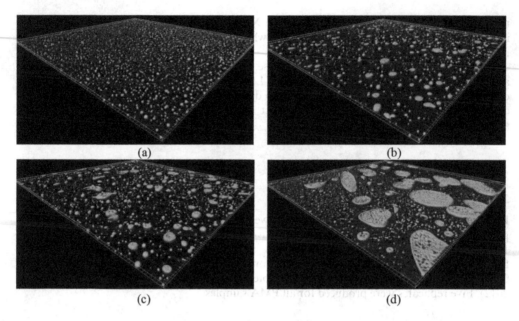

Figure 1.　3D morphology of polymer phase in PMA: (a) 1.5% SBS; (b) 3% SBS; (c) 4.5% SBS; (d) 6% SBS.

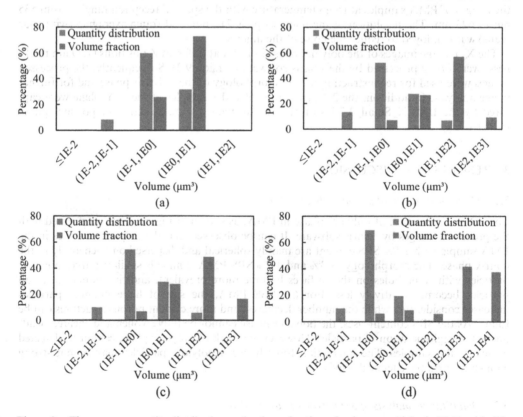

Figure 2.　The average quantity distribution and volume fraction of polymer particles in PMA with different concentrations: (a) 1.5% SBS; (b) 3% SBS; (c) 4.5% SBS; (d) 6% SBS.

high for samples with high polymer content. Meanwhile, the size of the maximum particle increases as the polymer content rises. The number of particles with size between 0.01 μm^3 and 10 μm^3 makes up over 90% for all PMA samples, but their volume fraction approximately shows a downward trend as the polymer content increases. Furthermore, the quantity percentage and volume fraction of particles larger than 100 μm^3 simultaneously increase with more polymer added.

To determine the swell effect of polymer particles in PMA, the volume fraction of polymer phase was calculated. As shown in Figure 3, a linear relationship can be established between the original and swollen volume fraction of polymer phase in PMA, and the volume of the SBS polymer particle apparently increases to about 2.5 times. Note that polymer particles smaller than the resolution of the FM are not counted. Consequently, the swell factor may be slightly higher than 2.5, but the number of extremely small polymer particles are believed to be limited.

3.3 Quantitative analysis of 2D polymer phase in PMA

Table 1 summarizes the area fraction and volume fraction of polymer phase in PMA (1.79 μm depth) obtained from 2D image and 3D reconstruction methods respectively. The area fraction of polymer phase in 2D image is greater than its volume fraction in 3D space, indicating the overestimation of the swell effect of polymer phase in PMA by 2D method. This is because the 2D images are formed by the projection of the 3D polymer on a 2D plane. Additionally, the fraction difference between 2D and 3D methods also increases for PMA with higher polymer content.

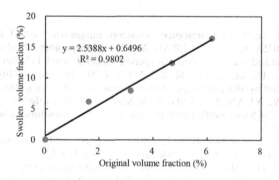

Figure 3. The relationship between the original and swollen volume fraction of polymer phase in PMA.

Table 1. The area fraction and volume fraction of polymer phase calculated by 2D and 3D methods.

Polymer content	2D area fraction	3D volume fraction	Fraction difference
1.5%	6.74%	6.19%	0.55%
3%	8.66%	7.96%	0.70%
4.5%	14.51%	12.44%	2.07%
6%	25.05%	16.44%	8.61%

4 CONCLUSIONS

To investigate the morphological difference of PMA with various polymer contents and the swell effect of polymer particles, this study reconstructed the polymer phase in PMA and analyzed the morphological characteristics and size distribution. The following conclusions can be drawn:

(1) The size of largest polymer particles increases with the rise of polymer content. Large polymer particles account for the higher volume fraction of polymer phase, and porous structure can be found on the surface of large particles.
(2) Connected polymeric network is not formed.
(3) A linear relationship exists between the volume fractions of the original and swollen polymer phase. The volume of SBS polymer particles likely expand to about 2.5 times in PMA.
(4) The traditional 2D image method would overestimate the proportion of swollen polymer phase in PMA. Therefore, 3D reconstruction method serves as a promising tool to study the morphological characteristics and volumetric properties of polymer phase in PMA.

ACKNOWLEDGEMENTS

This paper is based on a research project (Project No. 15204819) funded by the Research Grant Council of Hong Kong Special Administrative Region Government.

REFERENCES

BANKHEAD, P. (2014) Analyzing fluorescence microscopy images with ImageJ. *ImageJ*, 1, 195.
BEHNOOD, A. & MODIRI GHAREHVERAN, M. (2019) Morphology, rheology, and physical properties of polymer-modified asphalt binders. *European Polymer Journal*, 112, 766–791.
FU, H., XIE, L., DOU, D., LI, L., YU, M. & YAO, S. (2007) Storage stability and compatibility of asphalt binder modified by SBS graft copolymer. *Construction and Building Materials*, 21, 1528–1533.
HAO, G., HUANG, W., YUAN, J., TANG, N. & XIAO, F. (2017) Effect of aging on chemical and rheological properties of SBS modified asphalt with different compositions. *Construction and Building Materials*, 156, 902–910.
OLIVER, J., KHOO, K. Y. & WALDRON, K. (2012) The effect of SBS morphology on field performance and test results. *Road Materials and Pavement Design*, 13, 104–127.
YE, F., YIN, W. & LU, H. (2015) A model for the quantitative relationship between temperature and microstructure of Styrene-Butadiene-Styrene modified asphalt. *Construction and Building Materials*, 79, 397–401.
ZHU, J., BIRGISSON, B. & KRINGOS, N. (2014) Polymer modification of bitumen: Advances and challenges. *European Polymer Journal*, 54, 18–38.

Functional Pavements – Chen et al (eds)
© 2021 Taylor & Francis Group, London, ISBN 978-0-367-72610-2

Influence of the source on bitumen performance using a DSR

M. Rochlani, F. Wellner, G. Cannon Falla & S. Leischner
Institute of Pavement Engineering, TU Dresden, Dresden, Germany

ABSTRACT: Bitumen of penetration grade 50/70 is the most commonly used binder in asphalt pavements of Europe. This research attempts to differentiate between three bitumen of penetration grade 50/70 that have been acquired from three sources. The rheological properties, fatigue and rutting resistance were analyzed and compared using the results of Dynamic Shear Rheometer (DSR) tests. Additionally, to understand the changes in performance due to ageing, all the DSR tests were repeated for aged bitumen. Finally, the tests results were summarized in a performance diagram to aid in overall ranking of the bitumen taking into consideration relevant performance parameters.

1 INTRODUCTION

Owing to the large expenditure invested in construction and maintenance of road net-works, it has become crucial to take into account the quality considerations for bitumen. The quality of bitumen varies not only with the refining procedure used, but also with the crude oil source. Bitumen acquired from a particular refinery might meet the existing specifications, however, fail to pro-vide the required performance during the in-service period under heavy traffic and changing environmental conditions, indicating the insufficiency of the specifications that exist. As of now, conventional testing methods in Germany include needle penetration, softening point ring and ball etc.; belonging to the empirical (catalog) approach, these fail to define the rheological properties of the bitumen. In Germany the DSR temperature sweep between 30 to 90°C has been added recently as an additional procedure to gather information. The desired goal of this research was to verify the performance of three bitumen of the same grade but from three different sources using the results of DSR tests. Performance includes rheology, rutting and fatigue are to be tested to be able to compare and rank. Additionally, a method to appropriately rank all the materials using different parameters from the tests would be required.

2 LABORATORY TESTS

2.1 *Materials tested*

A total of three bitumen from different sources (1, 2 and 3) were tested, which include three unaged bitumen, three RTFOT aged and three PAV aged bitumen. In order to investigate the short-term and long-term performance of the bitumen, the bitumen were artificially aged in the laboratory. For short-term ageing, the materials were put in a RTFOT at 163 °C for 75 minutes. On the other hand, to replicate the short term + long term ageing effect was simulated using only PAV at 100 °C for 25 hours at 2.07 MPa.

2.2 *DSR tests*

An Anton Paar Modular Compact Rheometer 502 was used to evaluate the rheology, fatigue and rutting performance of the bitumen investigated. To study rheology, strain and frequency sweeps were conducted in the temperature range of -10 °C to 70 °C. The strain sweeps were firstly undertaken to evaluate the linear viscoelastic limit (LVE) strain, which are required to conduct the frequency sweeps. The strains were further reduced by 20 % to ensure the LVE region was maintained for the frequency sweeps. A frequency range of 0.0159 Hz to 47.7 Hz was used and the results acquired were superposed and approximated further using the time-temperature superposition principle. Secondly, for evaluating the rutting performance, single stress creep recovery (SSCR) tests were undertaken. In this test a sample is subjected to stress for one second, followed by nine seconds without stress, allowing the material to relax at a temperature of 60°C with the 25 mm plate. This step is repeated consecutively 10 times. From each cycle two variables were calculated and averaged:

$$\%R = \frac{\varepsilon_1 - \varepsilon_{10}}{\varepsilon_1} * 100[\%] \tag{1}$$

$$J_{nr} = \frac{\varepsilon_{10}}{100 * 3.2} [kPa^{-1}] \tag{2}$$

where ε_1 = strain increment after the first cycle; and ε_{10} = strain increment after the 10 second cycle. Lastly, fatigue performance was determined using stress-controlled tests on column cylindrical samples. The tests were carried out at a temperature of 20 °C and frequency 10 Hz for both unaged and aged samples. For each material, a total of nine samples were tested at three different stress levels. The stress levels were determined using pre-tests so as to attempt to keep the initial strain level of each material within the range of 0.5 to 2.0 to remain the LVE region.

3 TEST RESULTS

3.1 Frequency sweep tests

The frequency sweep test data were modeled using the time-temperature superposition is presented in terms of the master curves for dynamic shear modulus in Figure 1.

From Figure 1, the dynamic shear modulus of different bitumen at different aging conditions can be visually observed. It can be seen from the shear modulus master curve, for unaged materials, B1 and B3 tend to overlap, while B2 has slightly lower moduli values than the others at lower reduced frequencies (i.e. higher temperatures). If RTFOT aged bitumen are observed, B1 has the greatest visual shift, with an increase of approximately 600 % at higher temperatures to 3 % at lower temperatures (higher reduced frequencies), followed by B2 with an increase of 134% at higher temperatures and as low as 2% at lower temperatures while B3 shows almost overlapping results for unaged and RTFOT aged, with a numerical increase ranging from 20 % to 6.5 % as temperatures reduce. For PAV aged materials, all three bitumen, show a relatively higher increase, as expected. A better numerical evaluation of the effect of ageing could be observed using the aging index.

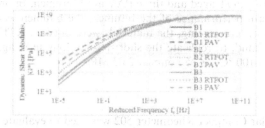

Figure 1. Master curves for the dynamic shear modulus.

3.2 Aging performance

The ageing of bitumen can be expressed numerically in terms of the ageing index A.I. (Equation 3). This index is calculated in terms of ratios corresponding to several physical grading tests and is given by the following equation:

$$A.I. = \frac{P_{aged}}{P_{unaged}} \qquad (3)$$

where, P_{unaged} = any physical parameter (e.g. viscosity) of the unaged bituminous materials; and P_{aged} = the same physical parameter as for an aged bituminous materials. The A.I. was calculated for dynamic shear moduli (bars) at 20°C and 60°C at a frequency of 10 Hz. The lower the ageing index value, lower is the ageing sensitivity of the material.

From the results in Figure 2, it can be concluded that B3 was most affected by RTFOT ageing, with B2 being the most susceptible and B3, the least susceptible to PAV ageing at higher temperatures.

3.3 Permanent deformation performance

SSCR tests were undertaken to evaluate the rutting performance. These tests were carried out on unaged and RTFOT aged materials. PAV aged materials were not tested as rutting is not critical in long term aged materials. Permanent deformation is hence usually tested in unaged and RTFOT state. In order to evaluate numerically the rutting performance, the Jnr and %R values were calculated and are presented in Figure 3. Unaged bitumen B2 has the highest recovery and lowest J_{nr} value, as desired, however, for RTFOT, it has the lowest recovery and

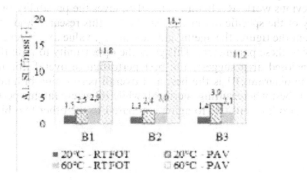

Figure 2. Aging index for the dynamic shear modulus.

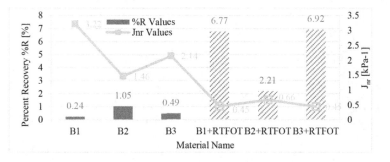

Figure 3. %R and J_{nr} values for unaged and RTFOT aged bitumen.

97

the highest J_{nr} value. With RTFOT ageing being more important, it can be concluded that bitumen B1 and B3 performs similar. It can be concluded that all three bitumen show significant different rutting performance - in particular bitumen B2 has much better rutting performance than the other two bitumen.

3.4 *Fatigue performance*

The fatigue test results were analyzed using the Dissipated Energy ratio approach. Figure 4 presents the results of the fatigue tests. From the figure, it can be observed that there is a clear distinction between the unaged, RTFOT and PAV aged bitumen curves, with the slope for all the materials other than B1 being relatively similar.

The ranking of the materials stays the same with respect to RTFOT and PAV aged materials. In RTFOT aged, the while bitumen B1 and B3 are almost overlapping, bitumen B2 has a much higher fatigue life than the other two, however for PAV aged materials, the differences between the bitumen are relatively equivalent with B1 again having the highest fatigue life. The better fatigue performance of B2 could be attributed to the higher stiffness of the bitumen.

4 CONCLUSIONS

Bitumen of grade 50/70 from three different sources and at different aging conditions were tested in terms of rheology, rutting and fatigue resistance using the DSR. It was found that despite the similar penetration grade, there were vast differences found in terms of specific performances. Using the characteristic parameters of the tests undertaken, the diagram presented in Figure 5 was constructed using to pictorially rank the admixtures based on their stiffness, ageing indexes, fatigue, and rutting behavior. This diagram, which was introduced by the authors in a previous work (Rochlani et al. 2019), gives the possibility of ranking bitumen on the basis of any of the specific parameters studied in this research and obtained via the DSR instrument. From the figure, the ageing indexes and J_{nr} value axis is reversed, as performance relates inversely to these parameters. This gives the opportunity to rank the materials in a way that a larger enclosed area suggests the best performance overall for the parameters used. According to the diagram, B1 is the best and overall performing bitumen, while for specific properties – B1 is best where rutting is critical, while B2 where fatigue performance is of value. This material is also less sensitive to aging compared to the other two bitumen at both aging conditions.

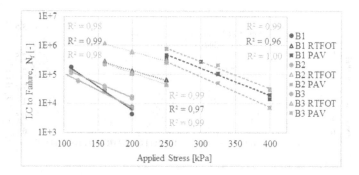

Figure 4.   Fatigue functions of the bitumen tested.

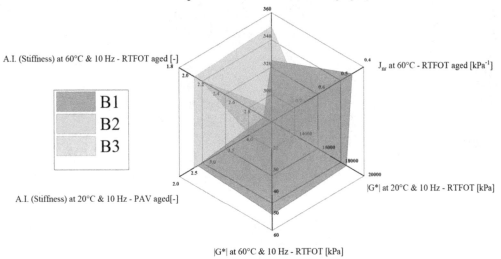

Figure 5. Performance diagram.

ACKNOWLEDGEMENTS

We hereby acknowledge the financial support of TU Dresden within the Scholarship Program for the Promotion of Early-Career Female Scientists for Ms. Rochlani. This paper is also based on the research project carried out at the request of the German Research Foundation (DFG), under research project No. WE 1642/11 and LE 3649/2 (FOR 2089). The authors are solely responsible for the content.

REFERENCES

Rochlani, M.; Leischner, S.; Falla, G.;Wang, D.; Caro, S.; and Wellner, F. 2019. Influence of Filler Properties on the Rheological Properties and Performance of Mastics. *Construction and Building Materials*. DOI: 10.1080/14680629.2020.1778508.

Asphalt mixture evaluation and performance

Functional Pavements – Chen et al (eds)
© 2021 Taylor & Francis Group, London, ISBN 978-0-367-72610-2

Laboratory investigation of epoxy resin based paving materials

Ali Raza Khan & Weiguang Zhang*
School of Transportation Engineering, Southeast University, Nanjing, China

Zia-Ur-Rehman
Department of Transportation Engineering and Management, University of Engineering and Technology, Lahore, Pakistan

ABSTRACT: With the advancement of construction industry, innovative materials are being used to be aesthetically pleasing in cities. Resin-based surfacing/paving is an innovative technique used for variety of pavement applications like footpath, parking lots, and bus stops etc. Before the construction of pavements, laboratory studies are important to provide evidence about the expecting physico-mechanical response of designed materials. Therefore, this study was conducted to measure the physical and mechanical properties of a resin-based surfacing/paving material. For this purpose, aggregates and an epoxy resin manufactured by Jiangsu SinoRoad Transportation Science and Technology Co., Ltd. were used. The aggregates have two size (1-2mm and 2-3mm) and two-part cold mix epoxy resin was selected to be used as binder. Results of dust and moisture content have shown that due to the limited amount of dust and moisture, aggregates can form strong bonds with the epoxy binder. Moh's hardness test also has shown that aggregates had high strength on hardness scale. Additionally, tensile strength, lap-shear strength, and butt joint strength tests depicted that the epoxy binder had high strength with short working (pot) life. When tested under low temperature (-10°C), results indicated that still the epoxy had enough elastic properties to resist cracks. Finally, combination of epoxy binder and aggregates for resin-based surfacing/paving applications hold improved properties and can be used for multiple applications. SinoRoad Transportation Science and Technology Co., Ltd. tested materials have comparable properties as other commercial products.

Keywords: Resin based surfacing/paving, Epoxy, Aggregates, Laboratory investigation, SinoRoad

1 INTRODUCTION

Epoxy paved colored aggregates have been used frequently to improve the aesthetic of pedestrian bridges, footpaths, running tracks, car parking, bus lanes, and road junctions. The resultant pavements have an extremely hard-wearing surface, capable to withstand both high and low temperatures with various range, do not fade in UV sunlight, and resistant to oil or petrol spillage (McCormack 2020). Before the application of such types of pavements, it is important to determine the physical and mechanical properties of both epoxy and aggregates. Mechanical properties were determined using the laboratory experiments to measure either material can obtain the desired properties or not. Epoxy resins are thermosetting polymers having good performance, the processing is easy, and have two epoxy groups per molecule

---

* Corresponding author

(Lokensgard 2013; Mano 1991). For the reason of having high reactivity, adhesive properties, rigidity, chemical resistance, and fluid nature before application make it suitable to be applicable in several mechanical applications (Deng et al. 2008; Hsieh et al. 2010). Additionally, the resin has also been used for a thin overlay of asphalt pavement on steel bridges (Zhang et al. 2019). The purpose of this study is to determine the physical and mechanical properties of epoxy and aggregate separately to ensure the quality of the material provided manufactured by Jiangsu SinoRoad Transportation Science and Technology Co., Ltd.

2 MATERIAL AND METHODS

2.1 *Materials*

Although, aggregates are available in different sizes and colors. According to the project specification, the epoxy bonded method was adopted with a layer thickness of 3mm. Reddish color aggregates of size 1-2mm and 2-3mm were used. Whereas, epoxy acts as a binding agent between the base and aggregate. For this study cold mix two-part (A & B) Jiangsu SinoRoad Transportation Science and Technology Co., Ltd. epoxy was used. The polymer matrix part A is epoxy resin and part B is hardener with a mixing ratio of 10:3. Part A is a viscous liquid, whereas, Part B viscosity is less as compared to Part A. Figure 1 shows the cross-sectional details of resin-based surfacing.

2.2 *Method*

2.2.1 *Aggregate*

Different experiments were performed including the appearance test to visually inspect the aggregates with the naked eye and with a magnifying glass. Dust content (Q_n) of aggregates was determined according to T0333 according to the wet method. Also, natural moisture content (w) was determined according to T0310.

Finally, Moh's hardness test was performed to measure the strength in numbers. These numbers were provided based on increasing strength with Talc (1) as the lowest strength and diamond (10) as the strongest mineral. According to ASTM C1895 minerals were scratched against the smooth surface of aggregate.

2.2.2 *Epoxy*

The appearance of both components was checked to determine the color, odor, and other physical properties. Mainly, mixture properties are the main concern that's why appearance test was also conducted for the mixture. The viscosity of the mixture was determined using Brookfield Viscometer according to GB/T 10247. According to the manufacturer mentioned ratio, weighted components were mixed and viscosity is monitored for continuously three (03) minutes at 25°C. The average of three (03) minutes is reported as the viscosity of the mixture. Working life is also called the pot life, was determined according to GB/T 7123 using the viscosity method at 23°C and 50% humidity.

Tensile strength (σ_t) and ductility (ε_t) of dumbbell-shaped samples were determined according to GB/T 2567. Whereas during the test, crosshead speed for the tensile test is 10mm/min and for

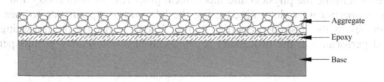

Figure 1. Aggregate for epoxy resin pavement.

elongation test, the speed of crossheads is 2mm/min. Lap-shear and butt joint strength are also checked in tension. Therefore, both test were performed according to GB/T 7124 and GB/T 6329.

Finally, pull off adhesion test and low-temperature bend tests were performed to measure the adhesion strength of the coating and cracking resistance at low temperature, respectively. Both these tests were performed according to the GB/T 5210 and ASTM D2136, respectively. Pull off adhesion test was performed at 25°C and 70°C, whereas, low-temperature bend test was performed at -10°C.

3 RESULTS AND DISCUSSION

3.1 *Aggregate*

Aggregates with size 1-2 and 2-3mm are visually inspected. Figures 2 (a & b) shows both samples side by side. The aggregate color is reddish, have angular edges, and has an almost equal proportion of both sizes as mentioned.

Table 1 shows the results of dust content (Q_n), moisture content (w), and Moh's hardness test. Aggregates of size 1-2mm have average dust content of 0.056%, whereas, 2-3mm size particles have a dust content of 0.18%. Also, both types of aggregates moisture content (w) averages are 0.00025% and 0.00017%, respectively. Overall, both dust and moisture contents are less as a result thin layer will not surround the particles and the bond will be stronger. Finally, Moh's hardness test results show that aggregate creates scratch on feldspar (6) and no scratch was observed on quartz (7). So, aggregates Moh's hardness number is 6.5.

3.2 *Epoxy Resin*

Figures 2 (c & d) represents both components (A and B) of epoxy and final hard resin. The viscosity of component A is higher, colorless, and odorless liquid as visible in Figure 2(c). Whereas, component B viscosity is less with a yellowish color shown in Figure 2(d), has a particular odor that creates irritation. After proper mixing and final setting, mixture converted to yellowish solid.

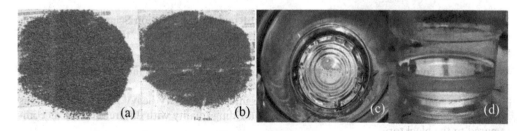

Figure 2.   (a) Aggregate of size 2-3mm (b) Aggregate of size 1-2mm (c) Component A (d) Component B.

Table 1.   Test results of aggregates.

Sr. No	Description of test	Unit	# of samples	Average result
1	Dust content (Q_n) of aggregate 1-2mm	%	03	0.056
2	Dust content (Q_n) of aggregate 2-3mm	%	03	0.056
3	Moisture content (w) of aggregate 1-2mm	%	03	0.00025
4	Moisture content (w) of aggregate 2-3mm	%	03	0.00016
5	Moh's hardness number of aggregate 1-2mm	No.	03	6.5
6	Moh's hardness number of aggregate 2-3mm	No.	03	6.5

Table 2. Test results of Epoxy.

Sr. No	Description of test	Unit	# of samples	Average result
1	Viscosity test	mPa.s	03	2.003
2	Working (pot) life	minutes	02	18
3	Tensile Strength	MPa	03	17.96
4	Ductility	%	04	7.18
5	Tensile lap shear strength	MPa	04	6.82
6	Tensile Butt joint strength	MPa	03	7.23
7	Pull off adhesion strength at 25°C	MPa	04	13.58
8	Pull off adhesion strength at 70°C	MPa	04	7.01
9	Low-Temperature bend strength test	-	-	No Cracks

Whereas, the results of different experiments were mentioned in Table 2. Mixture viscosity is 2.03mPa.s. Whereas, the mixture has the capability of accepting particles up to 18 minutes. Usually, the working life of cold mix epoxy is short and needs a quick response. Epoxy should have sufficient strength to withstand tensile loads. Three different types of tensile strength tests were performed including the tensile (dumbbell-shaped) specimen strength, lap shear strength in tension, and butt joint strength in tension. Dumbbell shaped specimens have a higher strength of 17.96MPa. Whereas, lap shear and butt joint tensile strengths are comparable to each other with 6.82 and 7.23MPa, respectively. Whereas, epoxy can elongate up to 7.18% of its original length. Moreover, the pull-off adhesion strength test was performed to measure either epoxy fails in adhesion or cohesion. The test was performed at 25°C and 70°C with pull-off strength of epoxy as 13.58 and 7.01MPa, respectively. Finally, to inspect cracking behavior al low temperature (-10°C) low-temperature bend test was performed on the specified samples. The result indicates that no cracks were observed on the bending surface even when observed through the magnifying glass. Eventually, all samples pass the test at low temperatures.

4   CONCLUSION

From the laboratory investigation of Jiangsu SinoRoad Transportation Science and Technology Co., Ltd. adhesives and aggregates for resin-based surfacing/paving, the following conclusions were drawn from this study.

– Aggregates color is consistent, having low moisture and dust content makes it suitable to make a good bond with the aggregates.
– Hardness and aggregate shape make it usable under temporary heavy load areas (i.e. bus stop) which provide sufficient friction and good riding quality with different colors as compared to the blacktop.
– Different physical and mechanical experiments suggest that epoxy contains good mechanical and physical properties.
– Tensile test (pull-off adhesion, lap-shear strength, tensile strength test) dictates that Sino-Road epoxy has higher tensile strength and shows good properties.
– Comparison of SinoRoad epoxy with other commercial epoxies suggests that SinoRoad epoxy resin have comparable properties, and can be used for different projects.

ACKNOWLEDGMENT

The authors would like to thank Jiangsu SinoRoad Transportation Science and Technology Co., Ltd. for every kind of support and providing the material.

REFERENCES

Deng, Shiqiang; Zhang, Jianing; Ye, Lin; Wu, Jingshen (2008): Toughening epoxies with halloysite nanotubes. In *Polymer* 49 (23), pp. 5119–5127.

Hsieh, T. H.; Kinloch, A. J.; Masania, K.; Taylor, A. C.; Sprenger, S. (2010): The mechanisms and mechanics of the toughening of epoxy polymers modified with silica nanoparticles. In *Polymer* 51 (26), pp. 6284–6294.

Lokensgard, Erik (2013): Plásticos Industriais: teoria e aplicações. In *Cengage Learning*.

Mano, Eloisa Biasotto (1991): Polímeros como materiais de engenharia: Editora Blucher.

McCormack, Tony (2020): Resin Based Paving. Available online at https://www.pavingexpert.com/resin, updated on 6/24/2020, checked on 6/29/2020.

Zhang, Hui; Zhou, Chengqi; Li, Kuan; Gao, Peiwei; Pan, Youqiang; Zhang, Zhixiang (2019): Material and Structural Properties of Fiber-Reinforced Resin Composites as Thin Overlay for Steel Bridge Deck Pavement. In *Advances in Materials* Science *and Engineering* 2019.

Functional Pavements – Chen et al (eds)
© 2021 Taylor & Francis Group, London, ISBN 978-0-367-72610-2

Experimental study on the performance of new ultra-thin overlay TOM-10 mixture utilizing a SBS/rubber composite modified asphalt binder

A.H. Liu, L.L. Liu, H.R. Zhu & M.M. Yu
Senior Engineer, JSTI Group, National Engineering Laboratory for advanced road materials, Nanjing, China

ABSTRACT: Asphalt binder has a significant impact on the properties of TOM-10 ultra-thin coating asphalt mixture. In this research, to evaluate and analysis the properties of asphalt binder and mixture, the SBS/rubber composite modified asphalt binder and high-viscosity modified asphalt binder were compared. In addition, SBS/rubber composite modified asphalt binder was obtained by high-speeding shearing and rolling with SBS modified asphalt and rubber powder which has been treated through some chemical processes. The results illustrated that the SBS/rubber composite modified asphalt has better viscosity and activation energy. Moreover, TOM-10 ultra-thin coating asphalt mixture with SBS/rubber asphalt binder exhibited higher high-temperature performance and water stability.

1 INTRODUCTION

As a new type of pavement preventive maintenance technology, ultra-thin coating is used to recover the function degradation of road surface, such as slight cracks, ruts and other distresses[1]. Most ultra-thin coating technology in China was to solve slight ruts, and road petroleum asphalt or SBS modified asphalt is commonly used. Therefore, the scope of application is relatively narrow which often with short service life. The typical of ultra-thin coating is from 1.5cm to 2cm, as well as that the asphalt binder and gradation have a significant impact on its performance, service life and scope of application [2][3].

To prolong the service life, as well as the scope of application, the excellent asphalt binder and gap gradation were utilized to develop TOM-10 ultra-thin coating mixture, which originally developed in Texas A&M Transportation Institute is adopted. The comparative study was carried out to make comparisons with SBS/rubber composite modified asphalt and high-viscosity asphalt. The analysis results are of great significance to promote the application of ultra-thin layer cover technology and similar engineering design.

2 RAW MATERIALS

The SBS/rubber composite modified asphalt is provided by Jiangsu Baoli company, which is produced by high-speeding shearing and rolling with SBS modified asphalt and rubber powder which has been treated through some chemical processes.

The high-viscosity asphalt is obtained by 70# asphalt produced by Shell company added with high viscosity additive PEA (Performance Enhancement Additive), The PEA mentioned here is developed by National Engineering Laboratory of new road materials.

3 PHYCIAL PROPERTIES OF ASPHALT BINDER

3.1 *Conventional properties*

The samples obtained were tested by ductility at 5°C, penetration at 25°C and softening point. The test methods were all in accordance with Test Rules for Asphalt and Asphalt Mixture for Highway Engineering (JTG E20-2011). The test results are shown in Table 1 respectively.

Table 1 shows the ductility of SBS/rubber composite modified asphalt is obvious higher than that of high-viscosity asphalt. So as to illustrate that the former would have better low-temperature performance. Moreover, the penetration of SBS/rubber composite modified asphalt is obvious lower than that of high-viscosity asphalt. Thus demonstrated excellent ability to resistance to shear failure.

3.2 *Rheological properties*

3.2.1 *Rotational viscosity test*

The viscosity test was carried out using Brookfield rotating viscometer. Previous work shows the viscous flow activation energy derived from the temperature dependence of the viscosity of asphalt binder can serve as the indicator of the high-temperature rheological property of asphalt. Therefore, the experimental temperature was set as 125°C, 135°C, 145°C and 155°C in this study. Figure 1 shows the temperature dependence of the viscosity of SBS/rubber composite modified asphalt and high-viscosity asphalt.

Studies reported that the asphalt with high activation energy exhibited excellent thermal stability [4]. Delmar Salomon et. al. established the empirical relationship by applying the Arrhenius equation which was showed below [5].

$$\eta = Ae^{\frac{E_f}{RT}} \tag{1}$$

Table 1. Test results of three indexes of asphalt.

Physical indicator	SBS/rubber composite modified asphalt	High-viscosity asphalt	Technical index requirement
Penetration 25°C(0.1mm)	62.3	71.3	60-80
Softening point (global method)°C	86	71	55
Ductility(5cm/min)	409	353	300

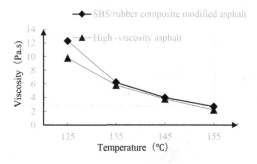

Figure 1. Viscosity curve of the two modified asphalt with temperature.

where η = viscosity; T = temperature, which unit is K; A = constant; E_f = activation energy; R=constant, the value is 8.314J.mol^{-1}.K^{-1}.

Equation (1) can be transformed into Equation (2) showed below for better understanding.

$$\ln \eta = \frac{E_f}{RT} + \ln A \qquad (2)$$

which shows a linear correlation between the $\ln\eta$ and the measurement temperature reciprocal with the activation energy over the ideal gas constant as the slope. Arrhenius curve of pitch rotational viscosity against temperature was shown in Figure 2. Then the activation energy can be derived as shown in the bar plot (Figure 3).

The result shows higher activation energy of SBS/rubber composite modified asphalt than that of high-viscosity modified asphalt indicating a better thermal stability of the SBS/rubber composite modified asphalt.

3.2.2 Bending beam rheological test

Low-temperature cracking of the asphalt pavements is a major pavement distress mechanism in cold regions. The permanent deformation was believed to form as the result of the non-viscous flow of the asphalt binder. By conducting the creep tests of asphalt at different temperatures, the service temperature of asphalt pavement can be determined. In this paper, the Bending Beam Rheometer test (BBR) was used to evaluate the rheological properties of SBS/rubber composite modified asphalt and high-viscosity modified asphalt at low temperature [6].

It can be seen from Figure 4 and Figure 5 that, at the same temperature, the bending stiffness of high rubber powder asphalt at low temperature is significantly lower than

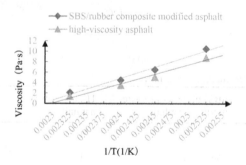

Figure 2.   Activation energy analysis diagram.

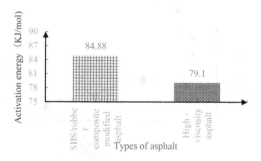

Figure 3.   Viscous flow activation energy results.

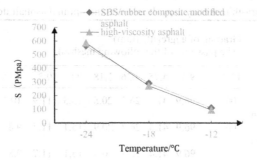

Figure 4.   Creep stiffness modulus S at low.

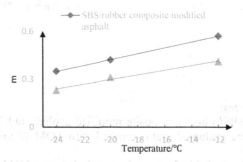

Figure 5.   Deflection rate change with temperature.

that of high viscosity asphalt, which is only half of the latter. The m value is greater than that of high-viscosity asphalt. The low-temperature performance of SBS/rubber composite modified asphalt is considered as the inherit of the good performance of rubber asphalt. According to AASHTO specification, when S is less than 300MPa and m is more than 0.3, it can be deemed to have sufficient resistance to low temperature damage at this temperature, and beyond this range, the asphalt mixture is judged to be in the cracking edge state at this time. At -24℃, the bending stiffness S of high viscosity asphalt at low temperature is 2 times of the determination index of 300MPa, and m is less than 0.3, indicating that the deformation resistance of high-viscosity asphalt at low temperature is weaker than that of SBS/rubber composite modified asphalt.

4   PERFORMANCE OF TOM-10 MIXTURE

4.1   *Mix design and the composition of raw materials*

A new type of ultra-thin overlay TOM-10 imported from the United States is adopted. the gradation type belongs to gap gradation. Since TOM-10 has a maximum nominal particle diameter of 9.5mm and is a fine-grained mixture, this paper uses 2.36mm as the dividing point of coarse and fine aggregates, and uses the coarse aggregate skeleton clearance ratio VCAmin (Voids in coarse aggregate of Asphalt mix)< VCA_{DRC}(Voids in coarse aggregate) to judge whether coarse aggregates contact to form a skeleton. According to the Marshall test volume index result, the design gradation is selected and the optimum asphalt dosage is determined. See Table 2.

Table 2.　Gradation of ultra-thin overlay asphalt mixture and optimal asphalt dosage.

Asphalt mixture type		Grading of aggregate (mm) (The pass rate of the following mesh%)									Optimum asphalt consumption
		13.2	9.5	4.75	2.36	1.18	0.6	0.3	0.15	0.075	
SRCMA* TOM-10	Synthetic gradation	100	96.9	41	26.3	20.8	15.3	11.9	10.1	8.8	6.5%
HVAM** TOM-10	Synthetic gradation	100	96.9	41.4	26.7	20.9	15.1	11.7	9.8	8.5	6.0%
MAM*** TOM-10	Synthetic gradation	100	96.9	41.4	26.7	20.9	15.1	11.7	9.8	8.5	6.0%
Gradation range	100	100	95~100	40~60	17~27	5~27	5~27	5~27	/	5~9	Min6

Notes:*SBS/rubber composite modified asphalt,**High-viscosity asphalt mixture,***SBS modified asphalt mixture.The same below.

4.2　Results and analysis

4.2.1　Rotational viscosity test

TOM-10 mixture using SBS/rubber composite modified asphalt and TOM-10 mixture using high-viscosity asphalt were evaluated and compared with TOM-10 mixture using SBS modified asphalt in rutting test.

　　Based on the test results in Table 3, it can be concluded that TOM-10 mixture using SBS/rubber composite modified asphalt and TOM-10 mixture using high-viscosity asphalt have better high-temperature stability than TOM-10 mixture using SBS modified asphalt. It is considered that the carbon nanoparticles released from the rubber powder in the SBS/rubber composite modified asphalt can enhance the flexibility of the mesh chain structure and thus increase the deformation resistance of the mixture. The addition of viscosifier PEA in the SBS modified asphaltt enhances the adhesion of SBS modified asphalt and the cohesion between asphalt and aggregate, thus improving the shear resistance of the mixture.

4.2.2　Water stability and low temperature cracking resistance

The water stability of asphalt mixture is essential to the service life of pavement[6]. The water stability of TOM-10 mixture using SBS/rubber composite modified asphalt and TOM-10 mixture using high-viscosity asphalt was evaluated by immersed Marshall test and freeze-thaw splitting test. Low temperature trabecular test was used to evaluate the low temperature performance of TOM-10 mixture using SBS/rubber composite modified asphalt and TOM-10 mixture using high-viscosity asphalt. The comparison was made against TOM-10 mixture using SBS modified asphalt. Test results of immersed Marshall, freeze-thaw splitting and low-temperature trabecula are shown in Table 4.

　　The test results in Table 4 show that:

(1) The splitting strength ratio of TOM-10 mixture using SBS/rubber composite modified asphalt and TOM-10 mixture using high-viscosity asphalt is much higher than that of

Table 3.　Rut test results of TOM ultra-thin mixture.

Indicators　Mixture type	SRCMA TOM-10	HVAM TOM-10	MAM TOM-10
Dynamic stability (cycles·mm^{-1})	11045	10843	6210

Table 4. Test results of TOM ultra-thin mixture immersed Marshall, freeze-thaw split and low temperature trabecula.

Mixture type Indicators	SRCMA TOM-10	HVSM TOM-10	MSM TOM-10
The residual stability (%)	91.3	89.4	84.3
The splitting strength ratio (%)	90.3	86.5	81.5
The maximum bending strain ($\mu\varepsilon$)	4100	3133	2604

TOM-10 mixture using SBS modified asphalt, indicating that significantly improved adhesion and overall stability of TOM-10 mixture by addition of rubber powder and viscosity modification. TOM-10 mixture using SBS/rubber composite modified asphalt and TOM-10 mixture using high-viscosity asphalt have better moisture damage resistance than TOM-10 mixture using SBS modified asphalt, and TOM-10 mixture using SBS/rubber composite modified asphalt has better water stability than TOM-10 mixture using high-viscosity asphalt.

(2) TOM-10 mixture using SBS/rubber composite modified asphalt and TOM-10 mixture using high-viscosity asphalt have better cracking resistance at low temperature than TOM-10 mixture using SBS modified asphalt. TOM-10 mixture using SBS/rubber composite modified asphalt has better cracking resistance than TOM-10 mixture using high-viscosity asphalt. This can be the evidence of the carbon black nanoparticles released from the rubber powder in the SBS/rubber composite modified asphalt can help to enhance the flexibility of the mesh chain structure and increase the bending strength of the mixture.

5 CONCLUSION

Through the study on the properties of SBS/rubber composite modified asphalt and high-viscosity modified asphalt, as well as the comparative study for the properties of TOM-10 mixture using SBS/rubber composite modified asphalt, TOM-10 mixture using high-viscosity asphalt and TOM-10 mixture using SBS modified asphalt, the following conclusions are mainly obtained:

(1) The high-temperature stability and low-temperature performance of SBS/rubber composite modified asphalt are better than that of high-viscosity asphalt, and the activation energy value of SBS/rubber composite modified asphalt is the highest. The bending stiffness at low temperature of SBS/rubber composite modified asphalt is lower than that of high-viscosity asphalt. The larger m value of high-viscosity asphalt indicates its rheological property at low temperature is better than that of high-viscosity asphalt.

(2) TOM-10 mixture using SBS/rubber composite modified asphalt has better high-temperature performance than TOM-10 mixture using high-viscosity asphalt, which is consistent with the result that the SBS/rubber composite modified asphalt has better high-temperature performance than the high-viscosity asphalt.

(3) The residual stability of TOM-10 mixture using SBS/rubber composite modified asphalt and TOM-10 mixture using high-viscosity asphalt was 91.3% and 89.4%, respectively, and the splitting strength ratio was 90.3% and 86.5%, respectively. The moisture damage resistance of TOM-10 mixture using SBS/rubber composite modified asphalt was better than that of TOM-10 mixture using high-viscosity asphalt.

(4) The low temperature performance of TOM-10 mixture using SBS/rubber composite modified asphalt is the best, which is consistent with the better rheological property at low temperature of SBS/rubber composite modified asphalt than high-viscosity asphalt.

Synthetic test results, due to the thin skin coat is located in the road surface, directly under vehicle load and the natural environment influence, pavement performance requirements are higher than ordinary roads, therefore, recommended for application of TOM-10 mixture using SBS/rubber composite modified asphalt ultra-thin layer of the actual project promotion.

REFERENCES

[1] He zhimin, 2018. Application of composite modified UTAC lamination in Beijing expressway [J]. Chinese & foreign highways, 38(2): 75–79.
[2] Sun zuwang, ren min, 2017. Practical technology for preventive maintenance of asphalt pavement [M]. Beijing: China building materials industry press.
[3] Dar Hao Chen, Moon Won, Xianhua Chen, Wujun Zhou, 2016. Design improvements to enhance the performance of thin and ultra-thin concrete overlays in Texas [J].Construction and Building Materials 116:1–14.
[4] Li peilong, ma lixia, feng zhengang, et al, 2017. Rheological properties of aging asphalt based on Arrhenius equation [J]. Journal of chang'an university (natural science edition), (05): 5-11+18.
[5] Wang f c, zhu J p, 2019. analysis on the influence of warm mixture on asphalt's thermal sensitivity [J]. New building materials, (7):150–153.
[6] Zhao z, dong p, 2019. study on properties of reactive terpolymer composite modified asphalt and its mixture [J]. New building materials, issue 11.

Functional Pavements – Chen et al (eds)
© 2021 Taylor & Francis Group, London, ISBN 978-0-367-72610-2

Evaluation of the effect of laboratory aging on fatigue performance of Hot in-place (HIR) mixtures

Jitong Ding, Fujian Ni, Zili Zhao & Qipeng Zhang
Southeast University, Nanjing, Jiangsu, China

ABSTRACT: Hot in-place (HIR) maintenance technology is expected to reduce aggregate mining, protect the environment, and economize construction costs. The prevention of cracking is an important concern for HIR pavement, because of the high content of reclaimed asphalt pavement (RAP), which exhibits a relatively high level of aging. This study aims to investigate the cracking performance degradation after long-term aging. Two recycling plans and three aging conditions (loose mix aging 4h 12h, and 24h) were considered. The cracking resistance were analyzed in terms of linear viscoelastic (LVE) characteristics, fracture energy and fatigue damge, respectively. Results show that the change LVE properties can be observed under 24h aging, indicating the modulus of the mixture is not sensitive to aging over a longer period. However, The fracture energy and apparent damage capacity indices can capture the degradation of cracking resistances. In addition, the SBR modifier can be applied to enhance the long-term anti-cracking properties.

1 INTRODUCTION

Hot in-place recycling (HIR) technology is a sustainable and economic technology for highway maintenance (Cao et al., 2017, Ma et al., 2010). Aged recycled pavement materials are brittle and more sensitive to thermal and fatigue cracking. Therefore, cracking is a common concern for the application of HIR pavements (Shu et al., 2008).

Asphalt pavement undergoes long-term aging during their service life. Many studies have focused on correlating the relationship between field and laboratory aging (Rahbar-Rastegar et al., 2018). For long-term aging simulation, the method recommended by AASHTO R30 is that compacted mixture aging in the oven for 5 days at 85°C. However, this method is time-consuming and results in the aging gradient (Carpenter and Shen, 2006, Zhang et al., 2019). Previous studies have recommended using a loose mix aged at 135°C for 24 hours to simulate 7-10 years of field aging (Mahmoud et al., 2014, Saha and Biligiri, 2016, Li et al., 2018). Although there is concern about the negative impact on compaction after loose mix aging, this aging method is an effective way.

To predict the cracking resistance of HIR asphalt pavement, it is critical to investigate pavement performance degradation under different aging levels (Mills-Beale et al., 2014, Yang et al., 2017). Therefore, the main objective of this study was to evaluate the linear viscoelastic characteristics, thermal and fatigue cracking resistance of HIR mixture through laboratory mixture performance tests in terms of different aging conditions. The loose mix aging method was conducted, aging 12h at 135°C to simulate 3-5 years aging level, and aging 24h at 135°C to simulate 7-10 years aging level. Two plans for recycling (rejuvenator, rejuvenator + SBR additive) were included in the study. A dense-graded HIR mixture was designed by incorporating 85% RAP with a nominal maximum aggregate size (NMAS) of 13.2mm. RAP was collected from the top layer of the Yanhai Highway in Jiangsu Province of China. One SBS modified binder meeting the Superpave PG 70-22 was selected for HIR mixture preparation.

2 MATERIALS AND SPECIMEN PREPARATION

2.1 *RAP*

To evaluate the performance of RAP used in this study, the RAP binder was extracted and recovered from the RAP mixes in terms of AASHTO T 164 and ASTM D 5404, respectively. the properties of the RAP binder are summarized in **Table 1**.

2.2 *Recycling additive*

One typical recycling agent containing tall-oil and its derivatives was selected, a dosage rate of 3% by the weight of the RAP binder was determined as the optimum additive quantity. SBR additive was utilized to modify the long-term performance of RAP. The SBR additive is a white liquid at room temperature, different content by the weight of RAP binder was added in RAP. The 4% SBR additive was designed to balance the performance and economy.

2.3 *Specimen preparation*

The process of fabricating HIR mixes was simulated construction of HIR mixtures, RAP heating temperature is 120°C for 2h, the virgin aggregate was heated to 180°C for 4h, for the scheme of SBR modified mixture, the SBR additive was added after RAP and rejuvenator mixed, the detailed process can be found in the previous report (Zhou et al., 2017). The mixes were fabricated with a Superpave gyratory compactor (SGC) to a height of 178 mm and a diameter of 150 mm. A loose mixed oven aging method was applied with different duration (4h, 12h, 24h) at 135°C for simulating short and long-term aging. The basic information of different HIR mixtures is shown in **Table 2**.

3 TEST METHODS

3.1 *Dynamic modulus test*

To identify the linear viscoelastic property of HIR mixtures, the uniaxial compression dynamic modulus test was conducted using small specimen in the temperature control chamber at -10°C,4.4°C, 21.1°C, 37.8°C, and 54.4°C. The specimens were cored and trimmed to

Table 1. Properties of RAP binder.

	High Temp.	Inter Temp.	Low Temp. S/MPa	m-slope	PG Grade
RAP binder	85.3	21.4	-16.1	-15.2	85-15

Table 2. Basic information of HIR mixtures.

Mixture	Additive	Aging time (h)	Air void (%)
NMS	Rejuvenator	4	4.43
NML1	Rejuvenator	12	4.36
NML2	Rejuvenator	24	4.54
MMS	Rejuvenator + SBR	4	4.25
MML1	Rejuvenator + SBR	12	4.64
MML2	Rejuvenator + SBR	24	4.41

a dimension of 110 mm in height and 38 mm in diameter (Kutay et al., 2009, Lee et al., 2017). Specific test specification reference AASHTO TP 79 requirements.

The mixture Glover-Rowe parameter was employed as an indicator of cracking potential (Rowe et al., 2014). This parameter is calculated through the dynamic modulus (E^*) and phase angle (δ) at a temperature of 15°C, and the frequency of 0.005 rad/s, which is defined by **Equation 1**.

$$G - R_m = |E^*|(\cos \delta)^2 / \sin \delta \tag{1}$$

3.2 *Semicircular bending (SCB) test*

The SCB test was carried out to evaluate the thermal cracking resistance of HIR mixtures at an intermediate temperature of 15°C. The test was set up for monotonic loading, specimens were loaded in a three-point bending setup at a constant rate of 50 mm/min until failure. The specimens were fabricated with a thickness of 50 mm and a diameter of 150 mm and non-notch treatment. Four parallel samples were conducted for each series. The tensile stress (σ_m) and fracture energy (G_f) of SCB specimens were calculated following **Equation 2 and 3**.

$$\sigma_m = 4.8P/Dt \tag{2}$$

$$G_f = 2W_f/Dt \tag{3}$$

Where P is the value of peak load in fracture failure process (N); W_f is the fracture work; t is the thickness of the specimen (mm); D is the diameter of the specimen (mm).

3.3 *Direct tension cyclic test*

The fatigue resistance of HIR mixtures was characterized by direct tension cyclic test. The Controlled-strain loading mode was conducted at the frequency of 10Hz and the test temperature of 18°C. The test data were analyzed using the simplified viscoelastic continuum damage (S-VECD) model. The damage characteristic curves (DCC) were developed following AASHTO TP107. The function can be fitted as a power function represented by **Equation 4**, where C_{11} and C_{12} are the model coefficients.

$$C = 1 - C_{11}S^{C_{12}} \tag{4}$$

The cracking index based on the S-VECD framework named apparent damage capacity (S_{app}) was utilized to evaluate the fatigue property of HIR mixtures (Wang et al., 2020).

4 RESULT AND DISCUSSION

4.1 *Linear viscoelastic characterization*

The time-temperature superposition principle (TTSP) was applied to develop the dynamic modulus master curves. A reference temperature of 21.1°C was used to construct the master curves. It has been proved that the reduced frequency ranges between 0.1-500Hz is the region most relevant to cracking resistance of mixtures in intermediate temperature (Cao et al., 2018).

Figure 1(a) compares the $|E^*|$ values of HIR mixtures under different laboratory aging times. The two groups of the mixtures (adding rejuvenator and rejuvenator + SBR) exhibit similar linear viscoelastic properties. Generally, the complex modulus of mixtures increases with the aging time increasing. Unexpectedly, the master curves of the mixtures do not change significantly with aging at 135°C for 12 hours. That's probably attributed to the HIR mixture

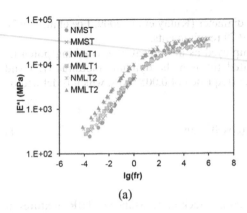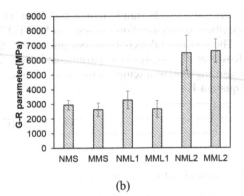

(a)   (b)

Figure 1. Linear viscoelastic characterization: (a) Dynamic modulus master curves; (b) The mixture G-R values.

containing high content of RAP that is under high levels of aging. However, the impact of 24h aging is distinctive, causing a significant difference to compare with 12h long-term aging.

Figure 1(b) shows the result of the G-R parameter for all the HIR mixtures. The higher aging levels result in the greater G-R value. Comparing the value of mixtures between short-term aging and 12h long-term aging, the magnitude of the increase is comparatively small. While the increase of G-R value is greater when raising the aging time from 12h to 24h. The addition of the SBR additive decreases the G-R value, indicating the improvement of cracking resistance.

4.2   SCB test results

Table 3 present the result of monotonic SCB tests for HIR mixtures. Tensile strengths increase whereas fracture energy decrease after subjected long-term aging. The addition of the SBR modifier has a negligible impact on the tensile stress of the HIR mixture, while there is a significant effect on the result of fracture energy. The post-peak fracture energy decreases as the aging levels increasing, meanwhile, the fracture pattern is manifested as the brittle fracture.

4.3   Direct tension cyclic test results

Figure 2(a) illustrates the damage characteristic curves (DCC) from the S-VECD model for all the HIR mixtures. The different aging conditions change the evolution of the damage curve. The position of the DCC becomes higher under higher levels of aging. The SBR modified HIR mixture follows a similar trend with the HIR mixture. The S_{app} parameter is a cracking

Table 3.   Summary of SCB test results.

Mixture	Tensilestrength		Fractureenergy		post-peakfractureenergy	
	Mean(MPa)	CV(%)	Mean(J/m2)	CV(%)	Mean(J/m2)	CV(%)
NMS	9.51	2.37	3254	8.78	950	11,78
MMS	9.54	3.13	3517	10.80	1013	12.52
NML1	9.81	4.35	3160	7.04	687	9.49
MML1	9.94	4.57	3347	9.52	739	13.80
NML2	10.50	5.02	2389	6.59	426	10.04
MML2	10.46	2.39	2749	9.48	303	13.74

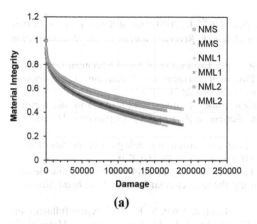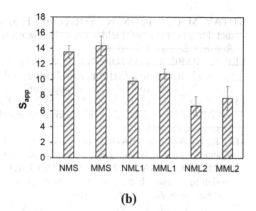

| (a) | (b) |

Figure 2.   Direct tension cyclic test results: (a) Damage characteristic curves; (b) The S_{app} values.

criterion from the S-VECD framework, which is based on the average reduction in pseudo stiffness and complex modulus of the mixture. **Figure 2(b)** shows the S_{app} value of the mixtures, the effect of short-term aging and 12h long-term aging on the values are similar. Also, aging decreases the performance enhancement of the SBR modifier.

5   CONCLUSIONS

A laboratory study was employed to investigate the linear viscoelastic characterization, thermal cracking, and fatigue resistance of the HIR mixture under different aging levels. Based on the analysis of tests results, the following summary and findings are offered:

The effect of 12h long-term aging on the linear viscoelastic properties of the HIR mixture is not significant compared with the short-term aged mixture, which could be caused by the high levels of aging and high content of RAP. Nevertheless, the post-peak fracture energy and S_{app} parameter could capture the degradation of fracture and fatigue performance, respectively.

Long-term aging (24h, 135°C) highly changes the linear viscoelastic characteristics and impairs the crack resistance of the HIR mixture. The fracture strength increases whereas the G_f and post-peak fracture energy decrease significantly, which is manifested as the brittle fracture. For fatigue behavior of HIR mixture, the pseudo stiffness becomes higher at the failure point and the S_{app} value shows a relatively high reduction, indicating the higher fatigue damage susceptibility.

The addition of the SBR modifier does not change the linear viscoelastic characteristic, while it improved fracture and fatigue properties. The simulated 7-10 year aging mixture showed significant performance degradation, in order to ensure the performance of the mixture, it is necessary to take appropriate modification measures.

REFERENCES

CAO, W., MOHAMMAD, L. N. & ELSEIFI, M. 2017. Assessing the effects of RAP, RAS, and warm-mix technologies on fatigue performance of asphalt mixtures and pavements using viscoelastic continuum damage approach. *Road Materials and Pavement Design*, 18, 353–371.

CAO, W., MOHAMMAD, L. N., ELSEIFI, M., COOPER, S. B. & SAADEH, S. 2018. Fatigue Performance Prediction of Asphalt Pavement Based on Semicircular Bending Test at Intermediate Temperature. *Journal of Materials in Civil Engineering*, 30.

CARPENTER, S. H. & SHEN, S. 2006. Dissipated energy approach to study hot-mix asphalt healing in fatigue. *Transportation Research Record: Journal of the Transportation Research Board*, 1970, 178–185.

KUTAY, M. E., GIBSON, N., YOUTCHEFF, J. & DONGRÉ, R. 2009. Use of Small Samples to Predict Fatigue Lives of Field Cores. *Transportation Research Record: Journal of the Transportation Research Board*, 2127, 90–97.

LEE, K., PAPE, S., CASTORENA, C. & KIM, Y. R. 2017. Evaluation of Small Specimen Geometries for Asphalt Mixture Performance Testing and Pavement Performance Prediction. *Transportation Research Record: Journal of the Transportation Research Board*, 2631, 74–82.

LI, Q., CHEN, X., LI, G. & ZHANG, S. 2018. Fatigue resistance investigation of warm-mix recycled asphalt binder, mastic, and fine aggregate matrix. *Fatigue & Fracture Of Engineering Materials & Structures*, 41, 400–411.

MA, T., HUANG, X., BAHIA, H. U. & ZHAO, Y. 2010. Estimation of rheological properties of RAP binder. *Journal of Wuhan University of Technology-Mater. Sci. Ed.*, 25, 866–870.

MAHMOUD, E., SAADEH, S., HAKIMELAHI, H. & HARVEY, J. 2014. Extended finite-element modelling of asphalt mixtures fracture properties using the semi-circular bending test. *Road Materials And Pavement Design*, 15, 153–166.

MILLS-BEALE, J., YOU, Z., FINI, E., ZADA, B., LEE, C. H. & YAP, Y. K. 2014. Aging Influence on Rheology Properties of Petroleum-Based Asphalt Modified with Biobinder. *Journal Of Materials In Civil Engineering*, 26, 358–366.

RAHBAR-RASTEGAR, R., DANIEL, J. S. & DAVE, E. V. 2018. Evaluation of Viscoelastic and Fracture Properties of Asphalt Mixtures with Long-Term Laboratory Conditioning. *Transportation Research Record: Journal of the Transportation Research Board*, 2672, 503–513.

ROWE, G. M., KING, G. & ANDERSON, M. 2014. The Influence of Binder Rheology on the Cracking of Asphalt Mixes in Airport and Highway Projects. *Journal of Testing and Evaluation*, 42.

SAHA, G. & BILIGIRI, K. P. 2016. Fracture properties of asphalt mixtures using semi-circular bending test: A state-of-the-art review and future research. *Construction And Building Materials*, 105, 103–112.

SHU, X., HUANG, B. & VUKOSAVLJEVIC, D. 2008. Laboratory evaluation of fatigue characteristics of recycled asphalt mixture. *Construction And Building Materials*, 22, 1323–1330.

WANG, Y. D., UNDERWOOD, B. S. & KIM, Y. R. 2020. Development of a fatigue index parameter, Sapp, for asphalt mixes using viscoelastic continuum damage theory. *International Journal of Pavement Engineering*, 1–15.

YANG, X., MILLS-BEALE, J. & YOU, Z. 2017. Chemical characterization and oxidative aging of bio-asphalt and its compatibility with petroleum asphalt. *Journal Of Cleaner Production*, 142, 1837–1847.

ZHANG, R., SIAS, J. E., DAVE, E. V. & RAHBAR-RASTEGAR, R. 2019. Impact of Aging on the Viscoelastic Properties and Cracking Behavior of Asphalt Mixtures. *Transportation Research Record: Journal of the Transportation Research Board*, 2673, 406–415.

ZHOU, Z., GU, X., NI, F., LI, Q. & MA, X. 2017. Cracking Resistance Characterization of Asphalt Concrete Containing Reclaimed Asphalt Pavement at Intermediate Temperatures. *Transportation Research Record: Journal of the Transportation Research Board*, 2633, 46–57.

Functional Pavements – Chen et al (eds)
© 2021 Taylor & Francis Group, London, ISBN 978-0-367-72610-2

Performance of SMA mixture added iron tailings coarse aggregate

Qingli Zhang
Hubei Provincial Road & Bridge Group Co. LTD

Yilong He
Shandong Jiaotong University

Kangming Wang, Changqing Jiao, Wen Zhu & Jiayu Wang
Hubei Provincial Road & Bridge Group Co. LTD

ABSTRACT: In view of the current situation that tailing resources have not been utilized effectively, the possibility of using iron tailings as coarse aggregate to SMA in pavement surface is discussed. In this study, the best gradation and the best asphalt content of SMA-13 are determined by material optimization, mix proportion optimization and performance verification. Rutting test, immersion Marshall test, freeze-thaw splitting Test, Schellenberg Binder Drainage test and Cantabro test were used to evaluate the performance of SMA. The results show that when the gradation is 10~15mm iron tailings: 5~10mm iron tailings: 0~5mm machine-made sand: mineral powder = 36:37:16:11, and the optimal asphalt-aggregate ratio is 6.3%, the test results of SMA-13 added iron tailings coarse aggregate can meet the requirements of current asphalt mixture technical specifications, and have good performance.

1   INTRODUCTION

With the vigorous development of infrastructure projects, the demand for mineral materials is increasing day by day. A large number of mineral resources have been developed and a large number of tailings have been produced. Therefore, the development and utilization of tailings have become a major problem to be solved in economic construction. If iron tailings can be effectively used in the new highway construction field, it can not only greatly reduce the cost, make up for the shortage of raw materials, but also solve the environmental pollution. Zhang, T. Z. added iron tailings into AC-13C and AC-16C asphalt mixtures. By testing their road performance, it is concluded that all performance indexes can meet the requirements, and some performances are better than ordinary asphalt mixture[1,2,3]. Pan, B. F studied the asphalt mixture added iron tailings sand by rutting test, low temperature bending and bending creep test, immersion Marshall test, freeze-thaw splitting test and paving test road, it is concluded that the cost is not significantly different from ordinary asphalt mixture[4,5,6]. In this paper, the basic properties of iron tailings, SMA mixture ratio design and road performance are studied in depth. This study discusses the possibility of SMA mixture added iron tailings coarse aggregate in low-grade pavement or pavement base.

2   RAW MATERIAL TEST

The raw materials are composed of coarse aggregate, fine aggregate, ore fines, asphalt and fiber. The coarse aggregate is iron tailings from Benxi City, Liaoning Province and basalt from Jinan City, Shandong Province, with specifications of 5~10mm and 10~15mm. The fine aggregate is limestone machine-made sand from Jinan City, Shandong Province, with the specification of 0~5mm. The mineral powder is fine limestone powder grinded and processed. The asphalt is SBS type I-D modified asphalt. The fiber is lignin fiber. The tests are based on *Test Methods of Aggregate for Highway Engineering (JTG E42—2005)* and *Standard Test Methods of Bitumen and Bituminous Mixtures for Highway Engineering (JTG E20—2011)*. The properties of raw materials were tested. Part of the test results of iron tailings and basalt coarse aggregate are shown in Table 1. Part of the test results of asphalt are shown in Table 2.

After a series of tests, the iron tailings and basalt coarse aggregate, limestone machine-made sand, mineral powder, SBS modified asphalt and lignin fiber all meet the technical requirements of raw materials in *Technical Specifications for Construction of Highway Asphalt Pavements (JTG F40-2004)*.

3 SMA MIXTURE DESIGN ADDED IRON TAILINGS COARSE AGGREGATE

There are two kinds of aggregate gradation design schemes: (1) Coarse aggregate is iron tailings (denoted by the letter M); (2) Coarse aggregate is basalt (denoted by letter B). In order to verify that the iron tailings coarse aggregate has good performance, the iron tailings coarse aggregate is used as the main part of the experiment to collect and process the basalt coarse aggregate. Sieving and washing each ore materials, the sieving test results of mineral aggregate are shown in Table 3.

According to the aggregate screening test results, combined with the upper and lower limit gradation range of SMA-13 in *Technical Specifications for Construction of Highway Asphalt Pavements (JTG F40-2004)* and the design experience in practical work. The coarse, medium and fine gradation are selected for optimization and comparison. After repeated calculation and comparison of mineral aggregate gradation, the optimal mineral aggregate ratio is 10~15mm (M, B): 5~10mm (M, B): 0~5mm machine-made sand: mineral powder = 36:37:16:11.

According to the mineral aggregate ratio and experience, the asphalt-aggregate ratio is selected to be 5.7%~6.9%, and 0.3% different intervals are used to produce different asphalt-aggregate ratio Marshall samples. According to the test results of different asphalt-aggregate

Table 1. Technical indices of iron tailings and basalt coarse aggregate.

Property		Unit	Freeway and first-class highway Surface layer	Other Layers	Test result Trailings	Basalt	Test method
Crushing value of stone, ≤		%	26	28	12.8	9.7	T 0316
Los Angeles abrasion loss, ≤		%	28	30	16.7	16.3	T 0317
Apparent relative	13.2~19mm	—	2.60	2.50	2.835	2.899	T 0304
density, ≥	9.5~13.2mm				2.868	2.931	
	4.75~9.5mm				2.896	2.936	
Water absorption, ≤		%	2.0	3.0	0.82	0.75	T 0304
Water washing method < 0.075mm particle content, ≤		%	1	1	0.3	0.2	T 0310

Table 2. Technical index of asphalt.

Property	Unit	SBS (I-D)	Test result	Test method
Penetration 25°C, 100g, 5s	0.1mm	40~60	51	T 0604
Ductility 5°C, 5cm/min, ≥	cm	20	36	T 0605
Softening point $T_{R\&B}$ ≥	°C	60	73	T 0606

Table 3. Screening test results of mineral aggregate.

Mineral specification	Pass rate(%)									
	16	13.2	9.5	4.75	2.36	1.18	0.6	0.3	0.15	0.075
10~15mm(M、B)	100	93.1	12.8	0.4	0.3	0.3	0.3	0.2	0.2	0.2
5~10mm(M、B)	100	100	97.7	0.6	0.2	0.2	0.2	0.2	0.2	0.1
0~5mm	100	100	100	99.6	61.3	37.8	23.1	14.2	10.1	7.7
Mineral powder	100	100	100	100	100	100	100	98.1	91.7	80.0

ratios, considering the heavy traffic sections and hot areas with high requirements for high temperature stability, the expected objective air void is 4%. When the air void is 4%, the asphalt-aggregate ratio of iron tailings is 6.3%, and the asphalt-aggregate ratio of basalt is 6.2%. The asphalt-aggregate ratio of the two schemes is almost the same, which is about 6.2% ~6.3%. In order to unify the asphalt-aggregate ratio and make the parameters of the mixture comparable, the asphalt-aggregate ratio is selected as 6.3%. When the gradation of SMA is 10~15mm (M, B): 5~10mm (M, B): 0~5 mm machine-made sand: mineral powder = 36:37:16:11, the best asphalt-aggregate ratio is 6.3%, and the lignin fiber is 0.3% of the total mass of asphalt mixture, the parameters of SMA can meet all the technical requirements of the current specifications.

4 ROAD PERFORMANCE OF SMA MIXTURE ADDED IRON TAILINGS COARSE AGGREGATE

4.1 *High temperature stability performance of asphalt mixture*

At present, high temperature stability is mainly measured by rutting test, and dynamic stability is used as evaluation index. The rutting test results are shown in Table 4.

According to the analysis of rutting test results, it is concluded that the high temperature stability of iron tailings coarse aggregate SMA is slightly lower than the basalt coarse aggregate SMA, and the reduction range is about 20%. The main reason for the difference of dynamic stability is that the different SMA aggregates lead to different compressive and crushing properties. Combined with the analysis of raw material test results, the crushing resistance of iron tailings is slightly worse than basalt, resulting in the breakage of coarse aggregate in the specimen, instability of skeleton structure, increase of rutting depth and decrease of dynamic stability. Although compared with basalt, the high temperature stability of iron tailings coarse aggregate SMA is slightly insufficient, but the dynamic stability value of iron tailings coarse aggregate SMA is more than 2 times of the specified requirements. Therefore, the high temperature stability greatly exceeds the requirements of the current asphalt mixture technical specifications and meets the service conditions.

4.2 *Water stability performance of asphalt mixture*

At present, the main research methods of water stability are immersion Marshall test and freeze-thaw splitting test. The residual stability of asphalt mixture is measured by Marshall

Table 4. Rutting test results.

SMA type	OAC (%)	DS (Times/mm)
M	6.3	6033
B	6.3	7273

Table 5. Immersion Marshall test and freeze-thaw splitting test results.

SMA type	Specimen	Mean value of stability (kN)	MS_0 (%)	Specimen	Mean value of splitting tensile strength (MPa)	TSR (%)
M	Immersion 0.5h	10.90	92.2	Freeze-thaw	0.989	96.5
	Immersion 48h	10.05		No freeze-thaw	1.025	
B	Immersion 0.5h	11.34	92.0	Freeze-thaw	1.026	97.5
	Immersion 48h	10.43		No freeze-thaw	1.052	

stability method, and the splitting tensile strength ratio TSR of asphalt mixture is measured by freeze-thaw splitting method. The test results are shown in Table 5.

According to the analysis of immersion Marshall test results, it is concluded that the stability of iron tailings coarse aggregate SMA decreases slightly after soaking for 0.5h and 48h than the basalt coarse aggregate SMA, but the residual stability of the two is almost the same. According to the analysis of freeze-thaw splitting test results, it is concluded that the splitting strength and freeze-thaw splitting strength ratio of iron tailings coarse aggregate SMA decreases slightly before and after freeze-thaw than the basalt coarse aggregate SMA. The difference between them is related to adhesion. Under the joint action of water and temperature, the adhesion between basalt coarse aggregate and asphalt is slightly better than that of iron tailings, and the asphalt on the surface of basalt coarse aggregate is more difficult to fall off. The density of iron tailings and basalt is also different, and the effective asphalt film thickness on the surface of ore is also different. Although the water stability is slightly less than the basalt, the residual stability and freeze-thaw splitting strength ratio of the iron tailings coarse aggregate SMA greatly exceed the requirements. Therefore, the water stability performance greatly exceeds the technical specifications of the current asphalt mixture and meets the service conditions.

5 CONCLUSION

(1) The strength of iron tailings is slightly less than the basalt, and the chemical composition of iron tailings and basalt is different, which makes the adhesion of iron tailings and asphalt slightly lower than the basalt, but the performance fully meets the current specification requirements.
(2) The results show that the best gradation of iron tailings coarse aggregate SMA is 10~15mm: 5~10mm: 0~5mm: mineral powder = 36:37:16:11, and the best asphalt-aggregate ratio is 6.3%. All parameters of SMA meet all the technical requirements of current specifications.
(3) Compared with the basalt coarse aggregate SMA, the high temperature stability of iron tailings coarse aggregate SMA is slightly lower, and the reduction range is about 20%. However, the dynamic stability value is more than 2 times of the specified requirements, and the high temperature stability greatly exceeds the current technical specifications requirements, and meets the service conditions.
(4) Compared with the basalt coarse aggregate SMA, the residual stability of iron tailings coarse aggregate SMA is almost equal, and the freeze-thaw splitting strength ratio is slightly decreased. However, the residual stability and freeze-thaw splitting strength ratio greatly exceed the specified requirements, However, the water stability greatly exceeds the requirements of the current asphalt mixture technical specifications, and meets the service conditions.

ACKNOWLEDGMENTS

This study was supported by Science and Technology Plan of Shandong Transportation Department (2019B63 and 2020B93).

REFERENCES

[1] Zhang, T. Z. 2010. RESEARCH ON FREEZING-THAWING CHARACTERISTICS OF INOR-GANIC BINDERS STABILIZED IRON TAILINGS. *Low Temperature Architecture Technology, 32(01):8–10.*

[2] Cui, Z. X. 2010. Iron Tailings Comprehensive Utilization at Home and Abroad. *Journal of Jilin Institute of Architecture & Civil Engineering, 27(04):22–26.*

[3] Jin, C. Z. 2012. Research on the Performances of Lime Fly Ash Stabilized Iron Tailing Gravel in Highway Application. *Journal of Wuhan University of Technology, 34(03):59–62.*

[4] Liu, T. J. 2007. Research of iron tailings application in the highway engineering. *Mining Engineering, (05):52-54.*

[5] Pan, B. F., & Chen, L. Y. 2011. Experimental Research on Dry Shrinkage Performance of Ferrous Mill Tailings Stabilized with Inorganic Binding Materials. *Advanced Materials Research, 1269:4166-4171.*

[6] Zhang, Z. 2010. Experiment of reinforced iron tailings used for road base. *Journal of University of Science and Technology Liaoning, 33(01):29–31.*

Functional Pavements – Chen et al (eds)
© 2021 Taylor & Francis Group, London, ISBN 978-0-367-72610-2

Effect of graphene on synthesized pavement performance of rubber asphalt mixture

Y. Lu, A.H. Liu, Y.L. Li & Q. Li
JSTI Group, National Engineering Laboratory for advanced road materials, Nanjing, China

ABSTRACT: Graphene has been used as a new material in materials science, energy biomedicine, drug delivery and other fields. However, the application of graphene in road engineering, especially the addition to rubber asphalt mixture, is still a frontier field. Therefore, the effect of graphene on rubber asphalt mixture was evaluated by laboratory tests to find the advantage of graphene modified rubber asphalt. The results show graphene can improve the resistance to water damage and high-temperature performance significantly. Besides, the Accelerated Pavement Testing verifies the durability of mixture can be extended by introducing graphene. But the conclusion of low-temperature performance exhibited a little lower by introducing graphene. Moreover, the field pilot was carried out to verify the advantage of rubber asphalt mixture in addition of graphene for the resistance to water damage and high-temperature performance.

1 INTRODUCTION

In recent years, graphene has been penetrated into the field of traffic engineering as a reinforcing phase. Research scholars at home and abroad have launched relevant research on the application of graphene in asphalt materials technology to improve the road performance of asphalt pavement [1]. Theoretically, graphene has a two-dimensional crystal structure, the extremely high specific surface area and similar structure to bitumen, which all make graphene and bitumen closely combined easily [2]. At present, some scholars have used graphene to modify rubber asphalt and found out that the addition of graphene increases the interfacial compatibility of the asphalt and rubber powder, reduces the construction temperature and dangerous gas emissions [3]. It is proved that graphene modified rubber asphalt mixture has excellent high dynamic stability, Marshall strength, and significant resistance to rutting and shearing [4]. By freeze-thaw split test, immersed Marshall test, Chinese rutting test, Hamburg rutting test, trabecular low temperature bending test and accelerated pavement loading test, this article has analyzed the effect of graphene on the comprehensive performance of rubber modified asphalt road [5–12]. Moreover, according to the project physical works, the graphene modified rubber asphalt mixture was practical.

2 EXPERIMENT AND TEST METHODOLOGY

2.1 *Raw materials*

(1) Graphene.
The graphene used in this work is provided by The Sixth Element (Changzhou) Materials Technology Co., Ltd. The physical properties of this graphene are shown in Table 1.

Table 1. Physical properties of grapheme.

appearance	pH	Tap density / (g/cm³)	Specific surface area / (m²/g)	Water quality score/%	Particle size / (D_{50}, μm)	carbon /%	oxygen /%	sulfur /%
Black powder	2.0~5.0	<0.1	180~280	<4.0	<10.0	75±5	16±3	<0.5

Table 2. Technical indicators of rubber asphalt.

Needle Penetration/ (25°C, 5s, 0.1mm)	Ductility/ (5°C, cm)	Softening Point/°C	density/ (15°C) (g/cm³)	Kinematic viscosity/ (135°C, Pa.s)	Elastic recovery/ (25°C) (%)
52	15	68.0	/	2.3(177°C)	82

(2) Rubber asphalt.

The particle size and dosage of crumb rubber used in this test are 40 mesh and 25% respectively. The relevant technical performance indicators of rubber asphalt used in this test are shown in Table 2.

(3) Aggregate.

The coarse aggregate in this experiment was collected from the construction of the stone yard in Chupulagou, Lintao County. Fine aggregates were collected from Yongdeng Xinhui Building Material Factory. The ore powder is collected from the diabase of Yongdeng Xinhui Building Material Factory. Basic technical performance indicators are shown in Table 3.

The aggregate gradation is shown in Table 4. Superpave method was selected for design of the mixture proportion. The optimal asphalt content was determined as 5.2% by the voidage of the mixture for all cases.

2.2 *Preparation of graphene modified rubber asphalt binder*

According to existing studies, Ultrasound can be used as a strong dispersing tool, which can effectively prevent the agglomeration of particles and make them fully dispersed. Therefore, the ultrasonic dispersion method is used to prepare graphene modified rubber asphalt binder.

Melted rubber bitumen was poured into a blended vessel and the temperature was kept at 182°C. Graphene accounted for 3‰ by mass of bitumen were poured into the vessel. Firstly, glass rods were stirred manually for 10min until no nanometer materials floated on the asphalt liquid level. Then, High shear emulsifying machine was used to shear the mix at 3500 r/min for 1 h and mechanical mixer was used to stir the mix for 0.5h. the temperature was kept at 177 ± 5°C. Finally, ultrasonic dispersing instrument was adopted to ultrasonic for 0.5h. After

Table 3. Basic performance indicators of aggregate.

Test projects	Aggregate identity				Aggregate source characteristics		
	Coarse aggregate angularity/%	Fine aggregate angularity/%	Flat, slender particles/%	Sand equivalent/%	Los Angeles wear loss/%	Ruggedness/%	Adhesion
Test value	100	46.2	7.6	72	22.9	3	5
technical standard	85/80	≥45%	≤15%	≥60%	≤28%	≤12%	≥4

Table 4. Gradation of hot asphalt mixture.

Sieve size	19mm	13.2 mm	9.5 mm	4.75 mm	2.36 mm	1.18 mm	0.60 mm	0.30 mm	0.15 mm	0.075 mm
Lower-upper limits	90~100	71~83	60~73	42~52	26~34	14~22	9~17	5~11	3~12	2~7
Passing percentage	100	98.6	79.9	46.2	29.9	21.8	14.7	8.3	5.3	4.4

Note: The figures in the table under the horizontal bar is strictly controlled, other parts are allowed to be different

the above-mentioned process, the graphene modified rubber asphalt binder was obtained. Rubber bitumen was handled as the above process to make the blank sample.

2.3 Testing methods and evaluation indicators

In order to study the effect of graphene as a reinforcement on the road performance of rubber-modified asphalt, in this paper, the graphene composite rubber modified asphalt mixture test piece with graphene content of 3‰ and asphalt content of 5.2% is adopted compared with test piece of rubber modified asphalt mixture with 5.2% asphalt. The two test pieces are analyzed and compared over water stability performance, high temperature performance, low temperature performance and durability performance by freeze-thaw split test, water immersion Marshall test, Chinese rut test, Hamburg rut test, trabecular low temperature bending test and small accelerated loading test [13].

2.3.1 Water stability test method
(1) Freeze-thaw split test (Figure 1)
Take splitting residual strength ratio as an evaluation indicator. The void ratio of the test piece used is required to be 7%±1.0% to form two sets of test pieces. After the first group of test pieces are saturated with water under vacuum, they are frozen at -18°C±2°C for 16h±1h, and then placed in a constant temperature water tank at 60°C±0.5°C for 24h. Finally, immerse the first and second groups of test pieces in a constant temperature water tank with a temperature of 25°C±0.5°C for not less than 2h. The results of the freeze-thaw split test of the first group of test pieces are the test results under the conditions, and the second group are the unconditional test results.

Figure 1. Splitting instrument.

(2) Water immersion Marshall test (Figure 2)
The residual stability is used to evaluate the water stability. The greater the ratio (residual stability), the better the water stability. The prepared two sets of test pieces were placed in a constant temperature water bath at 60°C±1°C for 48h and 30min respectively, and the test loading rate was 50±5mm/min. The Marshall stability was tested respectively.

2.3.2 Wheel tracking test

At present, the methods for evaluating the high temperature stability of asphalt mixture mainly include the rutting test in China, the rutting test in the United States, the rutting test in Hamburg and the rutting test in France. The Chinese rut test as shown in Figure 3 and the Hamburg rut test as shown in Figure 4 are used to evaluate the high temperature performance of graphene composite rubber asphalt mixture and rubber modified asphalt mixture in this paper.

(1) Rutting Test
Rutting is the most common disease at high temperature of asphalt pavement surface. Wheel tracking test was used to evaluate the anti-rutting performance of the unmodified and modified mixtures at high temperature on the basis of JTG E20. The test temperature was selected as 60°C and wheel load pressure was 0.7 MPa. Both unaged and long-aged mixtures of modified asphalt and unmodified asphalt were tested. Three replicates for each type of mixture were carried out. Dynamic stability (DS) calculated to evaluate the anti-rutting ability of asphalt mixtures is defined as follows:

Figure 2.   Marshall instrument.

Figure 3.   China Rut Instruments.

$$DS = \frac{(t_2 - t_1) \times N}{d_2 - d_1} \times C_1 \times C_2 \qquad (1)$$

where d_1 is the deformation at time t_1 (45 min), d_2 is the deformation at time t_2 (60 min), N refers to the rolling speed, C_1 and C_2 are testing machine coefficient and specimen coefficient respectively.

(2) Hamburg Rutting Test

The test obtains indicators such as rut depth, creep slope, spalling inflection point and spalling slope by analyzing rutting deformation curves, and uses creep slope and rut depth as indicators to evaluate high-temperature rutting resistance.

2.3.3 Low temperature indirect tensile test

Low temperature indirect tensile test is suitable for testing the mechanical properties of mixtures when the mixture is destroyed by indirect tensile at specified temperatures and loading rates. In this study, the low temperature indirect tensile test as shown in Figure 5 was performed on the long-term aged mixtures on the base of JTG E20. Four Marshall specimens were prepared for each asphalt. 10 °C test temperature was selected, and the specimens were insulated for 6 h before the test. Using the following equation to figure out indirect tensile strength (R_T) and indirect tensile failure strain (ε_T):

$$R_T = 0.006287 P_T / h \qquad (2)$$

$$\varepsilon_T = X_T \times (0.0307 + 0.0936\mu)/(1.35 + 5\mu) \qquad (3)$$

where P_T is the Maximum load, l is the Poisson's ratio, h is the sample thickness, and X_T is the total horizontal deformation under the maximum load.

2.3.4 Durability test method

A small accelerated loading test as shown in Figure 6 was used to evaluate the durability of the mixture. The specimen was placed in a 50°C water bath and a large amount of repetitive loading was applied in a short period of time. By recording the rutting depth and analysis of the specimen during different loading times over the loading process, the change rule of the rut depth of the test piece with the number of axle loads is to evaluate its resistance to rutting deformation and durability.

Figure 4.   Hamburg Rut Test Instrument.

Figure 5.   Trabecular low temperature bending test instrument.

Figure 6.   Miniature accelerated loading system (MMLS3).

3   RESULTS AND DISCUSSION

3.1   *Water stability*

Freeze-thaw splitting test results are shown in Figure 7. TSR is used to evaluate the water stability of the mixtures in this study. It can be observed in Figure 7 that splitting strength of both kinds of asphalt mixture treated by freezing and thawing decrease. The TSR value of asphalt mixture with graphene composite rubber is 0.9% higher than the splitting strength ratio (TSR) of rubber asphalt mixture, which shows that graphene composite rubber modified asphalt mixture has significantly improved water stability compared with rubber asphalt mixture. The high TSR means the material is not sensitive to water damage. Moreover, under both conditional and non-conditional conditions, cleavage strength values of graphene composite rubber modified asphalt mixtures were 10.8% and 11.9% higher than those of rubber modified asphalt mixtures, respectively, which shows that graphene composite rubber modified asphalt can improve the strength of rubber asphalt mixture. Nevertheless, according to Chinese standard, the TSR values of the modified asphalt mixtures can meet the construction requirements.

Water immersion Marshall test results are shown in Figure 8. MS_0 is used to evaluate the water stability of the mixtures in this study. It can be seen in Figure 8 that MS_0 of asphalt mixture with graphene composite rubber is above the other one, which indicates that graphene greatly improves the water failure resistance of asphalt mixture with rubber.

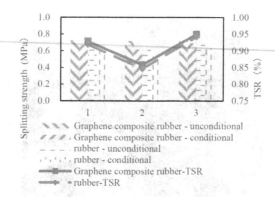

Figure 7.   Results of freeze-thaw split test.

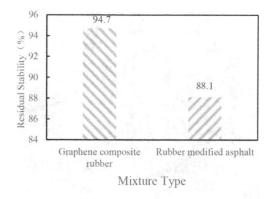

Figure 8.   Comparison of immersed Marshall test.

From the analysis of the modification mechanism[14–16], the incorporation of graphene can absorb the alkane groups and aromatic compounds in the rubber asphalt, strengthen the adhesion between the rubber and the asphalt, and thus improve the water stability of the modified asphalt, which is consistent with the results of the submerged Marshall test and the freeze-thaw split test.

3.2   High temperature stability analysis

With rutting test and Hamburg rutting test, the high temperature stability can be analyzed. The results are shown in Figures 9 and 10. Dynamic stability and rutting depth are used to evaluate high temperature stability of the mixtures respectively in this study.

From Figure 9 it can be seen that the dynamic stability of asphalt mixture with graphene composite rubber increase by 62.2% compared with asphalt mixture with rubber. Therefore graphene improves the high temperature stability resistance by increasing to permanent deformation resistance.

As seen in Figure 10, rutting depth increases with the number of rolling. Seen Table 5, when the number of rolling is 5000, 10000 and 20,000, the rut depth of asphalt mixture with graphene composite rubber is reduced by 44.67%, 43.91% and 44.44% compared with the asphalt mixture with rubber. And it meets the rut depth requirement of Modified asphalt mixture from PG70 to PG82 proposed by Hamburg, Texas. This phenomenon demonstrates graphene greatly increases high temperature stability of asphalt mixture with rubber.

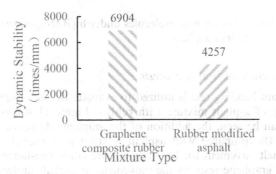

Figure 9. Comparison of rutting test results.

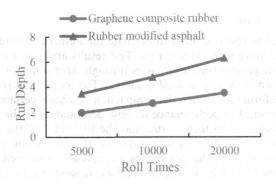

Figure 10. Comparison of Hamburg Rut Test Results.

Table 5. Field indoor est results of graphene-rubber modified asphalt mixture.

	Three index test			Volumetric test			
Test content	Needle penetration (25°C, 5s, 0.1mm)	Ductility (5°C, cm)	Softening point (°C)	The relative density of the specimen's gross volume	Voidage (%)	Voids in mineral aggregate (%)	Saturability (%)
Test result	54	27	70.5	2.546	4.1	14.4	71.5
Technical requirements	30~70	≮15	≮70	-	4	≥14	65~75

	Water stability test		High temperature stability test
Test content	Freeze-thaw split test/ TSR (%)	Immersion Marshall test/ Residual Stability ratio (%)	Rutting test/CV (%)
Test result	85.4	93.4	12.1
Technical requirements	80	85	≤20

From a microscopic point of view [14–16], graphene oxide has a large specific surface area and high surface activity, and is adsorbed on the surface of rubber particles. As a physical cross-linking point between rubber and asphalt, graphene promotes the combination between the two, limit the movement of asphalt molecules under high temperature conditions, thereby improving the stability of rubber asphalt.

3.3 *Low temperature crack resistance analysis*

Low temperature beam bending test is utilized to evaluate the influence of graphene on low temperature property of asphalt mixture with rubber. Figure 11 shows the results of beam bending test. As it can be seen, the addition of the graphene decreases the low temperature failure strain by 23.3% compared with asphalt mixture with rubber. But it can meet the requirements of asphalt pavement for low temperature crack resistance performance indicators. This is because graphene restricts the movement of asphalt molecules under high temperature conditions, increases the hardness of modified asphalt, and reduces its deformability under low temperature conditions, that is, low temperature performance is weakened [14–16].

3.4 *Durability analysis based on small accelerated loadings*

Small accelerated loading test is utilized to evaluate the influence of graphene on durability property of asphalt mixture with rubber. Figs. Test results are shown in Figures 12 and 13. The deformation of the general specimen passes through three stages of compaction deformation, shear flow deformation and instability failure. The initial load is mainly compaction, mainly compaction deformation; when the compaction reaches a certain degree, it begins to mainly flow deformation. The performance of flow deformation is that the position of the wheel track belt is depressed to form a rut, and the two sides of the test wheel track are formed uplift deformation due to lateral restraint. As the number of loading increases, the deformation develops more until the specimen is unstable and destroyed. It can be seen from the test results in Figure 12 and the appearance of the core sample in Figure 13 that the overall

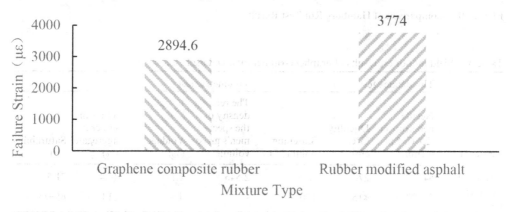

Figure 11.   Comparison of trabecular bending test results.

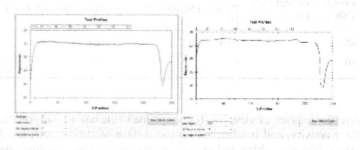

Figure 12.   Graph of small accelerated loading test results.

Figure 13.　Accelerated loading of test specimens.

(a) Field paving　　　　　　　　(b)Field rolling

Figure 14.　Paving and rolling on the test road.

degree of deformation after rolling is small, and there is no significant tire indentation. It shows that graphene composite rubber modified asphalt mixture has better rutting resistance, water damage resistance and fatigue resistance, which is consistent with the above water stability and high temperature stability tests.

4　FIELD PILOT APPLICATION

In order to better evaluate the pavement performance of graphene composite rubber modified asphalt and verify the actual road effect, physical engineering test road paving was carried out on the national Highway G309 modification section and the main line section of the Lanzhou South Ring Expressway, and the pavement process and effect were tested and evaluated.

The three major index tests, volume index test, water stability test and high temperature stability test were carried out on the graphene composite rubber modified asphalt mixture samples, and the test road was tested on site. The test contents included thickness, compaction and water penetration, structure depth, friction coefficient, flatness and splitting strength. The test results are shown in Table 5 and Table 6, indicating that they all meet the relevant technical documents and specifications.

Table 6. Field test results of graphene composite rubber modified asphalt mixture.

Test content	Thickness (cm)	Puddlability (%)	Water penetration testing (ml/ min)	Friction coefficient detection	Structural depth detection (cm)	Regularity inspection(mm)	Detection of splitting strength (Mpa)
Test result	4.2	93.9	23	72.6	0.75	0.54	1.365
Technical requirements	≥4.0	93~97	≤80	≥50	≥0.5	≤1.2	≥1.2

5 CONCLUSION

In this paper, effect of graphene on mechanical properties of asphalt mixture with rubber were investigated. Main conclusions can be drawn as follows:

(1) With the addition of graphene, the water stability, high temperature stability and durability of asphalt mixture with rubber are improved significantly, while the low temperature crack resistance is decreased. The results show that the splitting strength is increased by 10.8% and 11.9%, the splitting strength ratio (TSR) is increased by 0.9%, the dynamic stability is improved by 62.2%, and the rutting depth is reduced by 44.67%, 43.91% and 44.44% respectively when the rolling times are 5000, 10000 and 20000, respectively The crack resistance is reduced by 23.3%, but it can meet the requirements of low temperature anti cracking performance index of asphalt pavement.

(2) From the mechanism analysis [14–16] the incorporation of graphene can absorb the alkane groups and aromatic compounds in the rubber asphalt, strengthen the cohesion between rubber and asphalt binder, and thus improve the water stability of the modified asphalt, In addition, due to its large specific surface area, high surface activity, capable of acting with asphalt molecules and rubber, acting as a bridge between the two, making it more tightly bound, limiting the movement of asphalt molecules under high temperature conditions, that is, the high temperature stability of rubber asphalt is improved However, at the same time, the hardness of the modified asphalt is increased and the deformability under low temperature conditions is reduced, that is, the low temperature performance is weakened.

(3) According to the field test results of the physical engineering, all the test results meet the requirements of relevant technical documents, design and specifications, and show good high-temperature stability and water damage resistance.

REFERENCES

[1] Su, D.Y, Gao, Y. 2016. Research on recent graphene-enhanced composites projects in Europe and the United States. Jiangsu Science & Technology Information 12(34):55–59.
[2] Fang, C.Q, Yu.R, Liu S.L, Li Y. 2013. Nanomaterials applied in asphalt modification: a review. Journal of Materials Science & Technology 29(7): 589–594.
[3] 2018. The world's first graphene composite rubber modified asphalt pavement was built in nanning. Municipal technology (4).
[4] Sun, L, Wang, H.Y, Wang, S.X. 2012. Nano-modified asphalt and its road performance. Science press.
[5] Gao, J.Y, Zeng, M.L, Sun Z.L. 2019. Effects of nano-SiO$_2$ with modified surface on properties of recycled asphalt binder. China Civil Engineering Journal 52(03):120–128.
[6] Sun, L, Zhu, H.R, Xin, X.T, Wang, H.Y, Gu, W.J. Preparation of Nano-modified Asphalt and Its Road Performance Evaluation[J]. China Journal of Highway and Transport,2013,26(01):15–22.
[7] Chen, Y.K, Liu, J.Y, Bai, M.K. 2019. Road Performance Evaluation of Asphalt modified by Nano-ZnO. Technology and Economic Guide 27(25):69.
[8] Zhu, Q.P. 2019. Preparation and properties analysis of asphalt modified by nano-ZnO. Applied Chemical Industry 48(05):1031–1034.

[9] Chen, G.Z, Fan, L. 2018. Research on road technical properties of Nano TiO2 modified asphalt Shandong. Transportation Technology (05):6–9.

[10] Kang, A.H, Xiao, P, Zhou, X. 2010. Study on hot storage stability of nanometer ZnO/SBS modified asphalt. Journal of Jiangsu University (Natural Science Edition) 31(4): 412–417.

[11] Yu, R.E. 2016. Preparation and Properties of Asphalt Modified with Graphene Oxide/Polyurethane Composite. Doctoral dissertation, Xi'an: Xi'an University of Technology.

[12] Quan, X.D. 2011. The Research on Interface Mechanism and Performance for Crumb Rubber Modified Asphalt. Master Thesis, Wuhan: Wuhan University of Science and Technology.

[13] Li, P.L. 2012. Study on Pavement Performance and the Mix Proportion Design of Rubber Asphalt Mixture. Chang'an University.

[14] Zeng, Q. 2019. Preparation and Mechanism analysis of Graphene Oxide/Asphalt Composites. Shenyang Aerospace University.

[15] Guo, H.R, Meng Y J, Xu R G. 2019. The Rheological and Microscopic Properties of Graphene Rubber Composite Modified Asphalt. Journal of Building Materials1–9.

[16] Zeng, Q, Liu, Y, Liu, Q.C. 2020. Preparation and modification mechanism analysis of graphene oxide modified asphalts. Construction and Building Materials, 238:1–9.

Functional Pavements – Chen et al (eds)
© 2021 Taylor & Francis Group, London, ISBN 978-0-367-72610-2

Observation and research on performance of composite gussasphalt concrete steel deck pavement

K.Q. Zhang, J. Cao*, Z. Wu & R.J. Cao
JSTI GROUP, Nanjing, Jiangsu, China
National Engineering Laboratory for Advanced Road Materials, Nanjing, Jiangsu, China

ABSTRACT: The application of composite gussasphalt concrete steel deck pavement structure in the Fourth Nanjing Yangtze River Bridge provides a new idea for steel deck pavement of long-span bridges in China. In order to study the applicability of the pavement scheme to domestic high temperature, heavy load and hot rainy season, multi-functional inspection vehicle, laser profiler and friction coefficient detection vehicle were used to tracking detection of the pavement performance of the Fourth Nanjing Yangtze River Bridge from 2013 to 2020. The results show that after more than 7 years of use, its riding quality index, rutting depth index, skidding resistance index and surface condition index of the pavement are all excellent, and the overall road performance is good. The prediction results of the optimized rutting depth model show that the average rutting depth of the third two-way lane pavement will exceed 10 mm by 2022, and follow-up observation of this index should be strengthened. The research results are of great significance to the design optimization, promotion and application of the pavement scheme.

Keywords: Gussasphalt Concrete, Steel Deck Pavement, Pavement Performance, Prediction of Rutting Depth, Follow-up Observation

1 INTRODUCTION

The steel deck pavement is directly laid on the orthotropic steel deck, and provide driving comfort and safety, its mechanics response is much more complex than that of ordinary asphalt pavement. It has always been a hot and difficult point in academic research. After decades of research and development, several mainstream pavement schemes such as epoxy asphalt, gussasphalt and rigid flexible composite pavement have been formed. However, under different traffic and climate conditions, diseases still occur frequently in the pavement schemes above. Steel deck pavement is still a worldwide problem recognized by academic and engineering circles.

Gussasphalt steel deck pavement has been developed for decades in Germany, Britain and Japan, and the disease is not prominent in the early stage of application, and the overall application effect is good. However, with the rapid growth of traffic volume and the substantial increase of heavy load ratio, the diseases of gussasphalt pavement have been reported in Holland, Japan and other countries (Medani 2006, Himeno et al. 2003).

Since the introduction of gussasphalt pavement for the first time in Shengli Yellow River Bridge in Shandong Province in 2003, the application of "polymer modified gussasphalt mixture + modified asphalt SMA" in steel deck pavement in China has been opened. And it is further applied in Anqing Yangtze River Bridge, Caiyuanba Yangtze River Bridge and other river-spanning bridges due to its good water tightness, collaborative deformation ability with steel bridge deck and suitable engineering cost. However, under the harsh conditions of high temperature, heavy load and hot rainy season in China, the current situation of gussasphalt

---

* Corresponding author

steel deck pavement is not optimistic (Hao 2002, Xu 2014, Tang et al. 2018). In order to make full use of the advantages of gussasphalt pavement and improve its disadvantages, the Fourth Nanjing Yangtze River Bridge innovatively uses the composite gussasphalt concrete pavement of "gussasphalt mixture with TLA + dense built-in gradation mixture with high elastic modified asphalt" (Wu 2017). Its service status and long-term performance development trend are worthy of attention. Therefore, the long-term performance observation and research can lay a foundation for further optimization of the pavement scheme and its popularization and application in long-span steel structure bridges at home and abroad.

2 PAVEMENT SCHEME

2.1 Pavement structure

The pavement structure of the Fourth Nanjing Yangtze River Bridge is shown in Figure 1. The lower layer of pavement has good compactness and anti deformation ability due to the high content of mineral powder and asphalt. The results show that the asphalt binder formed by mixing hard bitumen and TLA can significantly improve its high temperature stability. The upper layer with high elasticity modified asphalt and dense built-in gradation can provide good rutting resistance and water tightness. The dynamic stability at 60°C can reach to 4000 times/mm. The dynamic stability of the composite structure at 60°C can reach to 4000 times/mm although it is only 350 times/mm of the lower layer, based on the innovative structure combination. Therefore, this pavement structure has good waterproof performance, cooperative deformation ability and high temperature stability.

2.2 Climate and traffic conditions

The Fourth Nanjing Yangtze River Bridge is an important channel for Nanjing Ring Expressway to cross the Yangtze River, and it is located in the southwest of Jiangsu Province, it belongs to the transitional climate zone from North subtropics to middle subtropics, and it is characterized by transition, monsoon and wettability. The weather is changeable in spring, hot and rainy in summer, clear in autumn and cold and dry in winter. The climatic and traffic conditions are shown in Table 1.

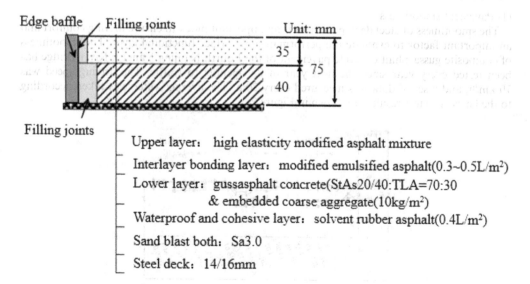

Figure 1.   Composite gussasphalt steel deck pavement structure.

Table 1. Climatic and traffic conditions of the Fourth Nanjing Yangtze River Bridge.

	Parameter category	Data
Climatic condition	Monthly average maximum temperature	32.5°C
	Extreme maximum temperature	39.7°C
	Extreme minimum temperature	-13.1°C
	Annual continuous high temperature days	51
	Rainfall	1106mm
Traffic condition	Traffic volume	35000 vehicles/day
	Truck proportion	35%

There's no obvious disease has occurred and no maintenance has been carried out in the Fourth Nanjing Yangtze River Bridge since it was opened to traffic in 2012. However, the high temperature stability of gussasphalt mixture is lower than that of ordinary asphalt pavement materials, and its dynamic stability at 60°C is only about 350 times per mm. Under the condition of rain and heat in the same season and long-term heavy load traffic conditions in China, there are risks of high-temperature rutting and fatigue cracking. Therefore, the tracking detection of it should be carried out, and find out the early disease of pavement in time and deal with it effectively in order to improve the service life of pavement.

3 LONG-TERM PERFORMANCE OBSERVATION METHOD AND DETECTION RESULTS

3.1 Testing and evaluation methods

In this study, the multi-functional inspection vehicle, laser profiler and friction coefficient detection vehicle were mainly used to track and detect the skid resistance, smoothness, rutting and damage status of steel deck pavement of the Fourth Nanjing Yangtze River Bridge. The test and calculation of each index refer to the Highway Performance Assessment Standards (JTG 5210-2018).

3.2 Results and analysis

(1) Pavement smoothness

The smoothness of steel deck pavement is an important index to ensure driving comfort and an important factor to evaluate the performance of surface functional layer. The smoothness of composite gussasphalt concrete pavement of the Fourth Nanjing Yangtze River Bridge has been tested every year since the next year of its opening to traffic. The driving speed was 70 km/h, and a set of data was measured every 20m. The average value was taken according to the lanes. The test results are shown in Figure 2.

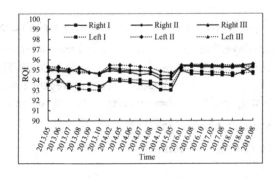

Figure 2. Change trend of riding quality index.

According to the test results of pavement smoothness, after 7 years of continuous driving load, high temperature in summer and low temperature in winter, the evaluation results of bridge deck pavement smoothness are greater than 90, and the evaluation grade is excellent, which indicates that all lanes still maintain good driving quality and service quality. By comparing the test results of riding quality index of each month horizontally, it can be found that the overall fluctuation of riding quality index of each lane per month is relatively small, indicating that the pavement has maintained stable performance during 7 consecutive years of use.

(2) Pavement rutting depth

The rutting depth of steel deck pavement is an important factor affecting the driving safety and comfort. At the same time, rutting is also a common early damage of gussasphalt concrete steel deck pavement. With the aggravation of rutting, the pavement will shift and slip, which eventually leads to the damage of pavement structure. Under the domestic high temperature and heavy load traffic conditions, this process develops very quickly, which will seriously reduce the performance and shorten the service life of pavement if not taken seriously.

In this study, a laser profiler was used to test the rutting depth of the two-way 6-lane of the Fourth Nanjing Yangtze River Bridge. The driving speed was 70 km/h, and a set of data was measured every 10m. The average value was taken according to the lanes. The test results are shown in Table 2.

According to the test results in the above table, as of 2019, the average rutting depth of each lane does not exceed 10 mm, and the rutting performance evaluation level is excellent. Especially the first two-way lane, the average rutting depth is less than 2 mm after 7 years of service, which have excellent anti rutting performance.

In order to study the development trend of rutting depth of each lane more intuitively, the third two-way lane with the largest rutting depth is divided into uphill and downhill sections according to the driving direction, and the test data over the years are plotted (see Figure 3).

Table 2. Test results of rutting depth over the years.

Year	Left lane width (mm)			Right lane width (mm)		
	First lane	Second lane	Third lane	First lane	Second lane	Third lane
2013	1.89	1.97	1.51	2.52	2.37	2.48
2014	2.53	3.00	3.03	3.17	3.45	3.89
2015	2.64	3.11	3.40	3.26	3.67	4.15
2016	2.26	2.74	2.96	1.96	3.26	4.44
2017	2.14	2.88	3.51	1.75	3.46	5.12
2018	1.78	2.97	4.43	1.90	3.79	5.93
2019	1.73	3.33	4.59	1.89	3.67	6.14

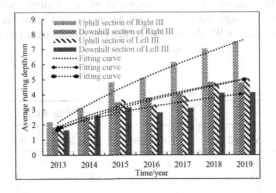

Figure 3. Rutting depth detection results.

By comparing the third lanes from the left and right in Figure 3, it can be seen that the rutting depths of the uphill and downhill sections of the third lane from the right are larger than those of the corresponding sections on the left, which is mainly due to the higher heavy load ratio of the right lane that of the left lane. In addition, by comparing the data of uphill section and downhill section, it is found that the left and right lanes have the same trend, that is, the rutting depth of uphill section is larger than that of downhill section, and the growth rate of uphill section is faster than that of downhill section. Because the driving speed of heavy-duty vehicle on uphill section is significantly lower than that on downhill section, which results in the action time of vehicle load on uphill section longer than that on downhill section. Asphalt has obvious viscoelastic characteristics. According to the time temperature equivalent principle of viscoelastic materials, prolonged time and elevated temperature are equivalent to molecular motion((Xu et al. 2013). Therefore, under the same load and action times, the prolonged load action time of the uphill section is equivalent to raising the service temperature of the pavement, which leads to the decrease of pavement modulus and increase of rutting depth.

(3)Anti-skid performance of pavement

The anti-skid performance of pavement is another important factor affecting the safety performance of driving. Due to the good water tightness, the surface friction coefficient and structural depth of gussasphalt pavement are smaller than those of ordinary asphalt concrete. The SFC of the Fourth Nanjing Yangtze River Bridge pavement with two-way six lanes was carried out by using friction coefficient test vehicle。 And the driving speed was 50 km/h, and a set of data was measured every 20m. The average value was taken according to the lanes. The test results are shown in Table 3.

It can be seen from the test results in Table 3 that with the increase of service life, the anti-skid performance of composite gussasphalt concrete steel deck pavement of the Fourth Nanjing Yangtze River Bridge shows a trend of obvious reduction in the early stage and tends to be stable in the later stage. This is because, in the early stage of operation, under the effect of wheel rolling, the gradation of high elastic modified asphalt mixture on the upper layer is further compacted, the macrostructure of pavement is reduced, which result in a decrease of the anti-skid performance. With the increase of wheel loads, the asphalt film on the road surface is basically polished, the compactness of mixture tends to be stable, and the macro structure reduction caused by the secondary compaction of mixture has a gradual weakening effect on the anti-skid performance of pavement. In addition, relevant research shows that the smaller the crushing value and abrasion value of aggregate and the greater the polishing value, the smaller the attenuation rate and amplitude of the friction coefficient of asphalt mixture, and the greater the final attenuation value (Kong et al. 2017, Zhu et al. 2018). And the coarse aggregate used for steel deck pavement of the Fourth Nanjing Yangtze River Bridge is hard basalt gravel with good wear resistance. Therefore, the final value of skidding resistance index can be maintained at a relatively high level.

(4)Pavement damage

The multi-functional inspection vehicle was used to test the damage of the steel deck pavement of the Fourth Nanjing Yangtze River Bridge. The test results are shown in Figure 4.

The results show that as of 2019, the pavement of the Fourth Nanjing Yangtze River Bridge has basically no damage, only a few parts have "serious" damages, and the PCI value of each two-way lane is above 95. According to the test results over the years, the damage of left lane is slightly lighter than that of right lane.

Table 3. Test results of sideway force coefficient (SFC).

Year	Left lane width (mm)			Right lane width (mm)		
	First lane	Second lane	Third lane	First lane	Second lane	Third lane
2013	71	69	64	71	67	65
2014	68	65	61	68	64	62
2019	/	67	/	/	65	/
2020	/	62	62	/	62	60

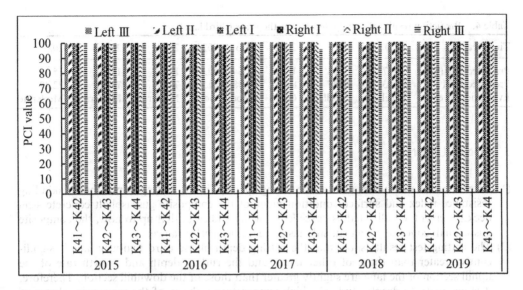

Figure 4. Pavement surface condition index from 2015 to 2019.

In addition, according to the results of manual walking inspection in March 2020, it is found that "serious" damages mainly occur in the third two-way lane, which is mainly caused by the falling of hard objects during the driving process of trucks and the damage by snow shovels during snow removal over the years. The damage is random, so there is no obvious pattern of damage distribution.

4 PREDICTION OF RUTTING DEPTH

According to the above long-term performance tracking and observation results, rutting is the main damage of composite gussasphalt steel deck pavement of the Fourth Nanjing Yangtze River Bridge. In order to further study the development trend of pavement rutting depth in the designed service life and formulate pavement maintenance planning scientifically, based on the actual traffic volume and the actual rutting depth over the years, the rutting prediction model of composite gussasphalt steel deck pavement proposed in the earlier stage is optimized in this study (Zhang 2013). The optimized model is as follows.

$$D = 0.6071 \times C \times C_t \times C_v \times C_w \times \sum_{i=1}^{n} \left(365 \times \frac{N_i}{DS}\right) + 0.5581 \quad (1)$$

Where: D-rutting depth; C-Coefficient of difference correction based on indoor test conditions and field rutting depth; C_t-temperature correction coefficient; C_v-driving speed correction coefficient; C_w-lane and wheel load correction coefficient; N_i-daily average equivalent wheel number in Year i (times/day); DS-dynamic stability (times/mm).

Based on the above prediction model, the prediction results of rutting depth over the remaining designed life of the Fourth Nanjing Yangtze River Bridge steel deck pavement can be obtained as shown in Table 4.

According to the above prediction results, the rutting depth of pavement increases significantly with the service life gradually entering the middle and late stages. By 2022, the rutting performance of pavement has been reduced to "good", so maintenance work should be carried out in time to avoid shear instability of pavement and prolong the service life of it.

Table 4. Prediction results of rutting depth over the years (third lane).

Time(year)	2020	2021	2022	2023	2024	2025	2026	2027
Predicted RD (mm)	8.47	9.53	10.58	11.64	12.69	13.74	14.80	15.85
Predicted RDI	91.53	90.47	88.26	85.08	81.93	78.78	75.60	72.42
Evaluation level	Excellent	Excellent	Good	Good	Good	Moderate	Moderate	Moderate

5 CONCLUSIONS

(1) After more than 7 years of use, the riding quality index, rutting depth index, skidding resistance index and surface condition index of the composite gussasphalt concrete steel deck pavement are all rated as "excellent", which indicates that up to now, the composite gussasphalt concrete pavement still has good service performance;

(2) The rutting test results show that the rutting depth of the third two-way lane is significantly greater than those of other lanes, and the rutting depth and growth rate of the uphill section of the lane are slightly greater than those of the downhill section. Therefore, it is necessary to deeply study the high-temperature stability of the pavement scheme to improve its high-temperature deformation resistance;

(3) The existing rutting prediction model is optimized based on the actual traffic volume and the measured rutting depth over the years. The optimized prediction results show that with the service life gradually entering the middle and later stages of the designed service life, the rutting depth of pavement increases significantly. By 2022, the rutting performance of pavement has been reduced to "good", so maintenance work should be carried out timely to extend the service life of pavement;

(4) With the secondary compaction of pavement during service, the anti-skid performance of pavement has been decreased to a certain extent. However, because the steel deck of the Fourth Nanjing Yangtze River Bridge uses hard basalt coarse aggregate with good wear resistance, the final value of skidding resistance index can be maintained at a relatively high level;

(5) The damage caused by the falling of hard objects from vehicles is the main damage type of steel deck pavement of the Fourth Nanjing Yangtze River Bridge at present, and the damage is random without obvious distribution rule. Therefore, it is necessary to strengthen the tracking and observation of pavement damage, carry out preventive maintenance in time, and delay the development trend of pavement damage.

REFERENCES

T.O.Medani. 2006. Design Principles of Surfacings on Orthotropic Steel Bridge Decks. Netherland: Civil Engineering and Geosciences of Delft University of Technology.

Kenji HIMENO & Tatsuo NISHIZAWA & Kitaro UCHIDA. 2003. Longitudinal surface cracking in asphalt pavement on steel bridge decks related to dissipated energy[C]. Japan-China 2th Workshop on Pavement Technologies, 293–302.

HAO ZH. 2012. Evaluation and Research on Road Performance of Long Span Steel Deck Pavement Structures. *Highway* 6: 103–108 DOI: 10.3969/j.issn.0451-0712.2012.06.021.

XU Y. 2014. The Performance Survey and Deterioration Analysis of Steel Bridge Deck Paving. Chongqing Jiaotong University.

JTG 5210. 2018. Highway Performance Assessment Standards.

TANG YJ & LIU SJ & FAN HQ. 2018. Comparative Study on in-situ Measur ed Perfor mance of Steel Deck Pavement. *Journal of Yancheng Institute of Technology(Natural Science Edition)* 31(3): 58–67 DOI: 10.16018/j.cnki.cn32-1650/n.201803011.

WU Z & ZHU L & XIONG WT. 2017. Research on Performance Observation of Composite Gussasphalt Steel Deck Pavement on Large Span Suspension Bridge. *Highway* 62(1): 32–36.

XU ZQ & LI C. 2013. Time-Temperature Superposition Principle Applied in Rutting Prediction. *Journal of Shandong Jiaotong University* 21(1): 65–70 DOI: 10.3969/j.issn.1672-0032.2013.01.014.

KONG LY & YING GG & LIN XW & TIAN QC. 2017. Anti-sliding performance attenuation model of asphalt pavement based on aggregate mechanical index. *Journal of Changsha University of Science and Technology(Natural Science)* 14(3): 13–20 DOI: 10.3969/j.issn.1672-9331.2017.03.003.

ZHU HZ & LIAO YY. 2018. Present Situations of Research on Anti-skid Property of Asphalt Pavement. *Highway* 63(1): 35–46.

ZHANG DJ. 2013. Study on steel deck pavement rutting model of composite guss asphalt concrete. *Engineering Science* 15(8): 63–69 DOI: 10.3969/j.issn.1009-1742.2013.08.011.

Functional Pavements – Chen et al (eds)
© 2021 Taylor & Francis Group, London, ISBN 978-0-367-72610-2

Impact of RAP classification on the road performance variability of recycled mixture

He. Zhan, Ning. Li, Wei. Tang & Xin. Yu
College of Civil and Transportation Engineering, Hohai University, China

ABSTRACT: In order to control the variability of recycled asphalt pavement (RAP) and ensure the road performance of recycled mixture. The coefficient of variability (CV) was used to describe the variability of RAP, and the classification standard of RAP was determined by controlling primary control sieve. The research results showed that asphalt aggregate ratio and gradation of fine RAP have great variability. The primary control sieve for the classification of coarse RAP, fine RAP are 12mm and 5mm, respectively. When the separation of RAP into three stockpiles, the variability of asphalt aggregate ratio and gradation of RAP have been controlled effectively, especially the CV of passing rate at 4.75mm primary control sieve is decreased by more than 80%. It will help minimize the variability in recycled mixtures and provide a constructed recycled mixture with better performance.

1 GENERAL INSTRUCTIONS

The maintenance of asphalt pavement is facing a severe environmental protection situation, as well as the multiple contradictions between the shortage of raw materials for infrastructure and the inefficient use of RAP. The efficient recycling of RAP has become an inevitable development trend for highway maintenance [1, 2]. Due to the lack of careful consideration of RAP variability during the production of recycled asphalt mixture, there is a large variability in recycled mixtures [3, 4], which affects the road performance of recycled mixture.

Hong et al. [5] have conducted researches that the material variability of RAP will lead to different degrees of variation in the performance of recycled pavement. Mcdaniel et al. [6] also found that controlling the variability of RAP is the key point to maintain the stability of the recycled mixture. Zhang et al. [7] have investigated the use of layered milling to reduce the variability of RAP, but the reduction ability was limited and affected the construction efficiency. However, there are few studies on the variability of SMA asphalt mixture. Therefore, this paper firstly analyzes the variability of RAP, and then the classification standard of RAP was determined by controlling primary control sieve. Finally, the effect of RAP classification on the road performance variability of recycled mixture was analyzed.

2 EXPERIMENTAL STUDY

2.1 Materials

In this study, RAP were recycled by cold milling from the SMA-13 pavement on the surface layer of a highway in Jiangsu Province, and its service life is about 11 years. The properties of RAP is presented in Table 1.

2.2 Reparation of recycled mixture

The content of RAP in the recycled mixture is 30%, the ratio of RAP 12~22mm, 5~12mm, 0~5mm is 3:5:2. RAP and recycling agent are heated to 130°C and 120°C, respectively. Then

Table 1.　Properties of RAP.

Materials	Items	Method	Results
RAP	Sand equivalent/(%)	JTJ/T 5521	64
	Moisture content/(%)	JTJ/T 5521	0.98
Recycled Asphalt	Penetration @25°C/(0.1mm)	ASTM D5	24.5
	Softening point/(°C)	ASTM D36	71.8
	Ductility@15°C/(cm)	ASTM D113	10.4
	PG	AASHTO M320-03	76-22

the recycling agent was sprayed on the surface of RAP, mixing at 175°C for 90s, and new aggregate and asphalt are added with the mixing time is 90s. The gradation and asphalt aggregate ratio of RAP and recycled mixtures were determined by centrifugal separation test.

2.3　Test and evaluation methods

According to the standard JTG E20-2011, the high-temperature, low-temperature and water damage resistant were evaluated by wheel tracking test, band test and freeze-thaw indirect tension test, respectively.

2.4　The coefficient of variability (CV)

The variability of RAP includes three parts: RAP gradation, RAP aggregate gradation and asphalt aggregate ratio. RAP gradation refers to the gradation determined by cold milling from pavement. The coefficient of variability (CV) was used to describe the variability of RAP, the equation of CV as follows.

$$S = \sqrt{\frac{1}{n-1}\sum_{i-1}^{n}(x_i - x)^2}, \; x = \frac{1}{n}\sum_{i-1}^{n}x_i \qquad (1)$$

$$CV = \frac{S}{x} \times 100\% \qquad (2)$$

Where S= the standard deviation; x= the sample average.

3　RESULTS AND DISCUSSION

3.1　The variability of RAP

Figure 1 showed the average aggregate passing rate and CV of RAP. From Figure 1, the sieve hole passing rate of RAP aggregate to exceed the upper limit of gradation specification, which indicates that the original pavement aggregate has been refined under the vehicle load and cold milling action, especially the biggest change in the passing rate of the 4.75mm sieve hole.

The CV of aggregate gradation increases first and then decreases with the increase of sieve size. The CV of fine aggregate is larger significantly than that of coarse aggregate, and the CV of 4.75mm sieve hole reaches the maximum value. 4.75mm is the primary control sieve of SMA asphalt mixture which supports skeleton structure. The CV of 4.75mm sieve hole is the largest, which indicates that the quality control of SMA is difficult and the fluctuation is large.The average value of asphalt-aggregate ratio of RAP is 6.7%, and the CV of asphalt-aggregate is 8.17%. The degree of variability of asphalt-aggregate ratio is much greater than that of aggregate gradation, which leads to the variability of RAP increase.

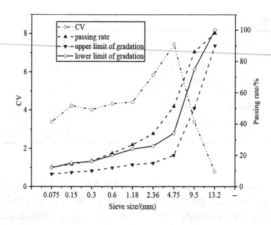

Figure 1. The average aggregate gradation value and CV of RAP.

3.2 Determination of primary control sieve for fine RAP

The primary control sieve of fine RAP is determined according to NCHRP SYNTHESIS-495, which due to that the report gives the variability control standards for RAP gradation and asphalt-aggregate ratio. It can be seen from Figure 1 that the variability of RAP aggregates in 2.36mm and 4.75mm sieve holes is large, so first assume that 3mm and 5mm are the key sieve holes for fine RAP material classification, and the RAP gradation passing rate and CV of 0-3mm and 0-5mm are shown in Table 2.

It can be seen from Table 2, the crushing and screening effect of 0-5mm RAP is good, and the passing rate of 4.75mm is over 97%, indicating that 0~5mm fine RAP can fully pass the 4.75mm sieve after sufficient sieving. The 0~3mm fine RAP has about 30% content on the 2.36mm sieve after sufficient sieving, which indicates the RAP crushing and sieving effect is not good. The main reason is that RAP with smaller particle size is easy to agglomerate in the sieve hole, so it is difficult for the shaker to screen fine RAP completely.

According to the CV of two different fine RAP gradation, the CV of 0~5mm fine RAP aggregate gradation and asphalt-aggregate ratio both meet the control standard of NCHRP report. The 0~3mm fine RAP is easy to agglomerate, resulting in the variability of asphalt-aggregate ratio is large. The key sieve hole of fine RAP classification is determined to be 5mm.

3.3 Determination of primary control sieve for coarse RAP

As we all known, the variability of coarse RAP gradation is small, so the primary control sieves for coarse RAP material classification are mainly determined according to the quality balance of RAP on each sieve hole. The passing rate of coarse RAP is shown in Table 3.

Table 2. Passing rate and CV of fine RAP.

Items		Passing rate/%		CV	
Fine RAP size/mm		0~3	0~5	0~3	0~5
Sieve size/mm	4.75	97.9	100	0.008	0
	2.36	71.3	70.5	0.027	0.074
	1.18	46.3	29.8	0.015	0.063
	0.6	26.1	16.5	0.021	0.074
	0.3	0	0	0	0
asphalt-aggregate ratio/%		7.4	7.8	0.046	0.105
NCHRP		-	-	≤0.05	≤0.1

148

Table 3. Passing rate of coarse RAP.

Sieve size/mm	Passing rate/%						Average/ %
	①	②	③	④	⑤	⑥	
16	100	100	100	100	100	100	100
13.2	92.4	93.1	92.1	94.5	97.1	89.2	93.1
9.5	71.6	72.3	71.3	73.1	75	68.3	71.9
4.75	30.2	31.1	29.7	33.2	35.1	28.1	31.2
2.36	15.3	16.3	14.2	17.9	18.5	13.6	16
1.18	10.6	11	9.9	12.4	14.9	13.4	12
0.6	4.9	5.7	4.4	6	6.5	4	5.3
0.3	0	0	0	0	0	0	0

Table 4. Road performance of recycled mixture before and after RAP classification.

Items	Before classification	CV	After classification	CV
DS/(cycle/mm)	8126	0.155	8656	0.063
The maximum tensile strain/(με)	3670	2.30	3849	2.33
TSR/%	81.9	1.432	84.7	0.944

Based on the principle of the quality balance of RAP on each sieve hole, the proportion of coarse material to the total amount should not be less than 30% in each sieve. From Table 4, it can be seen that the content of RAP coarse material on the sieve of 13.2mm is about 7%, and the content on the sieve of 9.5mm is about 28%. Therefore, the primary control sieve of coarse RAP classification should be distributed at 9.5mm-13.2mm. And considering the screening process of RAP, the primary control sieve of coarse RAP is determined to be 12mm. The RAP is divided into three stockpiles (Figure 2).

3.4 Impact of RAP classification on the variability of aggregate gradation and asphalt-aggregate ratio

The aggregate gradation, asphalt-aggregate ratio and their CV of RAP before and after classification are shown in Figure 3.

It can be seen from Figure 3(a,c), the volatility of coarse RAP after classification is relatively stable, while the volatility of fine RAP is still large. The effect of classification on reducing the variability of coarse RAP is significant. From the CV of aggregate passing rate and asphalt-aggregate in Figure 3(b,d), it can be seen that the CV decreases with the increasing of RAP classification sieve size, especially the CV at the 4.75mm sieve hole decreases significantly. After classification, the CV of aggregate at the 0-5mm, 5-12mm and 12-22mm sieve holes decreases by about 100%, 93% and 80%, respectively. Compared with non-classification,

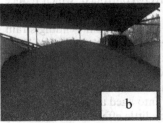

Figure 2. RAP with three stockpiles after classification.a=12~22mm;b=5~12mm;c=0~5mm.

149

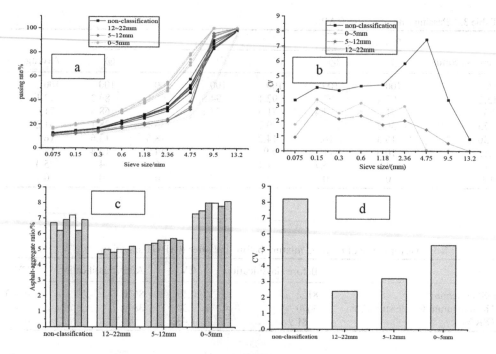

Figure 3. a=the aggregate gradation; b=the CV of the aggregate gradation; c=asphalt-aggregate ratio; d=the CV of asphalt-aggregate ratio.

the CV of 0-5mm, 5-12mm and 12-22mm aggregate decreases by 36%, 63% and 75%, respectively. However,the decrease degree of variability of asphalt-aggregate ratio is small, the main reason is that there are more fine aggregate and filler in fine RAP, which has larger specific surface area and could absorb more asphalt to produce agglomeration. On the whole, classification could minimize the variability of aggregate gradation and asphalt-aggregate ratio of RAP.

3.5 *Impact of RAP classification on road performance variability of recycled asphalt mixture*

The results of the impact of RAP classification on the variability of the road performance of the mixture are shown in Table 4.

The DS of the recycled mixture is far greater than the specification requirements, and the high temperature performance is great. After the RAP classification, the DS is increased by 6.5% and the CV is reduced by 59.3%. The high temperature performance of the recycled mixture is mainly affected by the excellent skeleton compact structure and the rheological properties of the aged asphalt. After classification, the variability of aggregate gradation and asphalt-aggregate ratio is effectively controlled, which reduce the variability of high temperature performance.

It can be seen from Table 4, the maximum bending strain of the recycled asphalt mixture after RAP classification is increased by 4.8%, and the CV is basically unchanged, which indicates that RAP classification has no obvious effect in the low temperature performance and variability control. It mainly due to that the low-temperature performance of recycled mixture is affected by asphalt, while the content of RAP in recycled mixture is only 30%. Adding recycling agent and new asphalt into aged asphalt can basically reach the level of new asphalt. Therefore, RAP classification has little effect on the low temperature performance of recycled mixture.

After RAP classification, the TSR of recycled mixture increased by 3.4%, and the CV decreased by 34%, which indicates that the classification can effectively reduce the variability of the water stability of recycled mixture. Aggregate gradation is the main factor affecting the water damage resistant of asphalt mixture. RAP classification reduces the variability of gradation and asphalt-aggregate ratio significantly, thereby the passing rate of key sieve holes such as 4.75mm and 9.5mm in recycled mixture has controlled, which makes the gradation of recycled mixture closer to the design value.

4 CONCLUSIONS

(1) The asphalt-aggregate ratio and aggregate gradation of fine RAP have great variability. Primary control sieve of coarse and fine RAP classification are 12mm and 5mm, respectively.
(2) When the separation of RAP into three stockpiles, the variability of asphalt aggregate ratio and gradation of RAP have been controlled effectively. It helps minimize the variability of high temperature and water damage resistant in recycled mixture.

ACKNOWLEDGEMENTS

This work was sponsored by the National Natural Science Foundation of China (NSFC) (51708178), the Natural Science Foundation of Jiangsu Province (BK20170886), and the Scientific research project of Jiangsu Communications Holding Co., Ltd (JETC-YF-2019-058).

REFERENCES

[1] L. Fang. Study on durability of SMA-16 mixture with high RAP content and plant mix temperature. Road engineering, 2016, 41 (04): 237–241.
[2] J. Wang, Y. C. Qin, S. C. Huang, et al. Variability of recycled asphalt mixture. Journal of Beijing University of technology, 2018, 44 (02): 244–250.
[3] A. Sreeram, Z. Leng. Variability of rap binder mobilisation in hot mix asphalt mixtures. Construction and Building Materials, 2019, 201(MAR.20):502–509.
[4] J. Montaez, S. Caro, D. Carrizosa, et al. Variability of the mechanical properties of Reclaimed Asphalt Pavement (RAP) obtained from different sources. Construction and Building Materials, 2020, 230:116968.
[5] F. Hong, R. Guo, F. Zhou. Impact of recycled asphalt pavement material variability on pavement performance.Road Materials and Pavement Design, 2014, 15(4):841–855.
[6] R. S. Mcdaniel, H. Soleymani, R. M. Anderson, et al. Recommended Use of Reclaimed Asphalt Pavement in the Superpave Mix Design Method. Nchrp Web Document, 2000.
[7] M. J. Zhang, W. Y. Qi. Influence of milling technology on RAP variability and grading. Petroleum asphalt, 2016, 30 (04): 12–17.

Functional Pavements – Chen et al (eds)
© 2021 Taylor & Francis Group, London, ISBN 978-0-367-72610-2

Experimental study on the initial and terminal performances of cold recycling emulsified asphalt mixtures

Yuanshuai Dong, Yun Hou, Chen Zhu & Yanhong Zhang
China Highway Engineering Consulting Corporation, Research and Development Center of Transport Industry of Technologies, Materials and equipments of Highway Construction and Maintenance Ministry of Transport, PRC, Beijing, China

ABSTRACT: When cold recycling emulsified asphalt mixtures are used in the overhaul projects of urban roads, the maintenance conditions by traffic closing always cannot be achieved. Therefore, great initial and terminal performances are required. In this paper, the initial performance of cold recycling emulsified asphalt mixture was evaluated by cohesion tests, sweeping tests and rutting tests. The terminal water stability and durability of cold recycling emulsified asphalt mixture were evaluated by splitting tests, Marshall tests, freeze-thaw splitting tests and fatigue tests. The effects of emulsifier type, gradation, cement dosage and RAP dosage on the initial and terminal performances of cold recycling emulsified asphalt mixtures were investigated for their applications in urban roads. The results showed that the abilities of cold recycling emulsified asphalt mixtures to resist early damages and to ensure good terminal performance can be enhanced by optimizing or adjusting the compositions of emulsifier, increasing the ratio of RAP, properly increasing the dosage of cement, and selecting finer gradation materials.

Keywords: cold recycling emulsified asphalt mixture, initial performance, terminal performance

1   INTRODUCTION

The highway construction of China is in a large-scale maintenance stage. The annual production of waste asphalt pavement materials in China reaches tens of millions of tons. If the waste asphalt pavement materials are discarded, a lot of land will be occupied, and the waste of resources and environmental pollution can be also caused [2]. Emulsified asphalt cold recycling technology can deal with recycled asphalt pavement materials (RAP) by 100%, and alleviate the shortage of raw materials for road construction in China [3–5].

After paving the pavements with cold recycling emulsified asphalt materials, the traffic needs to be closed for 2-7 days for curing, which undoubtedly limits the application of this technology [6,7]. The overhaul projects of urban roads often need to be constructed at night and be open to traffic during the day. When the cold recycling emulsified asphalt materials are used for urban roads, the traffic are required to be opened within 3-4 hours after the cold recycling layers are paved, resulting in the direct exposure of cold recycling emulsified asphalt materials to the traffic loading when the asphalt materials are of poor performance and unprotected. As a result, early damages are prone to take place [8]. Thus, good initial and terminal performances are required for the cold recycling emulsified asphalt materials. The initial performance refers to the ability to resist cracking, granulation and rutting during the open traffic curing, while the terminal performance refers to the performance after accelerated curing. In this paper, the initial and terminal performances of cold recycling emulsified asphalt materials were studied, and the methods and measures to improve the initial performance of cold recycling emulsified asphalt materials without reducing the terminal performance were analyzed. This paper is beneficial for the applications of cold recycling emulsified asphalt materials in urban roads.

2 EXPERIMENTAL PROCEDURES

In this paper, the compositions of cold recycling emulsified asphalt materials were designed. The content of large-sized RAP was 80%, denoted by C-80. The contents of medium-sized RAP were 80% and 100%, separately, denoted by Z-80 and Z-100. Then, the influences of emulsifier type, cement content, gradation type and RAP content on the initial and terminal performances of cold recycling emulsified asphalt materials were analyzed through cohesion tests, sweeping tests, rutting tests, splitting tests and fatigue tests. For the simulation of the initial performances of cold recycling emulsified asphalt materials under the practical engineering conditions, the curing conditions in the cohesion and sweeping tests were set as 20 °C, 90% of relative humidity, and 4 h [7]. The curing conditions in the terminal performance tests were set as 60 °C and 48 h.

3 MATERIALS

3.1 *Raw materials*

In this paper, four sorts of emulsifiers (a to d) and AH-70-type asphalt were selected for the preparation of emulsified asphalt, of which the performance met the requirements of the standard JTGT 5521-2019 "Technical Specifications for Highway Asphalt Pavement Recycling". Mineral materials could be well encapsulated in each of the emulsified asphalt materials. The old materials were divided into two types by their grain diameters: 0-10 mm and 10-20 mm. The new limestone aggregates were also divided into two types by their grain diameters: 10-15 mm and 10-20 mm. And, 42.5# regular portland cement was adopted. The quality of these selected raw materials was qualified after testing. Since the demulsification time of emulsified asphalt can be prolonged by increasing the dosage of emulsifier, herein, the minimum dosage of emulsifier was used for the preparation of emulsified asphalt.

3.2 *Design of proportioning*

According to the standard JTGT 5521-2019 "Technical Specifications for Highway Asphalt Pavement Recycling", the proportioning of cold recycling emulsified asphalt materials was designed. The gradation results of the mixtures are shown in Table 1, and the proportioning results are shown in Table 2.

4 INITIAL PERFORMANCE

4.1 *Cohesion tests*

According to the standard JTG E20-2011, 150-mm-sized specimens with a height of 8 ± 0.5 cm were prepared by rotary compaction [9]. After curing, the initial strength (resistance to cracking) of cold recycling emulsified asphalt mixture was evaluated by the Hveem method through cohesion tests. The strength was indicated by the cohesion value. In the operation of this instrument, the opening size of the steel ball feeding port was adjusted according to the lever loading principle, and the steel balls fell into a cylinder and were loaded on a specimen at a constant speed until the specimen was broken, when the steel ball feeding port was automatically closed. Then, the cohesion value was calculated according to the following formula by weighing the steel balls in the cylinder. The cohesion tester is illustrated in Figure 1.

$$C = \frac{M}{(0.031H + 0.00269H^2) \times D} \tag{1}$$

Table 1. Gradation results of the mixtures.

Mesh size (mm)	limestone 10-20mm	limestone 10-15mm	RAP 10-20mm	RAP 0-10mm	Mineral powder	Cement	C-80	Z-80	Z-100
26.5	100	100	100	100	100	100	100.0	100.0	100.0
19	19.2	100	99.3	100	100	100	83.7	99.9	99.7
16	5.3	66.5	75.7	100	100	100	76.2	88.9	91.3
13.2	2.5	55	49.7	100	100	100	70.4	81.9	81.9
9.5	1.2	8	9	99.2	100	100	61.6	64.7	66.7
4.75	0.9	0.6	0.5	64.7	100	100	39.6	41.2	42.5
2.36	0.9	0.6	0.5	45.8	100	100	28.6	30.0	30.8
1.18	0.9	0.6	0.5	25.6	100	100	16.8	17.9	18.4
0.6	0.9	0.6	0.5	10.7	100	100	8.0	9.1	9.3
0.3	0.8	0.6	0.5	4.6	100	100	4.5	5.4	5.5
0.15	0.8	0.6	0.5	2.5	90.1	100	3.2	4.1	4.1
0.075	0.7	0.6	0.5	0.5	76.6	100	2.0	2.8	2.8
C-80	20		20	58.5	0	1.5	100		
Z-80		20	18	59.5	1	1.5		100	
Z-100			36	61.5	1	1.5			100

Table 2. Proportioning results.

	C-80	Z-80	Z-100
Optimal emulsified asphalt content (%)	4.0	4.0	4.0
Optimum liquid content (%)	5.5	5.8	6.0
Theoretical maximum relative density	2.542	2.536	2.523
Relative density based on the gross volume	2.300	2.298	2.297

Figure 1. The cohesion tester.

C – cohesion value, g/cm^2;
M – mass of steel balls, g.
D – diameter of the specimen, cm.
H – height of the specimen, cm.

The results of cohesion tests are shown in Figure 2.The cohesion values of the mixtures containing emulsifiers a and b were significantly higher than those containing emulsifiers c and d under the same conditions of gradation materials, RAP dosage, and cement dosage. Especially, the cohesion value of the mixture containing emulsifier a was almost twice that containing emulsifier c. With the increase of cement dosage, the changes in the cohesion values of mixtures containing different emulsifiers were decreased slightly, and the cohesion values of mixtures containing different emulsifiers were ranked as follows: a > b > d > c, indicating that

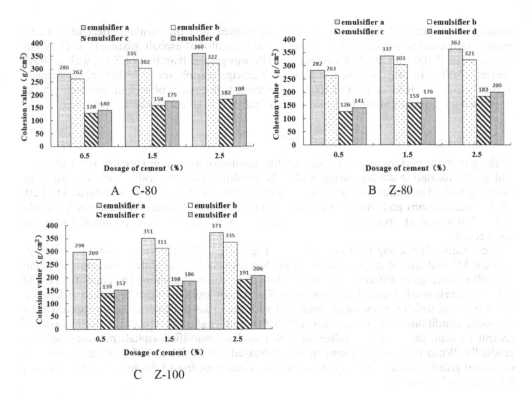

Figure 2.   Results of cohesion tests.

emulsifier type or demulsification speed had great influences on the initial strengths of cold recycling emulsified asphalt mixtures.

Under the same conditions of gradation materials, RAP dosage, and emulsifier, with the increase of cement content, the cohesion values of the cold recycling emulsified asphalt mixtures were increased gradually, but the changing trend was not linear: When the cement dosage was increased from 0.5% to 1.5%, the cohesion values were increased greatly. When the cement dosage was increased from 1.5% to 2.5%, the values were increased gradually. However, the dosage of cement should not be increased excessively to a level of > 1.8%. Otherwise, the cold recycling pavement is susceptible to cracking and other diseases [7–10].

According to the experience, a qualified cohesion value should be greater than 180 g/cm2. When the cement dosage was in the range of 0.5-1.5%, the cohesion values of the mixtures containing the emulsifiers c and d were smaller than 180 g/cm2. According to the results of cohesion tests, the emulsifiers a or b should be selected to produce emulsified asphalt and cold recycling emulsified asphalt mixtures for high initial strength. However, during the selection of emulsifiers, a high demulsification speed should not be the only criterion. Instead, enough working time is also necessary to prevent the deterioration of pavement quality caused by demulsification in the process of production and construction. When the numbers of ionic charges of different emulsifiers are the same and these emulsifiers have good compatibility, the emulsifiers can be properly mixed according to the demulsification speeds of emulsified asphalt materials containing different emulsifiers, so the formation speed of the strength of the mixture can be moderate. Thanks to that, enough working time and qualified initial strength can be achieved simultaneously. Namely, the initial strength of a cold recycling emulsified asphalt mixture can be improved by selecting an appropriate emulsifier or optimizing the composition of emulsifier.

After comparison, the cohesion values of C-80 and Z-80 mixtures are almost the same, which indicates that the gradation condition has slight influence on the initial strengths of the cold recycling emulsified asphalt mixtures. This is because both materials were prepared by continuous gradation and intercalation was not formed, and the bonding effect of a binder mainly contributes to the strength of a cold recycling emulsified asphalt mixture [11]. The cohesion value of Z-100 with an RAP content of 100% was greater than that of Z-80, with an RAP content of 80%. In other words, a higher RAP dosage contributes to higher initial strength. This is because the RAP content of 100% represents a high content of asphalt, and strength can be formed by the combination of demulsified asphalt and RAP encapsulating spent asphalt.

4.2 Sweeping tests

In this paper, the performances of cold recycling emulsified asphalt mixtures to resist abrasion and grains formation during curing under the condition of open traffic were evaluated by sweeping tests. The evaluation index was loss of weight. According to the standard JTG E20-2011 "Specifications and Testing Methods of Bitumen and Bituminous Mixtures for Highway", 150-mm-sized specimens with a height of 8 ± 0.5 cm were prepared by rotary compaction.

The results of sweeping tests are shown in Figure 3. Under the same conditions of gradation, RAP, and cement dosage, the sweeping test results of the mixtures containing different emulsifiers were quite different. The weight loss of mixtures containing the emulsifiers a and b was relatively small, lower than about 70%. The weight loss values can be ranked as follows: c > d > b > a, which is principally consistent with the order of demulsification speed. Under the same conditions of RAP content, emulsifier and gradation type, with the increase of cement content, the mass loss values of cold recycling emulsified asphalt mixtures declined gradually. When the dosage of cement was increased from 0.5% to 1.5%, the mass loss was decreased greatly. When the dosage of cement was increased from 1.5% to 2.5%, the mass loss was decreased gradually.

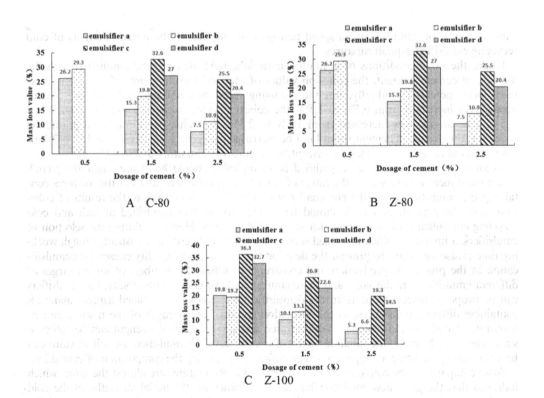

Figure 3. Results of sweeping tests.

156

The mass loss values of C-80 and Z-80 mixtures were compared, and it was showed that the value of the mixture containing medium-sized grains was smaller than that of the mixture containing large-sized grains. When the emulsifiers c and d were used in the C-80 mixture, the mixture was loosened directly after sweeping. This is because this mixture contained large-sized grains, and the surface of the mixture was worn by the rubber strips in the sweeping experiment. As a result, the large-sized grains were gradually exposed, and forces were easier to exert. In this way, the mixture around the large-sized grains was loosened. After a while, the mass loss of the mixture containing large-sized grains was increased. The mass loss values of Z-80 and Z-100 mixtures were compared, and the content of RAP had negligible impact on the mass loss of cold recycling emulsified asphalt mixtures. The results showed that the anti-loosening and anti-grain-losing performance of cold recycling emulsified asphalt mixtures during curing without traffic closing could be obviously improved by selecting an appropriate emulsifier, adding cement and reducing the content of large-sized aggregate. The relevant engineering application results showed that the mass loss values of the pavement were not as large as the test results during curing with open traffic. This is because confining pressure was not formed around the specimens in the sweeping experiment in the absence of other mixtures.

4.3 *Rutting tests*

Previous research results show that the dynamic stability of a cold recycling emulsified asphalt mixture reached 10000 times/mm after accelerated curing. Although the test results showed that ruts were not observed on the cold recycling layer, the relevant engineering applications results showed that the cold recycling layer had a large porosity, and compact rutting was possible. Therefore, the anti-rutting performance of a cold recycling emulsified asphalt mixture after accelerated curing cannot precisely reflect the actual situation. In this paper, the initial anti-rutting testing was conducted under the most adverse situation, i. e. the anti-rutting performance during curing with open traffic after the paving of the cold recycling layer, in simulated projects. According to the standard JTG E20-2011 "Specifications and Testing Methods of Bitumen and Bituminous Mixtures for Highway", the specimens with a size of 8cm × 30cm × 30cm for rutting tests were prepared. After curing, the rutting tests were carried out at 60 °C immediately for the evaluation of the initial anti-rutting performances of cold recycling emulsified asphalt mixtures, indicated by the dynamic stability.

The rutting testing results are shown in Figures 4 and 5. Under the same conditions of gradation and RAP content, the dynamic stability values of the mixtures containing different emulsifiers were close. The dynamic stability of the mixture containing the emulsifier a was the highest, 2120–2405 times/mm. The dynamic stability of the mixture containing the emulsifier c was the lowest, 2001–2236 times/mm. The dynamic stability values of the mixtures containing different emulsifiers were ranked as follows: a > b > d > c, principally consistent with the order of demulsification speed. Under the same conditions of emulsifier type and RAP content, the dynamic stability of C-80 containing large-sized grains was greater than that of Z-80 containing medium-sized grains. Although the two gradation schemes were continuous without embedded structures, the increasing of the mineral aggregate particle sizes was still beneficial to the rutting resistance of cold recycling emulsified asphalt mixtures. Under the same conditions of emulsifier type and gradation type, the dynamic stability of C-80 was slightly higher than that of Z-100, which is mainly due to the high asphalt content in the mixture containing 100% RAP and fine gradation materials. The relevant engineering applications experience showed that when the cold recycling layer of emulsified asphalt was cured in open traffic, cracks and abrasion were easy to occur, and ruts and loosening were rare. This is mainly because a certain proportion of water was contained in the cold recycling layer during the open traffic maintenance, and the water could cool the pavement. Although the surface temperature might be high, the inner temperature of the pavement was lower, which was beneficial to the overall resistance of the pavement to rutting. Moreover, when the cold recycling emulsified asphalt layer was loaded in the traffic, high dynamic water pressure was not easy to

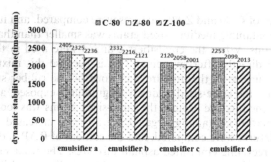

Figure 4.   Results of rutting tests.

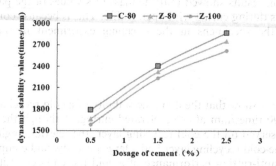

Figure 5.   Influence of cement dosage on the dynamic stability of cold recycling emulsified asphalt mixtures.

form inside the cold recycling layer. Because the surface water was dissipated very fast, the internal water continuously migrated upward to form connected pores, dissipating the water pressure in the pores. With the increase of cement dosage, the dynamic stability of cold recycling emulsified asphalt mixtures increased significantly. For example, the dosage of cement in the C-80 mixture was increased from 1.5% to 2.5%, and thus the dynamic stability was increased by *ca.* 20%. According to the standard JTGT 5521-2019, under the condition of the cement dosage of 1.5%, the dynamic stability of the three mixtures was greater than 2000 times/mm, meeting the requirements. Considering the resistance to rutting, these cold recycling emulsified asphalt mixtures can be cured in open traffic.

5   TERMINAL PERFORMANCE

5.1   *Water stability*

Water is one of the main factors resulting in damages to cold recycling emulsified asphalt pavements. Through freeze-thaw splitting tests, immersion splitting tests and immersion Marshall tests, the terminal water-damage resistance of cold recycling emulsified asphalt mixtures with different emulsifier types, gradation types and RAP contents was studied. The cement contents in the mixtures were 1.5%, and the testing results are shown in Table 3.

The testing results show that the terminal performances of the mixtures containing the emulsifier a and c were similar, indicating that the type of emulsifier had negligible influence on the terminal water stability of cold recycling emulsified asphalt mixtures. The testing results of C-80 and Z-80 mixtures showed that the Marshall stability and splitting strength of the mixtures containing large-sized grains were slightly smaller than those of the mixtures containing medium-sized grains. The testing results of Z-80 and Z-100 mixtures showed that with

Table 3. Water stability data of cold recycling emulsified asphalt mixtures.

Items		Emulsifier a			Emulsifier c		
		C-80	Z-80	Z-100	C-80	Z-80	Z-100
Water immersion Marshall test	wet	8.53	8.80	9.93	8.56	8.78	9.46
(40°C, kN)	dry	9.93	10.32	11.12	9.85	10.34	10.96
Residual Marshall stability (%)		85.9	85.3	89.3	86.9	84.9	86.3
Splitting test (15°C, MPa)	Wet	0.87	0.88	0.90	0.86	0.88	0.91
	dry	0.92	0.94	0.96	0.92	0.93	0.96
Ratio of dry splitting strength to wet splitting strength (%)		94.6	93.6	93.8	93.5	94.6	94.8
Freeze-thaw splitting test (MPa)	Freeze-thaw	0.25	0.24	0.22	0.25	0.26	0.24
	Without freeze-thaw	0.29	0.29	0.25	0.31	0.31	0.28
Freeze-thaw splitting strength ratio (TSR, %)		80.0	82.8	86.2	80.6	83.9	85.7

the increase of RAP contents, the splitting strengths with and without water immersion were gradually improved. This trend is different from those of the conventional hot mix and hot recycling asphalt mixtures. This is mainly because the strength of a cold recycling emulsified asphalt mixture is attributed to the bonding effect of the binder, and the asphalt mixture containing medium-sized grains contained many contact points, so the bonding strength was slightly larger. The emulsified asphalt could bond with the RAP encapsulating spent asphalt in a closer manner, so the indirect tensile strength was gradually increased. Therefore, Under the same conditions of emulsifier type, porosity, and other conditions, the terminal Marshall stability and splitting strength of the Z-100 mixture were the highest, followed by Z-80, and those of the C-80 mixture were the lowest. By comparing the residual Marshall stability, dry-wet splitting strength ratio, and freeze-thaw splitting strength ratio of the three mixtures, the water stability values of the three mixtures met the relevant requirements of the Specification, and the water stability the three mixtures was ranked as follows: Z-100 > Z-80 > C-80.

The terminal splitting strength of the cold recycling emulsified asphalt mixtures with different cement contents is shown in Figure 6. The results show that with the increase of cement contents, the splitting strength of cold recycling emulsified asphalt mixtures increased gradually. When the dosage of cement was increased from 1% to 1.5%, the splitting strength of C-80, Z-80 and z-100 was increased by 63%, 53% and 45%, respectively. When the dosage of cement was increased from 1.5% to 2%, the splitting strength of the three mixtures was increased by about 10%. Namely, the increments were reduced gradually.

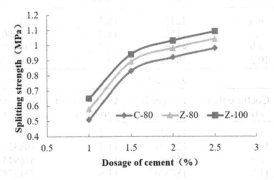

Figure 6. Influence of cement dosage on the splitting strengths of cold recycling emulsified asphalt mixtures.

In this paper, rutting specimens with the size of 450 mm × 300 mm × 80 mm were prepared with a wheel-rolling method. The specimens were directly placed into an oven at 60 °C for 48 hours. After cooling to room temperature, the specimens were demoulded and cut into beam-like specimens with the size of 380 mm × 63.5 mm × 5 mm, with an error of ± 5 mm. The four-point beam-bending fatigue tests were carried out at 15 °C ± 0.5 °C with a UTM tester. The loading was controlled by strains, and the waveform loaded was a partial sine wave with a loading frequency of 10 Hz. Previous research results showed that when the micro strain of a cold recycling material was greater than 400 $\mu\varepsilon$, the fatigue life was not only small but also highly variable, so the fitting of the fatigue equation was impossible. When the micro strain was smaller than 100 $\mu\varepsilon$, the stiffness modulus of the cold recycling emulsified asphalt mixture remained almost constant after more than 100000 times of cycles. Therefore, four strain levels of 150, 200, 250, and 300 $\mu\varepsilon$ were selected for fatigue tests. The ending criterion of the fatigue tests was that the stiffness modulus of a cold recycling asphalt mixture was decreased to 50% of the initial value. The mixtures were denoted by gradation type - RAP content - emulsifier type - cement content. For instance, C-80-a-1.5 represents a coarse-grained emulsified asphalt cold recycling mixture with an RAP content of 80% and cement content of 15%, containing emulsifier a. The fatigue testing results are shown in Table 4.

In this paper, the Wohler profile was fitted to the fatigue life and loading times for the prediction of the fatigue life of a cold recycling material.

$$\varepsilon = AN^{-b} \qquad (2)$$

wherein ε represents the amplitude corresponding to the strain that was repeatedly loaded; N represents the loading times when the specimen was damaged; A and b stand for the fitting parameters.

The fitting results are presented in Table 5, and the fatigue curves are shown in Figure 7.

The higher a fatigue curves is, the better the fatigue performance of the mixture is. The smaller the curves slope is, and the smaller the parameter b is, the greater the sensitivity of the mixture to the strain is. The testing results of Z-80-a-1.5 and Z-80-c-1.5 show that the type of emulsifier had negligible influence on the terminal fatigue performances of cold recycling emulsified asphalt mixtures. The testing results of C-80-a-1.5, Z-80-a-1.5, and Z-100-a-1.5 show that the fatigue performances of cold recycling mixtures containing medium-sized grains are better

Table 4. Fatigue testing results of the cold recycling emulsified asphalt mixtures.

Strain level	C-80-a-1.5		Z-80-a-1.5		Z-100-a-1.5	
	Fatigue life (times)	Coefficient of variation	Fatigue life (times)	Coefficient of variation	Fatigue life (times)	Coefficient of variation
300	5135	6.4	12357	10.5	26132	8.9
250	89363	7.8	112621	11.5	195686	10.3
200	313628	8.9	452901	13.6	715348	L9.5
150	1552502	10.5	1901928	13.3	2805236	10.8

Strain level	Z-80-a-0.5		Z-80-c-1.5		Z-80-a-2.5	
	Fatigue life (times)	Coefficient of variation	Fatigue life (times)	Coefficient of variation	Fatigue life (times)	Coefficient of variation
300	6892	8.5	11983	9.2	8493	10.9
250	83212	10.3	123281	11.7	108327	11.4
200	232493	8.9	443892	12.2	516943	12.6
150	1298403	11.2	1883247	11.6	2312987	12.1

Table 5. Fitting results for the fatigue life of cold recycling emulsified asphalt mixtures.

Mixture	Fatigue equation	R^2
C-80-a-1.5	$\varepsilon = 880.6N^{-0.119}$	0.9289
Z-80-a-1.5	$\varepsilon = 1137.7N^{-0.136}$	0.9563
Z-100-a-1.5	$\varepsilon = 1405.9N^{-0.147}$	0.9605
Z-80-a-0.5	$\varepsilon = 1021.4N^{-0.133}$	0.9457
Z-80-c-1.5	$\varepsilon = 1130.3N^{-0.136}$	0.9479
Z-80-a-2.5	$\varepsilon = 947.74N^{-0.121}$	0.9472

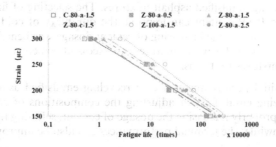

Figure 7. Fatigue curves of the cold recycling emulsified asphalt mixtures.

than those of mixtures containing large-sized grains. And, the fatigue performances of cold recycling emulsified asphalt mixtures containing 100% RAP are better than those of mixtures containing 80% RAP. This is mainly because a higher RAP content and finer gradation materials led to a higher total content of old and new asphalt in the mixture. In addition, the emulsified asphalt could be well bonded with the RAP encapsulating the old asphalt, and the bonding effect was stronger than that between emulsified asphalt and new aggregate. Thereby, the durability was better. Under the same conditions of emulsifier type, gradation and RAP content, when the dosage of cement was increased from 0.5% to 1.5%, the fatigue curves were elevated, and the fatigue performance was improved significantly. When the dosage of cement was increased from 1.5% to 2.5%, the fatigue life of the Z-80-a-1.5 mixture was longer when the strain was greater; The fatigue life of the Z-80-a-2.5 mixture was longer when the strain was smaller. The regression parameter b of the Z-80-a-2.5 mixture was smaller than that of the Z-80-a-1.5 mixture, indicating that the further increase of the dosage of cement would improve the sensitivity to strain. Therefore, comprehensively considering the terminal water stability, durability testing results, and economy, the dosage of cement should not be too high.

6 CONCLUSIONS

The cold recycling emulsified asphalt mixtures exhibited good initial and terminal performances, which are the crucial factors to determine whether the technology can be successfully applied, especially in urban roads. In this work, the influences of emulsifier type, gradation, RAP content and cement content on the initial and terminal performances of cold recycling emulsified asphalt mixtures were analyzed, and the methods to improve the initial performances of cold recycling emulsified asphalt mixtures were discussed. The main conclusions include the following:

1. In this paper, the initial strengths of cold recycling emulsified asphalt mixtures were evaluated via cohesion tests. The anti-abrasion and anti-grain-losing performances of mixtures

during curing with open traffic were evaluated by sweeping tests. The anti-rutting perform-ances of mixtures during curing with open traffic were evaluated by non-standard rutting tests. The results showed that through optimizing emulsifier, increasing cement content, selecting finer gradation materials and increasing RAP content, the initial performances of cold recycling emulsified asphalt mixtures could be improved. Although the selection of finer gradation materials and the increase of RAP content would reduce the initial anti-rutting performance, the reduction extent was small, and the test results still met the requirements of relevant specifications.

2. The type of emulsifier had negligible impact on the terminal water stability and durability of cold recycling emulsified asphalt mixtures. The increase of cement dosage, selecting of finer gradation materials and increasing RAP content can improve the terminal water sta-bility of cold recycling emulsified asphalt mixtures. The selecting of finer gradation mater-ials and increasing RAP content can improve the durability of cold recycling emulsified asphalt mixtures. Although the increasing of cement dosage was beneficial to the durability of cold recycling emulsified asphalt mixtures, an excess of cement would lead to increased sensitivity of the mixtures to strains.

To sum up, the initial performances of cold recycling emulsified asphalt mixtures can be improved by optimizing emulsifiers or adjusting the compositions of emulsifiers, increasing the content of RAP, properly increasing the dosage of cement, selecting finer gradation mater-ials, and so on. Meanwhile, the terminal performances can also be improved for their applica-tions in urban roads.

REFERENCES

[1] Wang H F, The Research on Asphalt Emulsion Cold Recycling Technology[D]. Changsha: Chang-sha University of Science and Technology, 2008.4.

[2] Wang W B. Study on Performance of Cold Recycling Mixture With Emulsified Asphalt and Cement[D]. Xi'an: Chang'an University, 2010.4.

[3] Wang Z, He L, Zhang J, Peng B. Research on Fatigue of Cold Recycling Mixtures with Emulsion Asphalt[J]. Highway. 2010,12. (12):160–163.

[4] Xing A X. Study on Performance of Cold Recycling Mixture With Emulsified(foamed) Asphalt[D]. Xi'an: Chang'an University, 2010.4.

[5] Li H X. Research on the Performance of Asphalt Cold Recycled Mixture and Pavement Structure[D]. Beijing: BEIJING JIAOTONG University, 2019.5.

[6] Peng B, Yin Z Q, Li L G. Study on Mix Design of Emulsified Asphalt Cold Recycled Mixture Based on Early Compression Strength[J]. Journal of Highway Transportation Research and Development. 2019,1:104–107.

[7] Wang Z, Li Z, Yang L Y. Methods Study on Improving the Early Performance of Emulsified Asphalt Cold Regeneration Mixture[J]. PETROLEUM ASPHALT. 2019,3:20–24.

[8] Sun J X, Liu L P, Sun L J. Effect of Early Strength Agent on Early Strength of Emulsified Asphalt Cold Recycled Mixture[J]. Journal of Wuhan University of Technology(Transportation Science and Engineering). 2017,6:1037–1040.

[9] Research Institute of Highway Ministry of Transport. JTG E20-2011 Standard Test Methods of Bitumen and Bituminous Mixtures for Highway Engineering [S]. Beijing: China Communications Press, 2011.

[10] Research Institute of Highway Ministry of Transport. JTGT 5521-2019 Technical Specifications for Highway Asphalt Pavement Recycling[S]. Beijing: China Communications Press. 2019.8.

[11] Wang Z, He L, Zhang J, Huang X M. Investigate of Mechanism and Rules of the Foamed Asphalt Mixture Strength[J]. International Chinese Conference of Transportation Professionals 2010.

[12] Wang Z, Hao L, Huang X M. Evaluation of Properties of Aged and Recycled Mixture without Extraction and Recovery[J]. Sustainable Construction Materials 2012. 2012, 287–302.

Functional Pavements – Chen et al (eds)
© 2021 Taylor & Francis Group, London, ISBN 978-0-367-72610-2

Experimental study on the properties of micro-surfacing mixtures containing RAP

Zhen Wang, Hailing Bu & Chuan Sha
Beijing Municipal Road & Bridge Building Material Group Co., Beijing, China

Tao Ma
Southeast University, Jiangsu, China

ABSTRACT: In this paper, the influences of RAP content on the mixing workability, cohesion, wear resistance, deformation resistance and noise reduction performance of micro-surfacing mixtures were studied, and the ranges of asphalt-aggregate ratio of micro-surfacing mixtures with different RAP contents were evaluated. The feasibility of adding RAP in micro-surfacing mixtures was explored. The results showed that with the increase of RAP content, the mixing time of micro-surfacing mixtures increased, curing rate decreased, abrasion resistance first decreased and then increased, deformation resistance decreased, and noise reduction performance was improved. The doping of RAP in micro-surfacing mixtures was feasible.

Keywords: micro-surfacing, RAP content, asphalt-aggregate ratio, noise reduction characteristic

1 INSTRUCTION

Currently, the asphalt pavements in China require large-scale maintenance and repair, along with the production of tens of millions of tons of waste asphalt pavement materials (RAP) every year [1]. RAP is generally treated by asphalt pavement recycling technologies, which can not only save a lot of raw materials and funds, but also avoid the occupation of lands due to the stacking of RAP, bringing obvious social and economic benefits [2,3]. With the continuous development and applications, the asphalt pavement recycling technologies in China have been relatively mature. However, with the increase of asphalt pavement recycling times, the refinement and aging degrees of mixtures will deteriorate, and the recycling of RAP will be more difficult.

As a preventive maintenance technology, micro-surfacing is widely used in the repair of pavement overlays and rutting in asphalt concrete pavements. The average service life of micro-surfacing pavement is in the range of 3 to 5 years, showing good road performance [4–6]. However, in the applications of micro-surfacing technologies in highway maintenance in recent years, it was found that the noise characteristics in micro-surfacing asphalt pavements were significantly different from those in normal pavements [7–9]. These noise characteristics seriously affected the comfort of drivers and passengers and their evaluation of road performance. To date, studies and applications of RAP in micro-surfacing asphalt mixtures are rare.

In this paper, the road performances of micro-surfacing mixtures containing different contents of RAP were measured, and the feasibility of adding RAP to micro-surfacing mixtures was explored. The effects of RAP contents on the mixing workability, cohesion, wear resistance, rutting resistance, and noise reduction performance of micro-surfacing mixtures were investigated. The application fields of micro-surfacing mixtures containing RAP were probed, and the appropriate RAP contents were recommended.

2 MATERIALS AND COMPOSITION

2.1 Materials

2.1.1 Modified emulsified asphalt
SBR modified emulsified asphalt was used as the binder, and the heavy traffic asphalt AH-70 was selected as the matrix asphalt in this work. Four slow-cracking and fast-setting cationic emulsifiers (denoted by 1, 2, 3, and 4), SBR latex, hydrochloric acid regulator and water were prepared. The technical indexes of the emulsified asphalt met the requirements of the relevant specifications.

During the preparation process, first, a certain amount of an emulsifier was added into water at 50–60 °C. Then, hydrochloric acid was added to adjust the pH to 2.0–2.2. The asphalt at 130–140 °C was mixed with a soap solution at 50–60 °C, and the mixture was treated with an asphalt gel mill at a high speed to prepare emulsified asphalt. When the emulsified asphalt was cooled to room temperature, a certain amount of latex was added, and the micro-surfacing emulsified asphalt was obtained.

The results of mixing experiments showed that though the surface of RAP was coated with aged asphalt, the superficial characteristics of mineral aggregate were improved, and the aggregate was easier to bond with asphalt. But, the compatibility between mineral aggregate (RAP and new aggregate) and emulsified asphalt was still poor. The mixing effect of emulsifier 4 on the mineral materials and old materials was poor. The emulsified asphalt prepared with emulsifier 3 could be well mixed with the micro-surfacing mixture containing 0% RAP. However, a "pseudo dilute slurry" state was observed in the mixing of emulsified asphalt prepared with emulsifier 3 with RAP-containing micro-surfacing mixtures. Namely, a large number of bubbles took place during the mixing process, and the slurry was not thickened. The mixing resistance was obvious, but the slurry was dilute. This is because of the poor compatibility between emulsified asphalt and old materials. Premature demulsification occurred in the mixing process, and the strength of the micro-surfacing mixture was low. An excess of emulsifier was present in the emulsified asphalt, and the demulsified micro-surfacing mixture was re-broken with the stirring of mechanical force, resulting in the formation of foams. The micro-surfacing mixture in the pseudo dilute slurry was filled with foams, resulting in many large voids and low bonding strength. The strength of the mixture was very low after curing. The mixing performances of different emulsifiers are shown in Figure 1. The modified emulsified asphalt produced with emulsifiers 1 and 2 could be readily mixed with the new mineral aggregate and RAP, so the emulsifiers 1 and 2 were mixed for using in this project. According to the demulsification rate, the ratio of both emulsifiers was set to be 1:1.

2.1.2 Aggregate, water and fibers
In this paper, basalts with grain sizes of 0–3, 3–5 and 5–8 mm were purchased from Yi County, Hebei Province and selected as the new mineral materials. The mineral powders

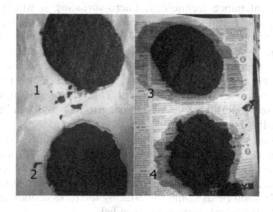

Figure 1. Mixing effects using different emulsifiers.

derived by grinding hydrophobic materials such as limestone or strong basic magmatic rock were selected as the filler. The physical and mechanical properties of these mineral materials were measured and met the relevant requirements. For the improving of the mixing workability of the emulsified asphalt and aggregate, in addition to the adjusting of the aggregate gradation, a certain proportion of cement should be added into the micro-surfacing mixtures to promote the stability of micro-surfacing mixtures and to adjust the demulsification rates of mixtures. In this paper, $42.5^{\#}$ regular Portland cement with a dosage of 1.5% was added. The water used in this paper was ordinary drinking water. Thanks to the adsorption, stability and multi-directional reinforcement effects of fibers, the gel structures of slurry mixtures could be improved and their comprehensive performances could be improved. The fibers used were glass fibers.

2.1.3 *Waste asphalt pavement materials*

In this paper, the surface layer of basalt in Beijing area was shredded and milled, and the grains obtained were graded according to their sizes (0-3, 3-5, and 5-10 mm). The RAP was used as the "black stone". The sand equivalent values of 0-3, 3-5, and 5-10 mm sizes RAP with basalts were measured to be 65%, 75% and 90%, respectively.

2.2 *Gradation*

The MS-3 gradation scheme was selected for the micro-surfacing mixtures. The contents of RAP in the mixtures A, B, C and D were set to be 0%, 40%, 60% and 94%, respectively. Because the RAP grains with the sizes of 0-3 mm were larger than the basalt grains with the sizes of 0-3 mm, a certain proportion of mineral powders were added to ensure the passing percentage of fine aggregate and to make the micro-surfacing regions contain enough mortar for the forming of strength. The RAP contents of the gradation schemes E and F were 100% and 60%, respectively, without mineral powders. The cement was added extra and excluded in the gradation schemes. The compositions of the mineral materials are shown in Table 1. The optimum contents of the MS-3 micro-surfacing fibers are in the range of 0-0.3%, and thereby the content in this experiment was set to be 0.2%.

Table 1. Compositions of the mineral materials.

Micro-surfacing mixture	RAP 0-3mm	RAP 3-5mm	RAP 5-10mm	Basalt 5-8mm	Basalt 3-5mm	Basalt 0-3mm	Mineral powder
A				20	25	55	
B	10	10	20		13	46	1
C	18	22	20			38	2
D	50	24	20				6
E	18	22	20			40	
F	58	22	20				

3   RESULTS AND DISCUSSION

3.1 *Mixing test*

The mixing test is primarily used to determine the mixing time and slurry state of micro-surfacing mixtures, for the optimization of asphalt-aggregate ratio [10]. The influence of RAP contents on the water contents added in the mixtures was studied by mixing tests. 10.0% of emulsified asphalt was mixed with 0%, 40%, 60% and 100% of RAP, separately. Water and cement were added through an external mixing method, and the contents were 8% and 1.5%, respectively. The results are shown in Figure 2.

The experimental results showed that under the same mixing conditions, with the increase of RAP content, the mixing time of micro-surfacing mixtures was increased, indicating that

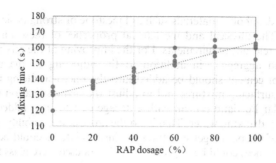

Figure 2. Effect of RAP content on the mixing time.

the addition of RAP could increase the workability of micro-surfacing mixtures and prolong the mixing time. The reason is that the surface of RAP was coated by aged asphalt mortar. The hydrophobicity of RAP surface was different from the hydrophilicity of fine aggregate, and the adsorption capacities for asphalt emulsifier and free water were also different [11]. Therefore, after mixing with emulsified asphalt, during the demulsification process, a layer of asphalt film would be formed on the surface of aggregate of the modified emulsified asphalt. The extruded water would be retained on the layer, and the demulsification times of different emulsified asphalt materials were different.

Therefore, in the micro-surfacing tests with the addition of RAP, the amount of water added should be adjusted by the mixing tests according to the dosage of RAP, to ensure the construction performances of the mixtures. The decrease of filler had no obvious influence on the mixing time.

3.2 Adhesion test

The adhesion index reflects the curing speed of a mixture and can be used to determine the open traffic time. For the same curing time, the higher the adhesion force is, the higher the molding rate is, and the greater the early strength is. The initial setting time refers to the time to when the cohesion of the slurry mixture reaches 1.2 N·m from the spreading. The open traffic time refers to the time to when the cohesion of the slurry mixture reaches 2 N · m from the spreading. The content of emulsified asphalt was 10%, and the cohesion testing results of materials with different RAP contents are shown in Figure 3.

The results show that under the same condition of emulsified asphalt content, with the increase of RAP content, the cohesion values of the mixtures gradually decreased. When the RAP content exceeded 60%, the cohesion force did not meet the requirements of the specifications. The cohesion values of mixtures C and D containing the filler were slightly lower than

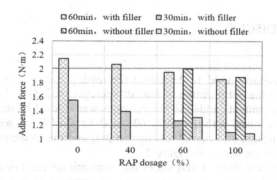

Figure 3. Testing results of micro-surfacing asphalt mixtures containing RAP.

166

those of mixtures E and F excluding the filler, and the values of mixtures C and D were equivalent. The emulsified asphalt can penetrate into the texture of aggregate surface, and certain chemical adsorption processes took place on the interface between asphalt and aggregate. The chemical adsorption processes could improve the resistance of asphalt layer adhered to the aggregate surface to water. However, the bonding between the emulsified asphalt and aged asphalt on the RAP surface was not strong enough in the presence of water, showing weak adhesion. Moreover, the acid-base neutralization reactions between the aggregate and emulsified asphalt slowed down, the early bonding strength of the mixtures decreased, and the curing and molding rates decreased. Therefore, when the micro-surfacing asphalt mixtures containing RAP were used, the open traffic time should be extended to a certain extent.

3.3 Wet abrasion test

A wet abrasion test (WTAT) can be used to measure the mass loss before and after the abrasion of a specimen, wear resistance performances of a mixture after molding, and optimum asphalt dosage. After the micro-surfacing asphalt concrete was molded for 1 h, the testing was performed. According to the testing results, the profiles regarding the relationships between asphalt dosage and wear loss were plotted, and the minimum asphalt dosage (Pbmin) was calculated according to the upper limit of allowable wear weight loss. In addition, the testing results reflect the compatibility of components in the mixture. If the compatibility is poor, a high-performance mixture cannot be obtained even if the quality of each component is good. Additionally, the wet abrasion value of the micro-surfacing mixture cannot meet the requirements. The wet abrasion test results of micro-surfacing mixtures with different RAP contents are shown in Figure 4.

The results show that with the increase of asphalt-aggregate ratio, the wet abrasion value of micro-surfacing mixture containing RAP gradually decreased after soaking for 1 day, which is due to the increase of asphalt-aggregate ratio increased the amount of mortar and reinforced the mixture, resulting in a lower abrasion value. When the content of RAP was smaller than 60%, with the increase of RAP content, the wear value of micro-surfacing mixture gradually increased. This principle was more obvious under the conditions of high asphalt-aggregate ratios. When the RAP content was further increased to 94%, the wear value decreased. This is due to the fact that when the proportions of the new and old materials were close, the contact surfaces between the old and new materials were increased, and these surfaces were subject to stress concentration, leading to the stripping of grains and elevation of the wear value. For the mixture E (RAP content = 60%) and F (RAP content = 100%) excluding the mineral powders, the wear values were greater than those of other materials under most circumstances. This is because the mineral powders had cementation and filling effects. The former improved the overall strength of the micro-surfacing mixture, and the latter made the surface of the mixture flatter and denser, thus reducing the wear values. According to the requirement that the abrasion value should be smaller than 540 g/m^2, the minimum emulsified asphalt dosage was calculated by interpolation (see Table 2).

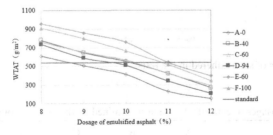

Figure 4. Wet abrasion testing results of micro-surfacing mixtures with different RAP contents and asphalt-aggregate ratios.

Table 2. Minimum emulsified asphalt dosages determined by the wet abrasion tests.

Gradation scheme	A	B	C	D	E	F
Dosage of emulsified asphalt (%)	8.7	10.1	10.2	9.6	11	10.8

3.4 Rutting test with a loaded wheel

The rutting tests with a loaded wheel were conducted 1000 times with a loading of 56.7 kg. The depths of tracks and lateral displacements of the specimens were measured for the detection of the contents of asphalt in the mixtures and controlling the upper limit of asphalt dosage, preventing oil spilling and displacement deformation due to the excessive dosage of asphalt. Generally speaking, the greater the asphalt dosage is, the greater the amount of sand adhered to the specimen is. When the amount of sand adhered reaches the required maximum value, the dosage of asphalt is the maximum value. The results of rutting tests are shown in Figure 5.

The testing results show that, with the increase of emulsified asphalt, the amount of sand adhered gradually increased. With the increase of RAP content, the amount of sand adhered gradually increased, but the increasing trend was moderate. With the addition of RAP, with the increase of emulsified asphalt content, the increasing trend of the amounts of sand adhered was obviously smaller than that of the micro-surfacing mixtures excluding RAP. This is because when the contents of emulsified asphalt in the mixtures excluding RAP were low, the oil films were thin and the amounts of sand adhered were small. After the addition of RAP, even if the dosages of emulsified asphalt were low, a certain amount of old asphalt was still present in the mixtures. The thickness of the regions with films was greater than that of the new materials, so the amounts of sand adhered were greater.

According to the technical indices of dilute slurry mixtures required by the "Technical Guides For Micro-Surfacing and Dilute Slurry Sealing Technologies", the maximum dosages of emulsified asphalt were determined under the condition of the amount of sand adhered reached 450 g/m^2. The results are shown in Table 3.

For the mixtures used to repair rutting, the transverse deformation percentage after rolling should not exceed 5%, which is called the rutting deformation test. For the micro-surfacing regions with small thickness values, the rutting deformation testing is not necessary, because the micro-surfacing regions are a thin layer and rutting will not be formed. The rutting testing results of micro-surfacing mixtures containing RAP are shown in Figures 6 and 7.

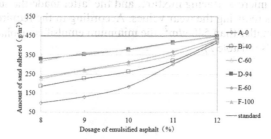

Figure 5. The amounts of sand adhered in the tests.

Table 3. Maximum dosages of emulsified asphalt determined by the amounts of sand adhered.

Gradation scheme	A	B	C	D	E	F
Dosage of emulsified asphalt (%)	12.3	12.3	12.3	12.1	12.2	12.1

168

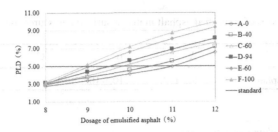

Figure 6.   Percentage of rutting lateral deformation (PLD) testing results.

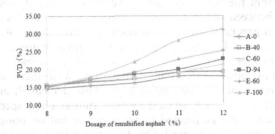

Figure 7.   Percentage of rutting depth (PVD) testing results.

The experimental results show that, with the increase of emulsified asphalt and RAP contents, both the percentages of rutting lateral deformation (PLD) and rutting depth (PVD) gradually increased. Namely, the anti-deformation performance was weakened. This is due to the increase of the overall asphalt content in the mixture by adding RAP. Because of the cementation and filling effects of the mineral powders, the percentages of rutting lateral deformation and rutting depth of mixtures E and F were greater than those of mixtures A, B, C and D. According to the requirements in the "Technical Guides For Micro-Surfacing and Dilute Slurry Sealing Technologies", the percentage of transverse deformation after rolling should not exceed 5%, and the maximum dosages of emulsified asphalt in the mixtures were calculated by interpolation. The results are shown in Table 4. When the filling with micro-surfacing mixtures was not necessary, the data may not be used for the evaluation and determination of the asphalt-aggregate ratio.

The afore-mentioned experimental results show that when the filling with micro-surfacing mixtures was not necessary, the maximum RAP dosage of micro-surfacing mixture was 100%, and the optimal emulsified asphalt dosages with different RAP contents are shown in Table 5. When the filling with micro-surfacing mixtures was necessary, the maximum RAP dosage of micro-surfacing mixture was 40%. The optimal dosages of emulsified asphalt in micro-surfacing mixtures containing RAP are shown in Table 6.

Table 4.   Maximum dosages of emulsified asphalt determined by the percentage of rutting lateral deformation.

Gradation scheme	A	B	C	D	E	F
Dosage of emulsified asphalt (%)	11.0	10.5	9.9	9.5	9.1	8.9

Table 5.   Optimal Dosages of emulsified asphalt in micro-surfacing mixtures containing RAP (filling is not necessary).

Gradation scheme	A	B	C	D	E	F
Dosage of emulsified asphalt (%)	8.7-12.3	10.1-12.3	10.2-13.3	9.6-12.1	11-12.2	10.8-12.1

Table 6. Optimal Dosages of emulsified asphalt in micro-surfacing mixtures containing RAP (filling is necessary).

Gradation scheme	A	B	C	D	E	F
Dosage of emulsified asphalt (%)	8.7-11.0	10.1-10.5	—	—	—	—

3.5 Noise reduction characteristics analysis

Generally, the road traffic noises can be divided into three categories according to the sources: 1. Noises generated from the power systems of vehicles; 2. Aerodynamic noises generated from the interactions between vehicles, external components of vehicles and the surrounding air; 3. Tire road noises generated from the interactions between the tires and road surfaces during the driving processes. The contributions of these three sources to the total noise are different under the different vehicle speed conditions. For the vehicles at a low speed, the noises generated from the power systems are dominant; with the increase of vehicle speed (> 50 km/h), the contribution of tire road noise is gradually increased; and the aerodynamic noise should be paid attention to only when the vehicle runs at a high speed.

The wet abrasion tests were performed to measure the indoor noise at micro-surfacing regions, and feasible results were obtained. Due to the high level of noise generated from the abrasion tester, the experimental results were interfered with. Herein, the sweeping tests of cold recycling asphalt mixtures were performed to measure the indoor noise at micro-surfacing regions. The precision of the equipment was high and noise generated from the equipment was weaker than that generated from the wet abrasion tester. Therefore, more accurate data could be obtained. In the sweeping tests, the surfaces of specimens were swept at different stirring rates to simulate the influence of driving speeds on the road noise at micro-surfacing regions. During the tests, the sound level meter should be placed 20 cm away from the specimen. In addition, the area of the test site should not exceed 20 m2 for the accuracy.

During the test, a shaped specimen was fixed on the testing platform. The A-weighted network and F-gear (fast gear) time weighting characteristics were adopted to measure the equivalent continuous sound pressure value (dB(A)(A)) in a period of time with the sound level meter. Before the test, the running conditions of the equipment should be checked, and the position of the sweeping pipe should be replaced after each test. During the test, the experimenters should keep quiet. The tester contained several gears, which could be adjusted to simulate different driving speeds, including 59, 107, 198, and 365 r/min. A test generally lasted for 30 s. In this project, the gear of 59 r/min was selected for testing. After the calibration, the noise level during idling was 49.5 dB(A)(A). The testing results of micro-surfacing mixtures containing RAP are shown in Table 7 and Figure 8.

The testing results show that with the increase of RAP content, the noise levels generated from the micro-surfacing asphalt mixtures gradually declined. This is because the RAP contained old asphalt, which increased the overall asphalt content in the micro-surfacing asphalt mixture, reducing the rigidity and increasing the flexibility of the mixture, thus reducing the vibration noise at the micro-surfacing pavement. According to the testing results, the noise level generated from the micro-surfacing mixture D (RAP content = 94%) was lower than that generated from the mixture A (RAP content = 0%) by 2.8 dB(A)(A).

Table 7. Noise levels during idling measured with the sweeping tester.

Gradation scheme	A	B	C	D
Noise dB(A)(A)	69.1	68.2	67.3	66.3
Difference dB(A)(A)	19.6	18.7	17.8	16.8

170

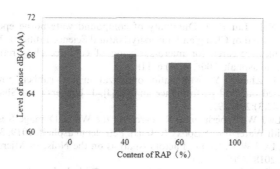

Figure 8.   Noise testing results of the micro-surfacing mixtures containing RAP.

4   CONCLUSIONS

In this paper, the performances of micro-surfacing mixtures with different RAP contents were tested. The effects of RAP on the mixing workability, adhesion, wear resistance, rutting resistance and noise reduction performances of micro-surfacing mixtures were studied.

(1) The results showed that the addition of RAP in micro-surfacing mixtures was feasible. With the increase of RAP content, the mixing time of micro-surfacing mixture increased, and the decrease of the content of filler had no obvious effect on the mixing time. With the increase of RAP content, the curing and molding rates of the mixtures decreased. When the RAP content exceeded 60%, the bonding strength did not meet the requirements of the specifications, and the open traffic time should be appropriately extended.
(2) With the increase of asphalt-aggregate ratio, the wet abrasion values of micro-surfacing mixtures containing RAP decreased gradually after soaking for 1 day. When the RAP content was lower than 60%, the wear weight loss values of the micro-surfacing mixtures gradually increased with the increase of RAP content. When the RAP content was further increased to 94%, the wear weight loss values decreased. The wear weight loss values of mixtures E and F excluding mineral powders were generally larger than those of other mixtures.
(3) With the increase of the contents of emulsified asphalt and RAP, the amounts of sand adhered at the micro-surfacing regions gradually increased, and both the percentages of rutting lateral deformation (PLD) and rutting depth (PVD) gradually increased. When the micro-surfacing treatment was not necessary, the maximum RAP content in the micro-surfacing mixtures was 100%. When the micro-surfacing treatment was necessary, the maximum RAP content in the micro-surfacing mixtures was 40%.
(4) The doping of RAP can effectively reduce the noise level generated from the micro-surfacing regions. With the increase of RAP content, the noise level was gradually decreased. The noise level at the micro-surfacing mixture D with an RAP content of 94% was lower than that generated from the mixture A with an RAP content of 0% by 2.8 dB(A)(A).

REFERENCES

[1] Fan S Z, Zhang Y J. Study on Properties of Micro-surfacing mixtures with RAP[J]. New Building Materials. 2019, 9:19–23.
[2] He P J. Construction Technology and Quality Control of Colored Emulsified Asphalt Micro-surfacing[J]. Technology of Highway and Transport. 2020, 36(1):22–27.
[3] Li T. Research on Technical Performance of Micro-surfacing Mixture With Adding Recycled Asphalt Material[J]. Highway Engineering. 2015, 40(5):170–175.
[4] Li W, Han S, Shao P K. Proportioning design and functional properties of super micro-surfacing[J]. Journal of Jiangsu University(Natural Science Edition). 2019, 40(3):366–372.

[5] Zheng M L, Fan X P, Liu F Q. Durability of compound waterborne epoxy emulsified asphalt microsurfacing[J]. Journal of Chang'an University(Natural Science Edition). 2020, 40(1):68–76.

[6] Wang H Y. Technical research for micro-surfacing of cold recycling retrieve asphalt mixtures planning[J]. Petroleum asphalt. 2008, 22(3):6–11.

[7] Wang C, Li X J, Zhu S Y. Mix ratio optimization for rubber powder noise reduction micro-surfacing based on road performance analysis[J]. J. University of Shanghai for Science and Technology. 2020, 42(3):291–297.

[8] Ye P P, Shi F Z, Lv J W. Experimental Research on the Water Damage Resistant Performance of Micro-surfacing with Water-borne Epoxy Resin[J]. Petroleum asphalt. 2019, 33(1):15–21.

[9] Zhu S Y, Shi F Z, Lv J W. Affecting Factors Analysis on the Noise of Micro-surfacing Mixture[J]. Petroleum asphalt. 2018, 32(6):43–50.

[10] Research Institute of Highway Ministry of Transport. Technical guide for microsurface and thin slurry seal[S]. Beijing: China Communications Press, 2011.

[11] Ren L N. Study on the application of recycled asphalt mixture in micro surfacing technology[D]. Xi'an: Chang'an University, 2017.4.

Functional Pavements – Chen et al (eds)
© 2021 Taylor & Francis Group, London, ISBN 978-0-367-72610-2

Comparative study of different density models in asphalt mixture density prediction

Tianjie Zhang, Haihang Han, Fengxia Chi & Yangyang Wang
Zhejiang Scientific Research Institute of Transport, Zhejiang Provincial Key Lab for Detection and Maintenance Technology of Road and bridge, Hang Zhou, Zhejiang Province, China

ABSTRACT: Ground Penetrating Radar (GPR) is widely used in predicting density of asphalt mixture based on composite dielectric models. In this study, the accuracy and stability of predicted density from six composite dielectric models were studied, including Linear model, CRIM, a modified CRIM, Bottcher model, ALL model and Rayleigh model. In addition, a BPNN-based pattern recognition was used to evaluate the density of asphalt mixture. The results showed that: Composite-dielectric-based models got a relatively low accuracy in density evaluation; Accuracy and stability of result from these models could not meet the requirement stipulated in standard; Results of BPNN-based recognition algorithm met the requirement stipulated in standard and its accuracy was increased with a larger database.

1 INTRODUCTION

In order to ensure an optimal lifetime and to avoid Hot Mix Asphalt (HMA) failures, a critical property of new asphalt pavement to control is the in-situ compactness (Araujo et al., 2017, LU et al., 2012). The compactness can be defined as the ratio between the bulk specific gravity and the theoretical maximum specific gravity. As the theoretical maximum specific gravity is given by the manufacturer, assessing the compactness or density is strictly equivalent.

Core drilling is the reference measurement for density evaluation. However, it is a destructive method (Dong et al., 2016). Also, the nuclear moisture density meter needs to be replaced due to its storage and cost issues as well as the risk incurred for users. Electromagnetic based methods like ground penetrating radar (GPR) is developed and widely used in road quality control and quality assurance (QC/QA) (Wang et al., 2018). GPR detects roads through the different dielectric constant between pavement layers. By analyzing dielectric constant of pavement materials, it could calculate the pavement density (LI et al., 2006).

Asphalt pavement is a three-phase mixture consisting of binder, aggregate, and air. Its density cannot be measured directly by GPR signals but could be evaluated by the composite dielectric model (Zhai et al., 2019). The composite dielectric model of asphalt mixtures can reveal the intrinsic relationship between the dielectric constant and the volume among each phase (Guo et al., 2018). Therefore, it is utilized when applying GPR to detect the density of asphalt pavement.

The composite dielectric model has been used in different fields. For example, the Rayleigh model has been widely used in petroleum exploration, and it played a significant role in predicting annual oil production (Qiang et al., 2003). The Lichtenecker-Rother model had a high consistency for saturated soil composed of soil particles, free water and air (Roth et al., 1990). In road engineering, researchers have applied the composite dielectric model in pavement detection. R. Mardeni produced asphalt concrete slabs (made of asphalt and single crushed stone) in lab, then used the Complex Refractive Index Model (CRIM) (Birchak et al., 1974) to predict road density (Roslee et al., 2010). Leng Zhen proposed the ALL model according to CRIM and Bottcher model, to improve accuracy of density prediction (Leng, 2012, Leng et al., 2012). In Illinois, USA, ALL model achieved higher accuracy than Rayleigh model and Bottcher model in real-time detection of pavement thickness and density (Wang et al., 2019).

Zhang Xiaoning found that the Rayleigh model (Rayleigh et al., 1892) performed well in the compaction control of roads in Guangzhou Province, China (LI et al., 2006). Wang Fuming applied a modified CRIM model in Henan Province, China, and it performed better than other models (Liu et al., 2009).

Compared with the composite dielectric model, a statistical-based model could avoid complicated derivation, and could continuously improve model accuracy by increasing the database. Research institutes have begun to use machine learning to analyze the data obtained from GPR response (Aydin and Erdem, 2019, Caorsi and Stasolla, 2008, Kafedziski et al., 2018, Li et al., 2012). For example, by analyzing the relationship between pavement density and data generated by gprMax, Shangguan Pengcheng used ANN-based pattern recognition to monitor the compaction of asphalt pavement; The final accuracy of density prediction reached approximately 95% (Wang et al., 2019).

Although various composite dielectric models have been applied in different scenarios, studies on comparing frequently-used models are few. In addition, AI-based statistical method is rarely used in density evaluation when utilizing GPR in QC/QA while statistical models could lower the coefficient of variation from materials. Therefore, evaluating the accuracy of composite dielectric models and proposing a new statistic model could effectively promote the application of density models, and expand the application of GPR.

This paper analyzed the accuracy of composite dielectric models in density evaluation of asphalt mixture using a database obtained from GPR response and materials. In addition, a BPNN-based recognition algorithm was proposed. First, test samples were formed in laboratory. Second, the relevant parameters of samples including height, diameter, dielectric constant, and radar response were obtained. Finally, bulk specific gravity of samples was calculated through different models according to the obtained parameters, and it was compared with the standard method (T 0705). Accuracy and stability of the models was validated by comparing actual and predicted density values.

2 MATERIALS

Two types of asphalt concrete samples, AC-13-Graded Asphalt Mixture, and SUP-20-Graded Asphalt Mixture, were produced in lab. They were made of Donghai 70 # matrix asphalt and basalt aggregate. Parameters of the mixture were shown in Table 1. The gradation of AC-13 and SUP-20 were shown in Figure 1.

In the process of making specimens, the bulk specific gravity of asphalt mixture was controlled by adjusting the mass of the mixture added to the container.

Specimens were preprocessed before testing in terms of budget and time. The procedure was shown in Figure 2. Firstly, specimens were cooled and demolded. Secondly, the demolded specimen was cut into 20 cm * 20 cm. Then, it was cut into two pieces (10 cm * 20 cm). After making test pieces, OKO-2 was used to obtain GPR response from the test specimens. OKO-2 is a ground penetrating radar developed by GEOTECH in 2015 which is designed for pavement QC/QA. Finally, height and bulk specific gravity of all test specimens, as well as the GPR response, were measured.

In order to improve accuracy of models, AS2855 Dielectric Constant Test was used to obtain the dielectric constant of asphalt and aggregate. AS2855 Dielectric Constant Test

Table 1. Parameters of asphalt mixture.

Type of Mixture	AC-13	SUP-20
Density of Aggregates	2.664	2.670
Theoretical Maximum Density	2.583	2.530
Density of Asphalt	1.15	1.15
Asphalt-Aggregate Ratio	0.049	0.043

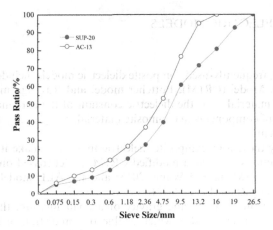

Figure 1.　Asphalt mixture gradation.

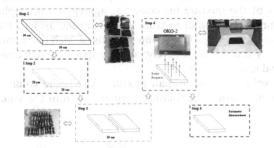

Figure 2.　Procedure of making and testing specimens.

consists of a test device (fixture), high-frequency Q meter (QBG-3E/QBG-3F/AS2853A), data acquisition system, and inductor (LKI-1). The dielectric constant of material (asphalt and aggregate) were obtained by the steps shown in Figure 3.

The dielectric constant of the tested samples was calculated by the following equation.

$$\varepsilon_r = \frac{D_1}{D_2} \tag{1}$$

where, ε_r is the dielectric constant of material, D_1 is the thickness of sample, and D_2 is the maximum resonance value.

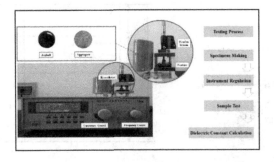

Figure 3.　Obtaining dielectric constant of asphalt and aggregate.

3 COMPOSITE DIELECTRIC MODELS

3.1 *Model building*

As shown in Figure 4, frequently-used composite dielectric models include linear model, Complex Refractive Index Model (CRIM), Bottcher model and Rayleigh model. Where ε is the dielectric constant of material; ε_0 is the dielectric constant of the core material; V_1, V_2, ...V_n are the volume ratios of components in composite material; ε_1, ε_2, ..., ε_n are the dielectric constant of each component.

Researchers modify these four composite dielectric models to make it applicable and suitable in different conditions, such as a modified CRIM model (based on CRIM model) proposed by Wang Fuming (Meng and Wang, 2013) and the ALL model (based on Bottcher model) proposed by Leng Zhen (Leng, 2012).

In order to evaluate the bulk specific gravity of asphalt mixture, these six models were selected to develop the density models which enable the prediction of bulk specific gravity from its dielectric constant. These six models were selected, because they had been successfully used in pavement QC/QA and their parameters are relatively easy to obtain.

When developing the density models in this study, the asphalt mixture was assumed dry. Thus, the components of the mixture included air, aggregate and asphalt binder. Figure 4 shows a phase diagram of the asphalt mixture describing the asphalt mixture's composition and parameters. The volumetric and mass contributions of each component to the entire mixture are represented by V and M, respectively; and the specific gravity and dielectric constant of each phase are G and ε, respectively.

Assuming the total volume of the asphalt mixture, $V_T = 1$, the volumes of air, binder and aggregate can be calculated using the following equations from the volumetric properties of the asphalt mixture.

$$V_a = 1 - \frac{G_{mb}}{G_{mm}} \tag{2}$$

$$V_b = \frac{M_b}{G_b} = \frac{G_{mb} P_b}{G_b} \tag{3}$$

$$V_{se} = \frac{M_s}{G_{se}} = \frac{G_{mb}(1 - P_b)}{G_{se}} \tag{4}$$

where: V_a is the porosity of asphalt mixture, G_{mb} is the bulk specific gravity of asphalt mixture, G_{mm} is the theoretical maximum specific gravity of asphalt mixture. V_b is the volume of asphalt, P_b is the asphalt-aggregate ratio, G_b is the density of asphalt. V_{se} is the volume of aggregate, M_s is the mass of aggregate, G_{se} is the density of aggregate.

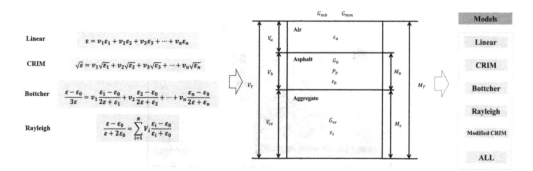

Figure 4.   Procedure of model building.

176

In general, the dielectric constant of air ε_a is assumed to be consistent with that of vacuum, which is 1.

Substituting equations (2) (3) and (4) in these six composite dielectric models mentioned above, and reorganizing these models gives the following modified models, as shown in Table 2.

3.2 *Accuracy analysis*

By substituting the measured material parameters and radar response into the six models mentioned above, the results of bulk specific gravity calculated by each model is shown in Figure 5. In order to evaluate the accuracy, bulk specific gravity was also measured by standard method (T 0705) stipulated in Standard *Test Methods of Bitumen and Bituminous Mixtures for Highway Engineering* (JTG E20-2011).

Table 2. Density models.

Model	Equation
Linear	$G_{mb} = \dfrac{\varepsilon_{AC}-1}{\frac{1}{G_{mm}} + \frac{1-P_b}{G_{se}}\varepsilon_s + \frac{P_b}{G_b}\varepsilon_b}$
CRIM	$G_{mb} = \dfrac{\sqrt{\varepsilon_{AC}}-1}{\frac{P_b}{G_b}\sqrt{\varepsilon_b} + \frac{(1-P_b)}{G_{se}}\sqrt{\varepsilon_s} - \frac{1}{G_{mm}}}$
Bottcher	$G_{mb} = \dfrac{\frac{\varepsilon_{AC}-\varepsilon_b}{3\varepsilon_{AC}}\ \frac{1-\varepsilon_b}{1+2\varepsilon_{AC}}}{\left(\frac{\varepsilon_s-\varepsilon_b}{\varepsilon_s+2\varepsilon_{AC}}\right)\left(\frac{1-P_b}{G_{se}}\right) - \left(\frac{1-\varepsilon_b}{1+2\varepsilon_{AC}}\right)\left(\frac{1}{G_{mm}}\right)}$
Rayleigh	$G_{mb} = \dfrac{\frac{\varepsilon_{AC}-1}{\varepsilon_{AC}+2}}{\frac{(1-P_b)}{G_{se}}\frac{\varepsilon_s-1}{\varepsilon_s+2} + \frac{P_b}{G_b}\frac{\varepsilon_b-1}{\varepsilon_b+2}}$
New CRIM	$G_{mb} = \dfrac{\sqrt{\frac{0.5853+\varepsilon_{AC}}{0.8366}}-1}{\frac{P_b}{G_b}\sqrt{\varepsilon_b} + \frac{(1-P_b)}{G_{se}}\sqrt{\varepsilon_s} - \frac{1}{G_{mm}}}$
ALL	$G_{mb} = \dfrac{\frac{\varepsilon_{AC}-\varepsilon_b}{3\varepsilon_{AC}-2.3\varepsilon_b}\ \frac{1-\varepsilon_b}{1-2.3\varepsilon_b+2\varepsilon_{AC}}}{\left(\frac{\varepsilon_s-\varepsilon_b}{\varepsilon_s-2.3\varepsilon_b+2\varepsilon_{AC}}\right)\left(\frac{1-P_b}{G_{se}}\right) - \left(\frac{1-\varepsilon_b}{1-2.3\varepsilon_b+2\varepsilon_{AC}}\right)\left(\frac{1}{G_{mm}}\right)}$

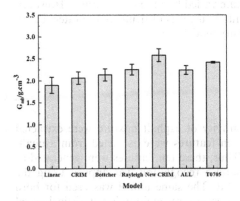

(a) AC-13

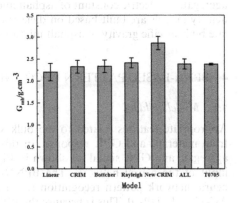

(b)SUP-20

Figure 5.　Average and Std of G_{mb} calculated by models.

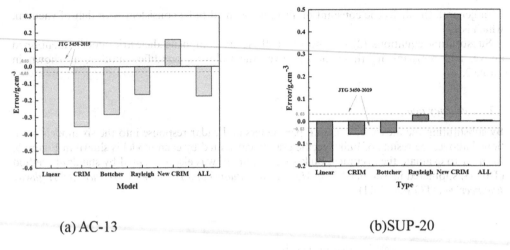

(a) AC-13	(b)SUP-20

Figure 6.   Error of G_{mb} calculated by models.

Standard deviation (Std) and coefficient of variation (Cv) for all six models were larger than that for standard method (T 0705). Std of bulk specific gravity required lower than 0.02 g/cm^3 stipulated in *Standard Test Methods of Bitumen and Bituminous Mixtures for Highway Engineering* (JTG E20-2011). However, Std of linear model was almost ten times than that of standard method (T 0705) in both AC-13 and SUP-20. ALL model had the lowest Std among the composite dielectric models, about 0.100 g/cm^3, in AC-13. Rayleigh model had the lowest Std among the composite dielectric models, larger than 0.100 g/cm^3, in SUP-20.

The prediction error of bulk specific gravity of density models is shown in Figure 6.

The allowable error for bulk specific gravity measurement was ±0.03 g/cm^3 stipulated in *Field Test Methods of Highway Subgrade and Pavement* (JTG 3450-2019). As shown in Figure 6(b), ALL model and Rayleigh model showed good accuracy in density prediction among density models. However, the error of ALL model in the AC-13 was larger than ±0.15 g/cm^3 while in the SUP-20 was lower than ±0.01 g/cm^3; The error of the Rayleigh model in the AC-13 was larger than ±0.15 g/cm^3 while in the SUP-20 was lower than ±0.03 g/cm^3. Only Rayleigh and ALL model (in SUP-20) reached the requirement stipulated in standard.

The Std and error of bulk specific gravity prediction of all six density models did not meet standard requirements. This was mainly because impurities were contained in asphalt and aggregate. Dielectric constant of asphalt and aggregate varied because of impurities. However, density models are built based on dielectric constant. Thus, it is difficult to accurately calculate bulk specific gravity of asphalt mixtures by density models.

4   BPNN-BASED PATTERN RECOGNITION

4.1   *Model building*

Appropriate features related to the bulk specific gravity of asphalt mixture were extracted from materials and GPR response. In this study, ten features were extracted from asphalt, aggregate and GPR signal, as shown in Figure 7. Thus, these ten features were generated as input of the neural network. The BPNN-based pattern recognition was applied using the neural network pattern recognition tool in MATLAB. The same model was used for both AC-13 and SUP-20. This is because the BPNN-based pattern recognition is a statistical-based model, and the material is a parameter for this model. The samples were divided randomly into three sets: 60% were used for training the network, 20% were used for validation, 20% were used for testing the network. Different number of hidden layers with different number of

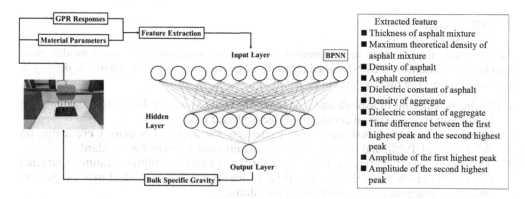

Figure 7.   Procedure of developing BPNN-based pattern recognition algorithm.

nodes was selected as the structure of BPNN. The structure of one hidden layer with seven nodes was found to have the best of performance. The gradient descent method was chosen as BPNN training algorithm. The learning rate lr was set to 0.01; The training goal was 0.001, and max training time was 5000.

4.2   *Accuracy analysis*

After training, the error of BPNN was lower than 5%, which has met the required accuracy. By analyzing the target value measured by T 0705 and the predicted value measured by BPNN of the testing samples, it can be found that the two values are similar, and the similarity of training is as high as 0.98, which indicates that the model is highly applied in the training sample. Error of testing samples (the first four samples was AC-13 and the others was SUP-20) were calculated. The results were shown in Figure 8.

As shown in Figure 8, error of all testing samples using BPNN meets the requirement (± 0.03 g/cm^3) stipulated in *Field Test Methods of Highway Subgrade and Pavement* (JTG 3450-2019). Compared with density models, the BP neural network model shows higher accuracy. This is mainly because BP neural network is a statistical-based model. It concerns data itself rather than properties of materials. Thus, it solves, to a great extent, the variation of dielectric constant of materials. In addition, accuracy of BP neural network would be increased with more data.

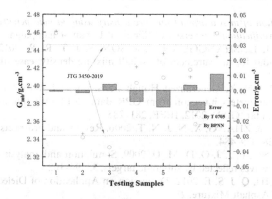

Figure 8.   Accuracy of BPNN-based pattern recognition algorithm.

5 CONCLUSION

In this study, the accuracy and applicability of six composite-dielectric-based models were developed. In addition, a BPNN based pattern recognition was used to evaluate the density of asphalt mixture. The findings are summarized as follows:

(1) Dielectric constant of asphalt and aggregate can be measured by Dielectric Constant Test to improve the accuracy of composite-dielectric-based models.
(2) Composite-dielectric-based models get a relatively low accuracy in density evaluation (in AC-13 and SUP-20), and it cannot meet the requirement stipulated in standard.
(3) The BPNN-based pattern recognition algorithm processes multiple features extracted from GPR signals while the traditional GPR density prediction method uses one feature only, the surface reflection amplitude in time domain.
(4) The BPNN-based recognition algorithm is a data-based model. It meets the requirement stipulated in standard and its accuracy would be increased with increasing data.

In terms of budget and time, the quantity of samples and types is less. In future research, more samples and types would be made to enlarge the database to improve the accuracy of models. In addition, the result of this study is not applied in in-site application. In future research, in-site samples would be obtained to validate the accuracy of models.

REFERENCES

ARAUJO, S., BEAUCAMP, B., DELBREILH, L., DARGENT, É., FAUCHARD, C. J. C. & MATERIALS, B. 2017. Compactness/density assessment of newly-paved highway containing recycled asphalt pavement by means of non-nuclear method. 154, 1151–1163.

AYDIN, E. & ERDEM, S. E. Y. Transfer and multitask learning using convolutional neural networks for buried wire detection from ground penetrating radar data. Detection and Sensing of Mines, Explosive Objects, and Obscured Targets XXIV, 2019. International Society for Optics and Photonics, 110120Y.

BIRCHAK, J. R., GARDNER, C. G., HIPP, J. E. & VICTOR, J. M. J. P. O. T. I. 1974. High dielectric constant microwave probes for sensing soil moisture. 62, 93–98.

CAORSI, S. & STASOLLA, M. J. A. I. G. 2008. ANN-based sub-surface monitoring technique exploiting electromagnetic features extracted by GPR signals. 19.

DONG, Z., YE, S., GAO, Y., FANG, G., ZHANG, X., XUE, Z. & ZHANG, T. J. S. 2016. Rapid detection methods for asphalt pavement thicknesses and defects by a vehicle-mounted ground penetrating radar (GPR) system. 16, 2067.

GUO, Y., XU, S., SHAN, W. J. C. R. S. & TECHNOLOGY 2018. Development of a frozen soil dielectric constant model and determination of dielectric constant variation during the soil freezing process. 151, 28–33.

KAFEDZISKI, V., PECOV, S. & TANEVSKI, D. Detection and classification of land mines from ground penetrating radar data using faster r-cnn. 2018 26th Telecommunications Forum (TELFOR), 2018. IEEE, 1–4.

LENG, Z. 2012. *Prediction of in-situ asphalt mixture density using ground penetrating radar: theoretical development and field verification.* University of Illinois at Urbana-Champaign.

LENG, Z., AL-QADI, I. L., SHANGGUAN, P. & SON, S. J. T. R. R. 2012. Field application of ground-penetrating radar for measurement of asphalt mixture density: case study of illinois route 72 overlay. 2304, 133–141.

LI, W., ZHOU, H. & WAN, X. Generalized Hough transform and ANN for subsurface cylindrical object location and parameters inversion from GPR data. 2012 14th International Conference on Ground Penetrating Radar (GPR), 2012. IEEE, 281–285.

LI, X.-T., WANG, D.-Y. & ZHANG, X.-N. J. N. T. 2006. Review and prospects of ground penetrating radar technology [J]. 28, 479–484.

LIU, J., CAI, Y. & WANG, F. J. J. O. D. M. U. 2009. Simulation and analysis on ground penetrating radar electromagnetic waves reflected by subgrade targets. 35, 99–102.

LU, C., LIU, H.-X. & LIU, Q. J. S. E. 2012. Discussion on Application of Dielectric Constant in Compactness Detection of Asphalt Mixture. 28.

MENG, M. & WANG, F. J. J. O. M. I. C. E. 2013. Theoretical analyses and experimental research on a cement concrete dielectric model. 25, 1959–1963.

QIANG, S., BING, D., LIMEI, M. J. P. G. & DAQING, O. D. I. 2003. Application of generalized Weng's model and rayleigh model in oil production prediction of polymer flooding [J]. 22, 58–59.

RAYLEIGH, L. J. T. L., EDINBURGH, MAGAZINE, D. P. & SCIENCE, J. O. 1892. LVI. On the influence of obstacles arranged in rectangular order upon the properties of a medium. 34, 481–502.

ROSLEE, M. B., RAJA ABDULLAH, R. S. A. & SHAFR, H. Z. B. M. J. P. I. E. R. 2010. Road pavement density analysis using a new non-destructive ground penetrating radar system. 21, 399–417.

ROTH, K., SCHULIN, R., FLüHLER, H. & ATTINGER, W. J. W. R. R. 1990. Calibration of time domain reflectometry for water content measurement using a composite dielectric approach. 26, 2267–2273.

WANG, S., ZHAO, S., AL-QADI, I. L. J. N. & INTERNATIONAL, E. 2018. Continuous real-time monitoring of flexible pavement layer density and thickness using ground penetrating radar. 100, 48–54.

WANG, S., ZHAO, S. & AL-QADI, I. L. J. T. R. R. 2019. Real-time monitoring of asphalt concrete pavement density during construction using ground penetrating radar: theory to practice. 2673, 329–338.

ZHAI, Y., ZHANG, B., WANG, F., ZHONG, Y. & LI, X. J. J. O. M. I. C. E. 2019. Composite Dielectric Model of Asphalt Mixtures Considering Mineral Aggregate Gradation. 31, 04019091.

Functional Pavements – Chen et al (eds)
© 2021 Taylor & Francis Group, London, ISBN 978-0-367-72610-2

The discrete and continuous retardation and relaxation spectrum of warm mix crumb rubber modified asphalt mixtures

F. Zhang
School of Science, Inner Mongolia University of Technology, Hohhot, Inner Mongolia, China

L. Wang
School of Civil Engineering, Inner Mongolia University of Technology, Hohhot, Inner Mongolia, China

Y.M. Xing
School of Science, Inner Mongolia University of Technology, Hohhot, Inner Mongolia, China

ABSTRACT: To study the linear viscoelasticity of crumb rubber modified asphalt mixtures before and after adding the warm mix additive. The methods of obtaining the discrete and continuous spectrum are presented. The discrete spectrum of asphalt mixtures can be obtained from complex modulus test results according to the generalized Maxwell and the generalized Kelvin model. Similarly, the continuous spectrum of asphalt mixtures can be obtained from the complex modulus test data by the inverse integral transformation. The results show that the discrete spectrum and the continuous spectrum have similar shapes, but the magnitude and position of the peak spectrum is different. The continuous spectrum can be considered as the limiting case of the discrete spectrum.

1 INTRODUCTION

The concept of spectrum distribution function plays a significant role in linear viscoelastic behavior theory (Alavi 2014), It independent on specific time system. The relaxation spectrum and the retardation spectrum are the distributions of the spectrum intensity to relaxation and retardation time. It is an essential characteristic of linear viscoelastic materials, from which all other viscoelastic functions can be obtained (Tschoegl 2012). The discrete relaxation spectrum can be determined by Prony coefficients from the generalized Maxwell (GM) model. In contrast, the Prony coefficients of the generalized Kelvin model can be used to determine the discrete retardation spectrum (Zhao, Ni et al. 2014). Schapery (Schapery 1962) uses the collocation method to solve the discrete spectrum. Cost and Becker (Cost and Becker 1970) proposed the multiple data method by applying the least-squares fitting technique to the Laplace transform domain. The recursive method proposed by Emri and Tschoegl (Emri and Tschoegl 1995) further improves the calculation method of the discrete spectrum. However, the accuracy of the discrete spectrum intensity depends on the spacing of the collocation points. The continuous spectrum using continuous time instead of the discrete time interval can avoid the ill-posed problem. Sun (Sun, Chen et al. 2015) constructed continuous relaxation and retardation spectrum based on the H-N complex modulus model. Liu (Liu, Luo et al. 2018) proposed a new method for determining the continuous relaxation spectrum. This study focused on obtained the discrete and continuous spectrum of crumb rubber modified asphalt mixture by Prony series and integral transform algorithm. the result was used to analyze the viscoelastic properties of crumb rubber modified asphalt mixtures before and after adding the warm mix additive.

2 MATERIALS AND METHODS

Four types of laboratory-produced mixtures were performed in this study. AC-16 was utilized in the following text. The performance grade of virgin asphalt is PG 64-22. Crumb rubber of 60 mesh and compounded mesh were added into virgin asphalt to produce crumb rubber modified asphalt binder. HMA was mixed and compacted at 180°C and 170°C, respectively. In contrast, WMA was mixed and compacted at 162°C and 152°C, respectively. The raw specimens were fabricated using a Superpave gyratory compactor and then cut and cored to standard dimensions (100mm in diameter and 150mm in height). The specimen air void content was controlled between 3.5% and 4.5%. The complex modulus test was conducted to measure dynamic modulus and phase angle for all specimens. The test was conducted in four different temperature (5°C 20°C, 35°C, 50°C) and seven different frequency (25Hz, 20Hz, 10Hz, 5Hz, 1Hz, 0.5Hz, 0.1Hz).

3 THEORETICAL BACKGROUND

The generalized sigmoidal model (Rowe, Baumgardner et al. 2016) has successfully characterized the mechanical behavior of asphalt mixtures, and this study uses the model to construct the viscoelastic function master curve that satisfies the K-K relations (Zhao, Chen et al. 2016).

3.1 The method of obtaining the master curve of linear Viscoelastic function

3.1.1 The master curve of dynamic modulus and phase angle

To obtain the master curve model parameters of the dynamic modulus and phase angle simultaneously, the error function ef_1 is introduced here. The expression of error function ef_1 is shown in Equation 3. The master curve of dynamic modulus and phase angle obtained from Equation 1 and 2. It can be solved by minimizing the error function, and the final results were shown in Table 1. The reduced frequency and shift factor calculated following Equations. 4, 5, respectively.

$$\lg|E^*| = \delta + \frac{\alpha}{1 + e^{\beta + \gamma(\log f_r)}} \tag{1}$$

$$\phi_1(f_r) = k\frac{\pi}{2}\frac{d(\lg|E^*|)}{d(\lg f_r)} = -\frac{\pi}{2}k\alpha\gamma\frac{e^{\beta+\gamma\lg f_r}}{(1 + \lambda e^{\beta+\gamma\lg f_r})^{1+\frac{1}{\lambda}}} \tag{2}$$

$$ef_1 = ef_{E^*} + ef_\phi = \frac{1}{N}\sqrt{\sum_{i=1}^{N}\left(\frac{|E^*|_{m,i} - |E^*|_{p,i}}{|E^*|_{m,i}}\right)^2} + \frac{1}{N}\sqrt{\sum_{i=1}^{N}\left(\frac{|\phi|_{m,i} - |\phi|_{p,i}}{|\phi|_{m,i}}\right)^2} \tag{3}$$

$$\lg f_r = \lg f + \lg \alpha_T \tag{4}$$

Table 1. Generalized sigmoidal model fitting parameters for dynamic modulus and phase angle.

Mixture type	Parameters								
	δ	A	β	γ	λ	C_1	C_2	k	Error/%
HMA-60	1.62	2.84	-0.60	-0.56	0.80	8.93	87.75	0.90	1.23
WMA-60	2.29	2.13	-0.14	-0.62	0.81	19.14	202.14	0.97	0.83
HMA-C	2.12	2.40	-0.54	-0.51	0.51	19.13	187.95	0.98	0.81
WMA-C	2.33	2.14	-0.31	-0.56	0.52	14.02	138.11	1.01	0.75

$$\lg \alpha_T = \frac{-C_1(T - T_r)}{C_2 + (T - T_r)} \tag{5}$$

where ef_1 is the error function of dynamic modulus and phase angle; N is the number of the measured data points, =28; $|E^*|_{m,i}$ is ith data point of the measured dynamic modulus, MPa; $|E^*|_{p,i}$ is ith data point of predicted by dynamic modulus master curve, MPa; $|\emptyset|_{m,i}$ is ith data point of the measured phase angle, °; $|\emptyset|_{p,i}$ is ith data point of predicted by phase angle master curve, °; k is a positive correction, it is to obtain potentially more accurate prediction.

3.1.2 The master curve of storage modulus and loss modulus

Similarly, the master curve models for the storage modulus and loss modulus can be determined regarding the method as mentioned above, and the final parameter results are shown in Table 2.

3.1.3 The master curve of storage compliance and loss compliance

In order to obtain the master curve model parameters of the storage compliance and loss compliance simultaneously, the error function ef_3 is introduced. The expression of error function ef_3 is shown in Equation 9. The master curve of storage compliance and loss compliance obtained from Equation 7 and 8. It can be solved by minimizing the error function, and the final results were shown in Table 3.

$$D^* = \frac{1}{E^*} = D' - iD'' = \frac{E'}{E'^2 + E''^2} - i\frac{E''}{E'^2 + E''^2} \tag{6}$$

$$\lg|D'| = \delta - \frac{\alpha}{1 + e^{\beta - \gamma(\log f_r)}} \tag{7}$$

$$D'' = -\frac{\pi}{2}kD'\frac{d(\log E')}{d(\log f_r)} = \frac{\pi}{2}kD'\alpha\gamma\frac{e^{\beta - \gamma \lg f_r}}{(1 + e^{\beta - \gamma \lg f_r})^2} \tag{8}$$

$$ef_3 = ef_{D'} + ef_{D''} = \frac{1}{N}\sqrt{\sum_{i=1}^{N}\left(\frac{|D'|_{m,i} - |D'|_{p,i}}{|D'|_{m,i}}\right)^2} + \frac{1}{N}\sqrt{\sum_{i=1}^{N}\left(\frac{|D''|_{m,i} - |D''|_{p,i}}{|D''|_{m,i}}\right)^2} \tag{9}$$

Table 2. Generalized sigmoidal model fitting parameters for storage modulus and loss modulus.

Mixture type	Parameters δ	A	β	γ	λ	C_1	C_2	k	Error /%
HMA-60	1.98	2.57	-0.36	-0.51	0.22	8.93	87.75	1.00	1.12
WMA-60	2.13	2.33	-0.12	-0.58	0.92	19.14	202.14	1.00	0.83
HMA-C	1.86	2.66	-0.54	-0.53	0.79	19.13	187.95	1.00	0.81
WMA-C	2.16	2.33	-0.28	-0.55	0.67	14.02	138.11	1.00	0.82

Table 3. Generalized sigmoidal model fitting parameters for storage compliance and loss compliance.

Mixture type	Parameters δ	A	β	γ	λ	C_1	C_2	k	Error /%
HMA-60	-1.70	2.81	-0.55	0.53	1.00	8.93	87.75	1.15	1.21
WMA-60	-2.16	2.26	-0.27	0.60	1.00	19.14	202.14	1.06	0.72
HMA-C	-2.01	2.46	-0.61	0.57	1.00	19.13	187.95	1.08	0.90
WMA-C	-2.20	2.23	-0.42	0.61	1.00	14.02	138.11	1.16	0.72

where ef_3 is the error function of storage compliance and loss compliance; N is the number of the measured data points, $=28$; $|D'|_{m,i}$ is ith data point of the measured storage compliance, MPa^{-1}; $|D'|_{p,i}$ is ith data point predicted by the storage compliance master curve, MPa^{-1}; $|D''|_{m,i}$ is ith data point of the measured loss compliance, MPa^{-1}; $|D''|_{m,i}$ is ith data point predicted by the loss compliance master curve, MPa^{-1}.

4 RESULTS AND DISCUSSION

4.1 *Determine the discrete and continuous spectrum*

Relaxation and retardation time spectrum are the most general functions that describe the dependence of material viscoelasticity on time or frequency. All viscoelastic functions can be combined through the time spectrum. By studying the relaxation and retardation time spectrum, we can obtain the distribution of spectrum and the contribution of various motion modes to the macroscopic viscoelasticity (Gurtovenko and Gotlib 2000), which opens up an effective way for the study of viscoelastic materials.

4.1.1 *Determine the discrete relaxation spectrum and retardation spectrum*
The generalized Maxwell model and the generalized Kelvin model are widely used to characterize the relaxation and retardation behavior of asphalt mixtures, and the Prony coefficients are used as the discrete relaxation and retardation spectrum. The detailed calculations can be found in the literature (Park and Schapery 1999), and the results are shown in Figure 1.

Figure 1 shows that bell-shaped discrete relaxation and retardation spectrum peak intensities correspond to relaxation and retardation times at around 10^{-3}s and $10^2 \sim 10^3$s, respectively. Compared warm-mix and hot-mix crumb rubber modified asphalt mixtures, it is found that after adding warm mix additive, the peak intensity of relaxation and retardation spectrum decreases, and the width of spectrum also decreases, the horizontal shift of the relaxation (retardation) spectrum to the left indicates that the warm mix asphalt mixture takes a short time to achieve relaxation (retardation) under load. It means that the elastic component of the mixture decreases and the viscous component increases. In other words, the rutting resistance of the crumb rubber modified asphalt mixture could be improved, and the cracking performance improved in low temperature.

4.1.2 *Determine the continuous relaxation spectrum and retardation spectrum*
Integral transform theory can be used to establish the relations between linear viscoelastic functions and continuous time spectrum. The continuous relaxation spectrum and the retardation spectrum can be derived from the corresponding storage modulus and storage compliance, respectively. The derivation of spectrum can be referred to the work of Liu

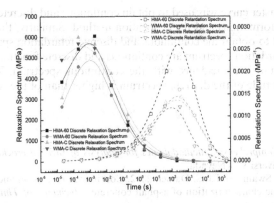

Figure 1.   The discrete relaxation spectrum and retardation spectrum of asphalt mixture.

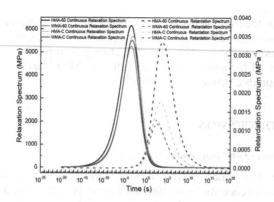

Figure 2.   The continuous relaxation spectrum and retardation spectrum of asphalt mixture.

(Liu, Luo et al. 2018) and Bhattacharjee (Bhattacharjee, Swamy et al. 2012). Finally, the results are shown in Figure 2.

Figure 2 shows the continuous relaxation and retardation spectrum also present the typical bell-shaped in a board time domain. Once warm mix additive was added, the width of the relaxation (retardation) spectrum narrowed, the peak intensity decreased, and the relaxation (retardation) time corresponding to the peak intensity shifted horizontally to the left. It indicates that the mixture's elastic component reduced and increased viscous component, so the mixture needs a shorter time to achieve the relaxation (retardation). It is consistent with the discrete spectrum. The faster relaxation mechanism is due to the reduced aging effect after adding the warm mix additive. From the molecular point, the reason for the above results can be attributed to the decrease in molecular weight and the concentration of polar functional groups in the asphalt binders (Yu, Gu et al. 2020).

5   CONCLUSION

Based on the complex modulus test, this paper uses the collocation method and integral transform theory to construct the discrete and continuous spectrum, respectively. The following conclusions can be drawn from the research.

According to the test data of complex modulus, the master curve model of dynamic modulus and phase angle is developed following the approximate K-K relations. These master curve models shared the same parameters and allowed for possible asymmetry. Similarly, the master curve model of storage modulus (compliance) and loss modulus (compliance) are also developed.

Storage modulus master curve was used to obtain continuous and discrete relaxation spectrum utilizing integral transformations and the collocation method. Similarly, the storage compliance master curve was used to obtain the continuous and discrete retardation spectrum. Finally, both the relaxation and retardation spectrum are continuous and discrete functions of the time.

The discrete spectrum is coincidence with the continuous spectrum. The most significant difference is the magnitude of the discrete spectrum is higher than of the continuous spectrum.

REFERENCES

Alavi, S. M. Z. 2014. *Comprehensive methodologies for analysis of thermal cracking in asphalt concrete pavements*, Reno: University of Nevada.
Bhattacharjee, S., A. K. Swamy and J. S. Daniel. 2012. Continuous relaxation and retardation spectrum method for viscoelastic characterization of asphalt concrete. *Mechanics of Time Dependent Materials* 16(3): 287–305.

Cost, T. L. and E. B. Becker. 1970. A multidata method of approximate Laplace transform inversion. *International journal for numerical methods in engineering* 2(2): 207–219.

Emri, I. and N. Tschoegl. 1995. Determination of mechanical spectra from experimental responses. *International journal of solids and structures* 32(6-7): 817–826.

Gurtovenko, A. A. and Y. Y. Gotlib. 2000. Viscoelastic dynamic properties of meshlike polymer networks: Contributions of intra-and interchain relaxation processes. *Macromolecules* 33(17): 6578–6587.

Liu, H., R. Luo and H. Lv. 2018. Establishing continuous relaxation spectrum based on complex modulus tests to construct relaxation modulus master curves in compliance with linear viscoelastic theory. *Construction and Building Materials* 165: 372–384.

Park, S. W. and R. A. Schapery. 1999. Methods of interconversion between linear viscoelastic material functions. Part I—a numerical method based on Prony series. *International Journal of Solids & Structures* 36(11): 1653–1675.

Rowe, G., G. Baumgardner and M. Sharrock. 2016. Functional forms for master curve analysis of bituminous materials. *Proceedings of the 7th international RILEM symposium ATCBM09 on advanced testing and characterization of bituminous materials.*

Schapery, R. A. 1962. A simple collocation method for fitting viscoelastic models to experimental data.

Sun, Y., J. Chen and B. Huang. 2015. Characterization of asphalt concrete linear viscoelastic behavior utilizing Havriliak–Negami complex modulus model. *Construction and Building Materials* 99: 226–234.

Tschoegl, N. W. 1989. *The phenomenological theory of linear viscoelastic behavior: an introduction.* New York: Springer.

Yu, D., Y. Gu and X. Yu. 2020. Rheological-microstructural evaluations of the short and long-term aged asphalt binders through relaxation spectra determination. *Fuel* 265: 116953.

Zhao, Y., P. Chen and D. Cao. 2016. Extension of modified Havriliak-Negami model to characterize linear viscoelastic properties of asphalt binders. *Journal of Materials in Civil Engineering* 28(5): 04015195.

Zhao, Y., Y. Ni and W. Zeng. 2014. A consistent approach for characterising asphalt concrete based on generalised Maxwell or Kelvin model. *Road Materials and Pavement Design* 15(3): 674–690.

Functional Pavements – Chen et al (eds)
© 2021 Taylor & Francis Group, London, ISBN 978-0-367-72610-2

Effect of short-term aging on the anti-cracking performance of stone matrix asphalt reinforced by basalt fibers

Changjiang Kou, Xing Wu, Aihong Kang & Peng Xiao
College of Civil Science and Engineering, Yangzhou University, Yangzhou, China

Zhao Fan
Jiangsu Expressway Company Limited, Nanjing, China

ABSTRACT: To investigate the anti-cracking performance of asphalt mixture with basalt fibers after the construction process, short-term aging procedure was conducted on basalt fiber reinforced stone matrix asphalt (BF-SMA) in a loose state and the anti-cracking indexes of unaged and short-term aged BF-SMA were tested through semi-circular bending test (SCB). Comparison was made between BF-SMA and lignin fiber reinforced stone matrix asphalt (LF-SMA). Results show that the fracture energy and flexibility index of BF-SMA are always higher than that of LF-SMA while the degradation extents of BF-SMA are much lower than that of LF-SMA, which implies that BF-SMA owns better and more stable anti-cracking performance. The same trend was also found in the rutting test and water stability test. The reduction coefficients are recommended to reevaluate the performance stability of asphalt mixtures before and after construction process and make the constructed road safer and more sustainable.

Keywords: stone matrix asphalt, basalt fibers, anti-cracking performance, semi-circular bending test, reduction coefficients

1 INTRODUCTION

Pavement damages such as cracking, rutting, raveling [1–3] etc. are more and more common on the asphalt pavements nowadays. For instance, there are still many cracks on the lignin fiber reinforced stone matrix asphalt (LF-SMA) pavements at the early stage of traffic opening. The reasons for this phenomenon can be explained in several ways. Firstly, the construction company might use the unqualified raw materials. Secondly, the overloading phenomena [4] are very common on the highways. Thirdly, the additives used is not good enough to improve the road performance. Last but not least, the performances of the raw materials will decrease in the construction process. The overloading phenomena becomes less owing to stricter regulations [5] and a more developed weight sensors buried in the roads and the advanced traffic police camera system. This paper focused on the reinforcement of paving materials by adding a kind of new additive, basalt fiber to replace the commonly used lignin fiber in the stone matrix asphalt (SMA) pavement, and followed the procedures presented by SHRP [6] to simulate the effect of short-term aging on the anti-cracking performance of stone matrix asphalt reinforced by basalt fibers. Immersion Marshall test was conducted to study the water stability. Rutting test was chosen to test the anti-rutting ability. Semi-circular bending test (SCB) was selected to study the anti-cracking performance [7]. The reduction coefficients can be used to modify the performance indexes provided by the laboratories to help to guide the design of the road. This paper has a certain significance to guide the design of the SMA pavements to make the road safer and more sustainable.

2 MATERIAL AND METHODS

2.1 *Raw materials*

SBS modified asphalt with penetration grade 70/100 and softening point 64 °C used in this study was provided by Jiangsu Tiannuo Road Materials Technology Co., Ltd. Coarse basalt aggregates and fine limestone aggregates were selected as the aggregates for SMA, the properties of which met the requirements of the *Test Specification for Aggregate of Highway Engineering* (JTG E42-2005). Chopped basalt fiber (6mm) was produced by Jiangsu Tianlong Basalt Continuous Fiber Co., Ltd. and lignin fiber by JRS company of Germany. Figure 1 presents the appearance of both fibers.

2.2 *Methods*

2.2.1 *Gradation design*

Figure 2 shows the gradation curve of the SMA with the nominal maximum aggregate size of 13.2 mm (SMA13). According to the previous researches about SMA13 from our group [8], the content of basalt fiber is chosen as 0.4% by the weight of asphalt mixture, and the content of lignin fiber is set as 0.3%. The fibers were mixed with the aggregates for 90 seconds firstly, and then the asphalt was added (mix for 90 seconds). The mineral powder was added in the end (mix for 90 seconds). The asphalt-aggregate ratio of the basalt fiber reinforced SMA13 is determined as 5.8%, and that of the lignin fiber reinforced SMA13 is set as 6.0%.

2.2.2 *Test methods*

The SMA13 mixtures under different aging degrees (unaged and short-term aged) were prepared according to the procedures provided by SHRP 1025 [6]. The mixtures are short-term aged at 135°C for 4 hours (under forced ventilation). Immersion Marshall test was used to study the

(a)   (b)

Figure 1.   Fibers: (a) Basalt fiber; (b) Lignin fiber.

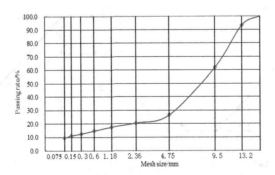

Figure 2.   Gradation curve.

189

water stability. Rutting test (using the wheel tracking test machine) was chosen to test the anti-rutting property (The wheel pressure is 0.7 MPa and the test temperature is 60 °C). SCB test was adopted to test the cracking resistance of the unaged and short-term aged SMA13 mixture, and was conducted according to the steps presented by AASHTO TP 105-13[9].

3   RESULTS AND DISCUSSION

3.1   *Volume parameters, residual stability and dynamic stability of the specimens*

The results of volume parameters, residual stability and dynamic stability of the unaged basalt fiber reinforced SMA13 (BF-SMA13) and lignin fiber reinforced SMA13 (LF-SMA13) were listed in Table 1. The water stability and the anti-rutting ability of the BF-SMA13 is better than that of the LF-SMA13 (aged and unaged). The degradation percentage of the residual stability of the BF-SMA13 is lower than that of LF-SMA13.

3.2   *SCB test*

In the SCB test, fracture energy (G_f) is used to evaluate the cracking resistance of the mixtures. Higher G_f value means that the mixture has better anti-cracking ability. Flexibility index (*FI*) is used to assess the growth rate of the cracking. Lower *FI* value means quicker cracking growth rate and worse cracking resistance. The test is carried out at 25°C.

As shown in Figure 3, the G_f of BF-SMA13 decreases by about 7.29% and the *FI* value is 10.58% lower after being short-term aged. The corresponding decrease percentages of the G_f and the *FI* value of LF-SMA13 is 14.61% and 18.87%. The results show that the anti-cracking ability of the mixture will be affected by the short-term aging (the construction process). The degradation extents of the G_f and *FI* value of BF-SMA13 is much lower than that of LF-SMA13, and the G_f and FI value of BF-SMA13 are always bigger than that of LF-SMA13 (aged and unaged), which means that basalt fiber can better improve the anti-cracking performance and the anti-aging ability of SMA than lignin fiber.

Table 1.   Test results of volume parameters, immersion Marshall test and wheel tracking test.

Index	VV/%	VMA/%	VFA/%	Residual stability/%		Dynamic stability/ (times/mm)	
				unaged	aged	unaged	aged
BF-SMA13	4.3	16.8	74.4	90.84	90.17	5250	6000
LF-SMA13	3.8	17.2	78.0	90.11	88.83	4200	3938

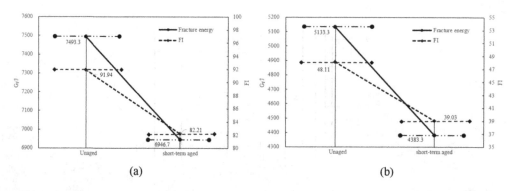

(a)   (b)

Figure 3.   SCB test results: (a) BF- SMA13; (b) LF-SMA13.

Table 2. Reduction coefficients.

Index	Residual stability	Dynamic stability	G_f (SCB)	FI (SCB)
Reduction coefficient	0.993	1.143	0.927	0.894

3.3 Reduction coefficients

The reduction coefficients of basalt fiber reinforced SMA13 defined as the ratio of the index value of the short-term aged and the unaged mixture are listed in Table 2.

This paper just takes the residual stability, dynamic stability, and the SCB test indexes of BF-SMA13 as an example. They can be used to modify the test results given by the laboratories. Therefore, it is safer and more accurate if designers use the test results modified by the reduction coefficients to judge the mixture used on the road construction, and use the modified performance indexes to choose the pavement materials that meet the needs of Party A. By doing this, the road will be more sustainable and safer. Otherwise, the performance of the constructed pavement might not meet the standards in the contract, which could result in the early damages of the pavement.

4 CONCLUSIONS

Comparison study was carried out to evaluate the effect of short-term aging process on the anti-cracking performances of basalt fiber reinforced SMA13 and lignin fiber reinforced SMA13. In this study, basalt fiber behaves better in improving the water stability, anti-rutting and anti-cracking performance of SMA before or after aging than lignin fiber. And the degradation extents of the anti-cracking indexes, G_f and FI value of basalt fiber reinforced SMA13 is much lower than that of lignin fiber reinforced SMA13, implying that BF-SMA owns more stable anti-cracking performance during construction or even service process.

It should be mentioned that this paper just takes the anti-cracking performance of the mixture as an example to show the influence of the construction process on the properties of fiber reinforced asphalt mixtures. Other performances will be tested in the future studies.

REFERENCES

[1] Norouzi A, Richard Kim Y. Mechanistic evaluation of fatigue cracking in asphalt pavements[J]. International Journal of Pavement Engineering, 2017, 18(6): 530–546.
[2] Gong H, Sun Y, Mei Z, et al. Improving accuracy of rutting prediction for mechanistic-empirical pavement design guide with deep neural networks[J]. Construction and Building Materials, 2018, 190: 710–718.
[3] Bendtsen H, Kohler E, Lu Q, et al. Californian and Danish Study on Acoustic Aging of Road Pavements[C]//Compendium of Papers, TRB 89th Annual Meeting, Washington, DC, 2010.
[4] Chen Y, Wang K, Zhang Y, et al. Investigating factors affecting road freight overloading through the integrated use of BLR and CART: a case study in China[J]. Transport, 2020, 35(3): 236–246.
[5] Pervaiz M, Panthapulakkal S, Sain M, et al. Emerging trends in automotive light weighting through novel composite materials[J]. Materials Sciences and Applications, 2016, 7(01): 26.
[6] Standard Practice for Short Term Aging of Asphalt Concrete Mixtures, SHRP 1025, 1992.
[7] Zhang J, Tan H, Pei J, et al. Evaluating crack resistance of asphalt mixture based on essential fracture energy and fracture toughness[J]. International Journal of Geomechanics, 2019, 19(4): 06019005.
[8] Lu P, Xiao P, LV Y. Rutting performance of basalt fiber asphalt mixture based on dynamic creep test[J]. Journal of Jiangsu University (Natural Science Edition), 2015, 36(04): 480–484. (In Chinese).
[9] AASHTO TP 105-13. Standard Method of Test for Determining the Fracture Energy of Asphalt Mixtures Using the Semicircular Bend Geometry (SCB), American Association of State and Highway Transportation Officials, 2013.

Functional Pavements – Chen et al (eds)
© 2021 Taylor & Francis Group, London, ISBN 978-0-367-72610-2

Performance prediction and evaluation for steel bridge deck pavement

L.L. Chen & G. Liu
Intelligent Transportation System Research Center, Southeast University, Nanjing, Jiangsu, China

H.T. Wu
Kunshan Transportation Bureau, Suzhou, Jiangsu, China

Z.D. Qian
Intelligent Transportation System Research Center, Southeast University, Nanjing, Jiangsu, China

ABSTRACT: To evaluate the performance of steel bridge deck pavement (SBDP) reasonably and accurately, an evaluation model of SBDP was established in this paper. Firstly, through typical distress investigation, the effect of different distresses on the SBDP performance was studied. Then, evaluation indexes of SBDP, were proposed based on their effects and existing evaluation indexes of highway pavement. Next, computing models for each index were established and their weights were determined through Analytic Hierarchy Process (AHP) and Delphi Method. Lastly, the Wuhan Baishazhou Bridge was selected as case study to evaluate the established model. Result shows that this evaluation model can assess performance of SBDP efficiently and provide reasonable maintenance guidance for the management department.

Keywords: steel bridge deck pavement, performance prediction and evaluation, evaluation model, Analytic Hierarchy Process, Delphi method

1 INTRODUCTION

Steel bridge deck pavement (SBDP) is the asphalt layer placed on the top of bridge deck and acts as an important structural layer of a steel bridge. It not only provides a smooth and skid-resistant ride surface for traffic, but also shelters the underlying steel deck from damages such as corrosion. However, many investigations found that distress in SBDP are much more common than those in ordinary highway asphalt pavement, because of the special support conditions and strict usage environment of SBDP (Huang, 2006; Qian, et al. 2011; Chen, et al. 2020). These distresses may significantly reduce the serviceability and the service life of a steel deck bridge. In order to better serve the operation and maintenance of SBDP, the performance of SBDP needs to be evaluated and predicted.

Evaluation on pavement performance was mainly conducted for ordinary asphalt pavement (Chinese Standard, 2018; Ma, et al, 2019; Abdulmawjoud, 2020). The first pavement performance index is the Present Serviceability Rating (PSR), developed by American Association of State Highway Officials (AASHO) (Carey and Irick, 1960). Ralph et al (1994) put forward the concepts of structural and functional condition. The pavement structural condition refers to its ability to support the current and future traffic, whereas the functional condition refers to its ability to provide a safe, smooth, and a quiet riding surface for public traveling. Nejad and Zakeri (2011) established an expert system which could classify pavement distress. Many evaluation methods have also been proposed. The fuzzy logic method, which could analyse linguistic or non-crisp data and the uncertainty of data, was utilized to evaluate the pavement condition through considering the severity, densities and weighting factors of different distress (Qin, et al, 2013). A model also came out to describe the deterioration of pavement skidding resistance (Vittorio et al, 2020). Some researches have also been carried out on the evaluation of SBDP performance. Chen (2006) proposed some evaluation indexes of SBDP qualitatively, but did not give quantitative computation method. Huang (2008) utilized the fuzzy neural network model to assess SBDP performance and proposed evaluation index system, mainly including crack, rutting, roughness and anti-slip performance. Qian et al (2006) explored the influence of many factors on mechanical responses of SBDP and found that pavement material, thickness, load position and some relevant design parameters of steel box girder have great influence on SBDP performance.

Through above investigations, it is clear that most researches in this field mainly focused on evaluation of ordinary asphalt pavement. Some distress of SBDP is also similar to that of ordinary asphalt pavement. However, ordinary asphalt pavement and SBDP are quite different in many aspects, such as pavement structure, materials, functional effects, service conditions, distress forms, damage mechanisms and severity. Thus, evaluation methods for ordinary asphalt pavement can not be applied to evaluating SBDP performance directly. Although some attention has been paid to the evaluation of SBDP performance, they only assess SBDP performance qualitatively and consider SBDP distresses from some aspects. Therefore, a quantitative and comprehensive evaluation model suitable for SBDP was needed.

The objective of this paper is to build an evaluation model for assessing performance of SBDP reasonably and accurately. Firstly, based on the evaluation method used to evaluate the performance of ordinary asphalt pavement, evaluation indexes are proposed to meet the functional requirements of SBDP firstly. Then, computing models for each index are established and their weights are also determined through theoretical analysis. Lastly, an engineering case study is carried out as an example to test this model.

2 EVADUATION INDEXES

The existing evaluation system for ordinary asphalt pavement mainly consists of five technical indexes, including pavement damage, unevenness, rutting, anti-skidding performance and structural strength (Chinese Standard, 2018). The evaluation for the performance of steel bridge deck pavements is the investigation and assessment of its distress influence. Therefore, the evaluation model of steel bridge deck pavements should be based on different distress indicators. However, there are great differences on many aspects between SBDP and ordinary asphalt pavement, such as main distress forms, distress impact and criteria for severity. Several evaluation indexes for ordinary asphalt pavement are of little significance for SBDP, and need to be ignored. On the contrary, some evaluation indexes are not included in existing specification for ordinary asphalt pavement. But they are significant and need to be added to assess critical distress forms for SBDP. Based on typical distress of SBDP and existing evaluation system for ordinary asphalt pavement, indexes for SBDP performance were determined, including Steel Deck Pavement Condition Index(SDPCI), Steel Deck Pavement Skidding Resistance Index(SDSRI), Steel Deck Pavement Adhesive Condition Index(SDACI) and Steel Deck Pavement Patching Condition Index(SDPPCI). Besides, Delphi method, as an evaluation method which considers expert's opinions anonymously, is commonly employed for the estimation of driving safety. Scores given by experts are summarized, analyzed, and classified (Chen, et al. 2020). Based on the evaluation method for ordinary asphalt pavement, the computing model was constructed and the weight values of distress severity were determined through Delphi Method.

2.1 Steel Deck Pavement Condition Index (SDPCI)

The damage forms of SBDP include cracking, map cracking and potholes. The distress development can be divided into two stags according to the severity. At the beginning, the damage only affects driving safety and comfort. Then it would develop quickly. When the damage penetrates and reaches steel deck, the steel deck would rust rapidly owing to the invalidation of waterproof layer. That would do harm to the structural safety of steel bridge. Thus, the damage condition of SBDP should be selected as critical index of SBDP performance. Description of the damaged severity of different pavement types and its influence weight can also be determined. The computing equation for damage distribution ratio (DDR) is as follows.

$$DDR = 100 \times \frac{\sum_{i=1}^{i_0} W_i A_i}{A} \tag{1}$$

Where DDR= damage distribution ratio (%); A= area of pavement (m^2); A_i= area of the i-th damage on pavement (m^2); W_i= weight of the i-th damage on pavement; i= the i-th damage on pavement considering damaged severity; i_0= total number of damage types.

Rutting depth index(RDI), which is used to evaluate vertical deformation in evaluation standard of ordinary asphalt pavement. However, rutting rarely occurs in SBDP. Because SBDP possesses thin layers and high support strength of steel deck. Thus, vertical deformation index does not need to be adopted and could be considered independently for Guss asphalt (GA) pavement.

In order to obtain the relationship equation between SDPCI and damage distribution ratio-(DDR), Delphi Method was used through regression analysis. The equation is defined in Equation (2).

$$SDPCI = 100 - a_0 \times DDR^{a_1} \tag{2}$$

Where SDPCI= Steel Deck Pavement Condition Index; DDR= damage distribution ratio (%); a_0, a_1 = regression coefficients, different when type of pavement changes.

2.2 Steel Deck Pavement Skidding Resistance Index (SDSRI)

Similar to pavement skidding resistance in ordinary asphalt pavement, steel deck pavement skidding resistance is determined by the friction value between the surface of pavement and tire. Therefore, the evaluation method used in ordinary asphalt pavement is selected to compute skidding resistance index in this paper. However, there are several differences between SBDP and ordinary asphalt pavement in pavement material. The evaluation criterion needs to be adjusted based on SBDP material.

Steel deck pavement skidding resistance could be evaluated through side-way force coefficient (SFC). Then Delphi Method was adopted to regress the evaluation outcome of experts. Finally, the corresponding computing equation is obtained as follows.

$$SDSRI = \frac{100 - SRI_{min}}{1 + a_0 \exp(a_1 SFC)} + SRI_{min} \tag{3}$$

Where SFC= side-way force coefficient; SRI_{min} = calibration parameters; a_0, a_1 = regression coefficients, different when type of pavement changes.

2.3 Steel Deck Pavement Adhesive Condition Index (SDACI)

Because the SBDP works with steel deck through the adhesive layer. The adhesive layer functions as the force transmission zone and becomes the weak part for SBDP. The invalidation of the adhesive layer also become the unique distress form for SBDP. In order to evaluate its effect reasonably, a new evaluation index needs to be proposed. There are two forms of invalidation for adhesive layer. The first one is that the adhesive layer is corrupted by the top-down permeation water in damaged pavement. The steel

194

deck would contact the atmosphere and rust may occur. The second one is that the pavement is separated from adhesive layer, but the pavement is still complete. It is milder distress form compared with the first one. Furthermore, different weight should be given to these two distress form. Through Delphi method, the influence weight of the invalidation of adhesive layer could be determined.

SDACI is used to evaluate the failure degree of adhesive layer. Its severity is determined through invalidation ratio of adhesive layer(ALR). In addition, ALR is related to the area of invalid adhesive layer. The computing equation for ALR is shown as Equation 4.

$$ALR = 100 \times \frac{\sum_{i=1}^{2} \omega_i A_i}{A} \tag{4}$$

Where ALR= Invalidation ratio of adhesive layer (%); A= Area of pavement; A_i= Invalidation area of the i-th adhesive layer (m^2); ω_i= the weight of the i-th invalid adhesive layer; i= the i-th invalid adhesive layer according to its severity.

Through regressing the evaluation outcome of experts, the relationship between SDACI and ALR can be acquired, as presented in Equation 5.

$$SDACI = 100 - a_0 \times ALR^{a_1} \tag{5}$$

Where SDACI= Steel Deck Pavement Adhesive Condition Index; ALR= Invalidation ratio of adhesive layer; a_0, a_1= regression coefficients, different when type of pavement changes.

2.4 Steel Deck Pavement Patching Condition Index (SDPPCI)

Patching distress includes various distress which have been repaired. The condition of patching distress has great influence on the whole perform of SBDP. Thus it is of great necessity to separate patching distress as an independent index. Its severity is determined through pavement patching ratio (PPR). PPR is related to the pavement patching area. The computing equation for PPR is shown as Equation 6. Through regressing the evaluation outcome of experts, the relationship between SDPPCI and PPR can be acquired, as presented in Equation 7.

$$PPR = 100 \times \frac{\sum_{i=1}^{n} A_i}{A} \tag{6}$$

$$SDPPCI = 100 - a_0 PPR^{a_1} \tag{7}$$

Where SDPPCI= Steel Deck Pavement Patching Condition Index; PPR= Pavement patching ratio (%); A_i= Patching area of the i-th pavement distress (m^2); a_0, a_1= regression coefficients, different when type of pavement changes.

3 EVALUATION MODEL

This section is based on the established indexes for performance of SBDP. Then, based on AHP method and Delphi method, the weight values for evaluation indexes were acquired. Finally, an evaluation model for the performance of SBDP is established.

3.1 Establishment of hierarchical model for the performance of SBDP

According to the relationship between evaluation indexes and pavement performance, a hierarchical structure was founded, including target layer, rule layer and index layer, as illustrated in Figure 1.

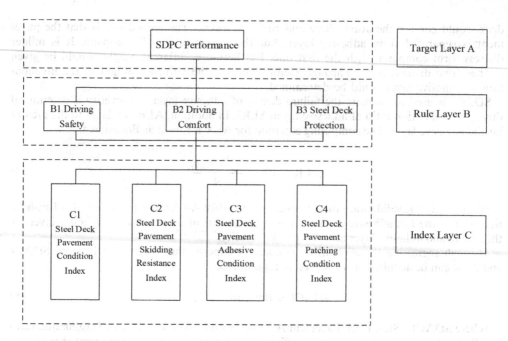

Figure 1. Hierarchical structure of SDPC performance evaluation.

3.2 Establishment of judgment matrix

Judgment matrix is determined through comparing every two factors. Its value is obtained by scale method. Then, the selected experts provide the judgment matrix of B ~ A and C ~ B. The judgment matrix of B ~ A given by the expert is as follows.

$$A = \begin{bmatrix} 1 & 1/3 & 1/2 \\ 3 & 1 & 2 \\ 2 & 1/2 & 1 \end{bmatrix} \tag{8}$$

3.3 Calculation of relative importance degree and consistency check

The relative importance degree of every factors in a layer is determined through judgment matrix. The column vectors of judgment matrix are normalized; $\bar{a}_{ij} = a_{ij} / \sum_{k=1}^{n} a_{kj} (i = 1, 2, \ldots, n)$, the matrix $\bar{A} = (\bar{a}_{ij})_{n \times n}$ relative to $\bar{a}_{ij}$ can be obtained:

$$\bar{A} = \begin{bmatrix} 0.16 & 0.18 & 0.14 \\ 0.50 & 0.55 & 0.57 \\ 0.33 & 0.27 & 0.29 \end{bmatrix} \tag{9}$$

Summing the acquired judgment matrix through normalizing the column vectors by row, $\overline{W}_1$ can be acquired., $\overline{W}_i = \sum_{j=1}^{n} \bar{a}_{ij}, i = 1, \ldots, n$ thus it was obtained that $\overline{W}_1 = 0.69$, $\overline{W}_2 = 0.22$, $\overline{W}_3 = 2.09$.

Through normalizing vector $\bar{W} = [\overline{W}_1, \overline{W}_2, \overline{W}_3]$, the feature vector, which refers to the weight coefficient of the scores, can be obtained, $W_1 = \overline{W}_1 / \sum_{i=1}^{3} W_i = [0.23, 0.07, 0.7]^T$. Computing maximum eigenvalue, the equation of λ_{max} is $\lambda_{max} = \sum_{i=1}^{n} (AW)_i / (nW_i)$, in which $(AW)_i$ refers to the i-th component of vector AW.

196

Examining uniformity is aimed to prevent logic mistakes in the established judgment matrix. Firstly, C_I was computed, $C_I = \frac{\lambda_{max}-n}{n-1} = \frac{3.0092-3}{3-1} = 0.0046$. Then, value of random consistence index R_I is averaged according to the quantity of elements in this layer. R_I is the average value of C_I of 1000 random sample matrix.

The consistency ratio of judgment matrix was obtained: $C_R=C_I/R_I=0.0046$. Generally, uniformity of judgment matrix is reasonable when C_I is less than 0.10. This means that the selection of the importance ratio between elements of matrix is reasonable.

3.4 Calculation and test of comprehensive importance degree

After obtaining the weight of elements in bottom layer relative to element in middle layer, the weight under single criterion needs to be composed from top to bottom. In this research, rule layer (B layer) includes three elements: B_1, B_2 and B_3. Index layer (C layer) includes 4 evaluation indexes: C_1, C_2, C_3, C_4. Ranking weights of B layer relative to target layer(A layer) are b_1, b_2 and b_3 respectively, and ranking weights of C layer relative to B layer are c_{1j}, c_{2j}, c_{3j}, c_{4j} respectively. Thus, weight coefficients of elements in C layer relative to A layer are C_1, C_2, C_3, C_4 respectively. The coefficients mentioned could be given through Equation 10.

$$C_i = \sum\nolimits_{j=1}^{3} C_{ij}b_j, i = 1, \cdots, 4 \tag{10}$$

When the judgment matrix of B_j relative to C layer has passed consistency test, consistent indexes based on single rank C_{1j}, j=1, ..., 3 and the relative consistent indexes based on average random R_{1j} could be determined. According to this, consistent ratio of C layer could be determined as follows:

$$C_R = \sum\nolimits_{i=1}^{3} C_{ij}b_i / \sum\nolimits_{1}^{3} R_{ij}b_i \tag{11}$$

3.5 The determination of weight

$$\bar{f_i} = \sum\nolimits_{i=1}^{m} f_i/m \tag{12}$$

Where $\bar{f_i}$= average weight of evaluation indexes; f_i= weights derived from the judgment matrix of experts; m= number of experts.

The evaluation results were obtained from management department, bridge user and research department. To eliminate the personal subjective factors, the scoring result from experts was averaged after computing. The equation for evaluating the comprehensive performance of SBDP was established as follows.

$$SDPQI = 0.37 \times SDPCCI + 0.23 \times SDPACI + 0.17 \times SDPSRI + 0.23 \times SDPRCI \tag{13}$$

4 CASE STUDY

The Wuhan Baishazhou Bridge is a highway and railway combined bridge. Its main bridge is 618m long. The panel of main girder is orthotropic steel plate. Epoxy asphalt concrete of double layer was selected. According to the patrol data of this bridge, its main distress is crack and pothole. Therefore, the SDPCI was adopted as the main performance index to investigate the performance development of SBDP. Based on the patrol data of Baishazhou Bridge from September 2015 to March 2016 and the suitable standard prediction decay equation of pavement performance in road engineering, the equation could be determined as follows.

$$SDPCI = SDPCI_0 \left\{ 1 - \exp\left[-\left(\frac{\alpha}{t}\right)^{\beta} \right] \right\} \qquad (14)$$

Where t= Time (month) from the date of the performance indicator starting to the predicted day; $SDPCI_0$= Initial value of performance index; SDPCI= Predicted performance index; α, β= Fitting parameters, α is related to residual life of pavement, β represents deterioration monde.

Through fitting analysis, the fitting parameters were obtained: α=21.15, β=0.3282 (R^2=0.9590, SSE=0.035). The result indicated that standard decay equation could forecast the pavement distress accurately through adopting evaluation model proposed in this paper.

5 CONCLUSION

In this paper, an evaluation model for assessing performance of SBDP was established based on the evaluation method for ordinary asphalt pavement. Firstly, according to the representative distress and functional requirements of SBDP, the relative criteria for severity division and computing method for indexes were proposed. Then, the evaluation model for performance of SBDP was established based on AHP and Delphi method. Lastly, the evaluation method in this paper was tested through an engineering case study. The following conclusions can be drawn from the study:

(1) The evaluation indexes were determined based on the analysis of main distress types and functional requirements of SBDP, including Steel Deck Pavement Condition Index-(SDPCI), Steel Deck Pavement Skidding Resistance Index(SDSRI), Steel Deck Pavement Adhesive Condition Index(SDACI) and Steel Deck Pavement Patching Condition Index-(SDPPCI). The selected indexes improve the accuracy and rationality of evaluation model for performance of SBDP.

(2) Through AHP method and Delphi method, the weight values for evaluation indexes were acquired and the computing model was constructed. Then, an evaluation model for the performance of SBDP was established to guide the assessment of SBDP performance.

(3) A predictive model for performance of SBDP was employed to test this evaluation model through an engineering case study. Result shows that this evaluation model can assess performance of SBDP efficiently and provide reasonable maintenance guidance for the management department.

ACKNOWLEDGEMENTS

The authors gratefully appreciate the financial support for this research from the National Key R&D Program of China (Nos.2018YFB1600300 and 2018YFB1600304).

REFERENCES

Abdulmawjoud, A. A. 2020. Performance evaluation of hot mixture reclaimed asphalt pavement. *Journal of Engineering ence and Technology*, 15(1), 477–492.

Carey J. and Irick, P. E. 1960. The pavement serviceability-performance concept. *Highway Research Board Bulletin*.

Chen, T. J. Research on Crack Behavior of Long-span Steel Bridges Deck Expoxy Asphalt Pavement. *Doctoral thesis. Southeast University, 2006.*

Chen, L. L., et al. 2016. Multiscale numerical modeling of steel bridge deck pavements considering vehicle-pavement interaction. *International Journal of Geomechanics*, 16(1): B4015002.

Chen, L. L., et al. 2020. Feasibility evaluation of a long-life asphalt pavement for steel bridge deck. *Advances in Civil Engineering*, 2020(2020):1–8.

Chen, L. L., et al. 2020. Determination of Allowable Rutting Depth Based on Driving Safety Analysis. *Journal of Transportation Engineering, Part B: Pavements*, 2020, 146(2).

Chinese Standard. JTG 5210–2018, *Highway Performance Assessment Standards*.

Guo, D. J., et al. 2018. Application of uncertainty analytic hierarchy process method for asphalt pavement construction quality control in china. *Transportation Research Record*.

Huang, W. 2006 *Theory and method of deck paving design for long-span bridge*. Beijing: China Architecture & Building Press.

Huang, H. T. Study on service performance evaluation and maintenance decision of steel deck pavement. *Master's thesis, Southeast University,2008*.

Ma, L. et al. 2019. Evaluation of transverse cracks for semi-rigid asphalt pavements using deflection basin parameters. *Transportation Research Record Journal of the Transportation Research Board*.

Nejad, F. M. and Zakeri, H. 2011. An expert system based on wavelet transform and radon neural network for pavement distress classification. *Expert Systems with Applications*, 38(6), 7088–7101.

Qian, Z.D., Chen L.L., Jiang C.L., et al. 2011. Performance evaluation of a lightweight epoxy asphalt mixture for steel bridge deck pavements. *Construction and building material*, 3117–3122.

Qian, Z.D., Huang, W., Du, X, et al. 2006. Research on Effects of Shape of Long-span Cable-supported Bridge on Mechanical Analysis of Surfacing under Vehicular Load. *Engineering ence, 2006*.

Qin, Z. B., et al. 2013. Multi-objective comprehensive evaluation for performance of asphalt concrete pavement based on fuzzy entropy. *Journal of Central South University*, 44(8), 3474–3478.

Ralph C. G. H., et al. 1994. Modern pavement management. Krieger Publishing.

Vittorio N., et al. 2020. Cumulated frictional dissipated energy and pavement skid deterioration: evaluation and correlation. *Construction and Building Materials*, 263 (2020) 120020.

Functional Pavements – Chen et al (eds)
© 2021 Taylor & Francis Group, London, ISBN 978-0-367-72610-2

Feasibility study of lightweight asphalt concrete in large flexibility steel bridge deck pavement

C.C. Zhang
Key Laboratory of Safety and Risk Management on Transport Infrastructures, Ministry of Transport, Southeast University, Nanjing, Jiangsu, China
China Iconic Technology Company Limited, Hefei, China

L.L. Chen
lligent Transportation System Research Center, Southeast University, Nanjing, Jiangsu, China

C.L. Jiang
Chongqing Municipal Design and Research Institute, Chongqing, China

G. Liu
Intelligent Transportation System Research Center, Southeast University, Nanjing, Jiangsu, China

ABSTRACT: This paper presents a feasibility study of lightweight asphalt concrete in large flexibility steel bridge deck pavement (SBDP). Firstly, considering effect of global bridge structure, the most critical response of large flexibility steel bridge was simulated through multiscale analysis. Then, lightweight epoxy asphalt concrete was designed by replacing part of the basalt aggregate with ceramsite. After that, the performance of the designed lightweight asphalt concrete were tested to find whether it can meet the requirements of the large flexibility SBDP. Results show that splitting tensile strength of lightweight epoxy asphalt concrete with different ceramsite mixed ratios can meet the requirements. Meanwhile, temperature stability and water stability can meet the requirements of specifications for epoxy asphalt concrete. Based on the performance of lightweight epoxy asphalt concrete with different ceramsite addition, lightweight epoxy asphalt concrete with 40% (mass percentage) ceramsite addition is recommended as material of the large flexibility SBDP. Nearly 1/4 of the mass can be saved compared to that of normal design.

Keywords: steel bridge deck pavement, flexibility steel bridge, lightweight asphalt concrete, multiscale analysis

1 INTRODUCTION

With the continuous development of bridge construction, span of coast-crossing steel bridge is constantly increasing. Various types of bridges continue to appear, including suspension bridge with long span, cable-stayed bridge and steel truss bridge for highway and railway. In general, the flexibility of bridge increases constantly with the increase of span. Thus, new challenge is brought for SBDP. Steel bridges with large flexibility have long span, and have high requirements for dead load and deformation resistance of pavement materials. It is an effective approach to solve the problem by selecting a kind of durable and lightweight material with high deformation resistance (Huang 2006; Wang et.al. 2005; Chen et al. 2020). On the other hand, sand and gravel aggregate used in asphalt concrete is mainly obtained by mountains exploiting and channel excavating, which causes unrecoverable destruction for environment.

Currently, exploitation of sand and gravel aggregate has been limited in multiple places. Therefore, it is urgent to find substitute of asphalt concrete.

As a lightweight material, ceramsite has been applied in civil engineering widely. In road engineering, some researches have been made on ceramsite cement concrete and ceramsite asphalt concrete. The application mainly focuses on bituminous surface treatment and maintenance (EPPS et al. 1981; Texas. 1981). Existing research results show that ceramsite substituted aggregate can improve performance for asphalt concrete on skid-resistance, rutting resistance performance and moisture stability. Also, it can reduce the weight of asphalt concrete to provide a new choice for paving engineering (Mallick et.al. 2004; Chen et al. 2020). In bridge deck pavement, exploratory researches have been conducted on the use of ceramsite asphalt concrete pavement on special bridge, such as bascule bridge (Qian et.al. 2011; Liu et al. 2012; Liu et al. 2018). However, application of ceramsite asphalt concrete on long-span bridges, especially large flexibility steel bridges has not been studied.

Under this background, existing long-span bridges were selected as project case in this research. Then the performance requirements of large flexibility steel bridges for paving materials were determined through numerical simulation. Subsequently, optimization design and examination of performance of lightweight ceramsite asphalt concrete were made through laboratory tests. And on this basis, the feasibility of lightweight ceramsite asphalt concrete as paving material of large flexibility steel bridge was analyzed. The research results can provide a reference for design and construction of long-span SBDP.

2 NUMERICAL ANALYSIS

Multi-tower suspension span bridge is a new bridge structure type that has been widely used in recent years. Consecutive layout of multiple main spans is achieved by means of installing one or more main tower. It improves spanning capacity of bridge substantially while increasing flexibility of suspended-cable structure. Taizhou Yangtze River Highway Bridge, regarded as a super-span bridge with large flexibility in China, is the first long-span steel bridge that applied three-tower and two-span suspended-cable structure in the world. The main span is $2 \times 1080m$, and ratio of deflection to span is nearly 1/250. Taking the case of Taizhou Yangtze River Bridge, the performance requirements of large flexibility steel bridges for paving materials were analyzed.

2.1 Establishment methods and ideas of numerical model

In order to take the influence of various adverse stress states of the whole bridge structure on the pavement layer into consideration, Taizhou bridge should be simulated and analyzed precisely. The size of Taizhou bridge is of thousand-meter scale. While the detail structure, such as seam of slab, position of stiffener, is of centimeter scale or smaller. If bridge model size is analyzed according to detail structure, it requires much computing power. Conventional computer is incapable to deal with yet. Large-scale special computers lacking special computing software tend to have low efficiency in programming calculation. Analytical size is too large to get detailed stress currently. And the analysis problem caused by span and size has a great influence to analyze stress of bridge deck pavement. In this context, the multiscale method of "whole bridge-beam of local bridge-orthotropic composite structure" was used to analyze the stress of bridge deck pavement (Chen et al. 2016; Song and Zhou. 2019; Qian et.al. 2016).

2.2 Overall bridge model

This paper mainly studies the performance of bridge deck pavement, and pays attention to the stress results of pavement. Pavement is built after completion of bridge during construction. Thus, it does not participate in the overall stress of the bridge in the early stage (such as self-weigh of structure and secondary dead load). The main load originates from vehicle load, temperature load and wind load after completion of bridge. Deck pavement and main beam work together while main beam is equivalent to tension-bending component in the mechanical response analysis. Consequently, vertical bending moment is considered main internal force and vehicle load is considered main load. The most unfavorable beam segment of SBDP under live load should be determined. The calculation model and analysis result are shown in Figure 1.

According to the calculation results of bridge, the maximum negative bending moment under the random traffic load is -118448 $kN·m$.. It is located in middle tower. The maximum bending moment is 205133 $kN·m$.. It is located in position about 1/12 away from side span. For deck pavement system, position that has maximum negative bending moment is the most unfavorable. Although position of steel box girder in middle tower has been strengthened, it still has too large negative bending moment and makes its mechanical response larger than that of standard beam.

2.3   The most critical steel box girder deck system model

64-meter-long steel box girder near middle tower of suspension bridge is regarded as the most unfavorable position. Internal force results (bending moment, torque and axial force) from left and right sides of bridge model are selected as boundary condition of local beam model in next step. Parameters of model are shown in Table 1. Epoxy asphalt concrete is used in Taizhou Yangtze River Bridge as pavement material. In normal temperature, dynamic modulus of the material is 9000MPa, and Poisson's ratio is 0.25 (Qian et al. 2016).

According to analysis results, choosing 32 meters on both sides of middle tower, a total of 64 meters local beam is analyzed and simulated by mixed finite element method (Su et al. 2005). The model, including four standard beams, is 64m long and 39.1m wide. Models of top slabs, webs, bottom slabs, stiffeners of steel box girder are simulated in shell63 elements, and deck pavement is simulated in solid45 elements. Models of plates and pavement are simulated in actual size and connection. Mises-stress distribution of local beam under load is shown in Figure 2.

2.4   Orthotropic composite system model

In order to improve the accuracy of model and obtain force details of local area, orthotropic plate of crucial force position of SBDP is intercepted on the basis of local beam model analysis. Simultaneously, larger-scale finite element analysis is considered. Orthotropic plate that is 9.6m long (contains 4 diaphragm) and 4.2m wide (width of 7 U-stiffened) is chosen in local beam model.

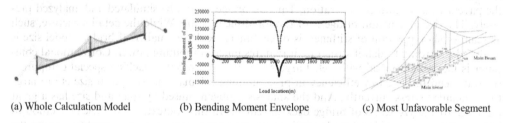

(a) Whole Calculation Model   (b) Bending Moment Envelope   (c) Most Unfavorable Segment

Figure 1.   Diagrams of the most critical response of bridge deck pavement system.

Table 1.   Geometric dimensions and material parameters of steel deck.

Thickness of steel deck /mm	U Plate Size/mm	U Plate Spacing/mm	Diaphragm thickness /mm	Diaphragm spacing/mm	Poisson's ratio	Elastic modulus/MPa
14	300×280×6	600	12	3.2	0.3	210000

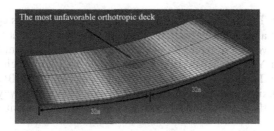

Figure 2.   Stress distribution of the most unfavorable box girder.

Model is established in the way of sub-model. As initial stress and boundary condition, calculation results of local beam model are included in the model by incision boundary. Thus, orthotropic plate model not only considers the influence of global bridge effect to deck pavement, but also reflects force details of pavement system and steel deck. Bridge deck, diaphragm and stiffener are simulated in shell element, and deck pavement is simulated in solid element. Value under the most unfavorable load is shown in Table 2. For comparison, traditional method of simplifying constraints is applied in the paper, directly using the orthotropic plate model for analysis and calculation. The results are calculated in Table 2.

As shown in Table 2, bridge structure has great effect on deck pavement. The maximum stress and strain of deck pavement under the influence of global bridge structure, are about 17% greater than the value which is calculated in local beam model simplified constraints. It is considered that influence coefficient of Taizhou Bridge structure on local stress of deck pavement is 1.17. Besides, the maximum tensile stress and tensile strain of pavement surface are 0.919 MPa and 835.4 $\mu\varepsilon$ repectively.

3   EXPERIMENTAL PROGRAM

As a typical large flexibility steel bridge, deflection span ratio of Taizhou Yangtze River Highway Bridge is far greater than that of other types of steel bridges. Under the most unfavorable condition, it also has larger mechanical responses. Compared with ordinary asphalt concrete, lightweight asphalt concrete has smaller self-weight, which is beneficial to load condition on large flexibility steel bridge. However, whether mechanical performance of lightweight asphalt concrete can satisfy requirements of large flexibility steel bridge should be researched deeply. Under the back ground, grapholith ceramsite is selected as lightweight aggregate on the basis of investigation and analysis, then lightweight asphalt concrete is designed, and performance is also studied.

3.1   *Performance of lightweight asphalt concrete material*

Epoxy asphalt concrete is selected as deck pavement material in Taizhou Yangtze River Bridge. For comparison, epoxy asphalt concrete is selected as binding material of lightweight

Table 2.   Most critical response of steel deck pavement.

Constraint condition	Tensile stress of pavement surface/MPa		Tensile strain of pavement surface/$\mu\varepsilon$		Maximum deflection span ratio between U-stiffened
	Horizontal	Vertical	Horizontal	Vertical	
Internal force of bridge as boundary condition	0.919	0.556	835.4	423.1	1/1063
Simplified condition	0.787	0.506	721.0	407.8	1/1169

203

concrete in this paper. The lightweight epoxy asphalt concrete contains epoxy asphalt, ceramsite and regular basalt aggregate. Particle size of shale ceramsite used is 5-10mm. Shale aggregate and mineral powder are from the same manufacturer with material used in deck pavement of Taizhou Yangtze River Bridge. According to Lightweight Aggregates and Its Test Methods PartII: Test Methods for Lightweight Aggregates (GB/T 17431.2-1998) and Specifications For Design and Construction of Pavement on Highway Steel Deck Bridge (JTG/T3364-02-2019), the main technical parameters of each component are listed in Table 3.

3.2 Performance test of lightweight epoxy asphalt concrete

Original aggregate is substituted by partial lightweight aggregate in researches of lightweight concrete at home and abroad. Refer to related researches (Zhong et al. 2010; Liu et.al. 2007), AC-10 was adopted in the design of lightweight epoxy asphalt concrete, and partial coarse aggregate was substituted by ceramsite in this paper. For comparison, control groups with mass substitution ratio of 0%, 5%, 15%, 25% and 40% were set up in experiment. Among them, reference group with 0% had no lightweight aggregate. Based on Marshall test, as shown in Table 4, optimum asphalt content of each group materials were determined.

In order to confirm applicability of lightweight epoxy asphalt concrete used in large flexibility steel bridge, several aspects should be considered as followes. Firstly, mechanical performance of lightweight asphalt concrete needs to satisfy requirement of response under the most unfavorable condition. Therefore, splitting test and bending test were made to confirm tensile and compression strength and deformation resistance of the materials. Concurrently, its high temperature stability and water stability were tested in this paper to ensure materials can satisfy the requirements of steel deck pavement. Test parameters and analysis, as well as test results, are shown in Table 5.

3.2.1 Splitting test

The experiment was carried out according to *Splitting Test of Asphalt Mixture (T0716-2011)* from *Test Methods of Asphalt and Asphalt Mixtures for Highway Engineering (JTG E20-2011)*. Splitting strength of asphalt mixtures with different contents was studied at 25°C. Standard Marshall specimens were prepared and solidified completely at 120°C. Direction of

Table 3. Main technical parameters of light epoxy asphalt concrete.

Asphalt type	Technical parameter	Test value	Technical requirements[1]
Epoxy asphalt	Tensile strength/23°C, MPa	7.56	≥6.0
	Elongation at break/23°C, %	325	≥270
Ceramsite	Density grade	890	600-900
	Cylinder compressive strength/MPa	7.32	≥6.5
	Content of mud and friable particle	1.25	≤2
Shale aggregate	Apparent density (/cm^3)	2.91-2.95	≥2.65
	Compressive strength/MPa	140	≥120
	Los Angeles wear value/%/	11.5	≤22.0
	Crushing value/%	9.4	≤12
Limestone powder	Density (g/cm^3)	2.715	≥2.5
	Plasticity index/%	3.2	4.0

Table 4. Optimum oil stone ratio of different ceramsite mixture.

Ceramsite mixed ratio/%	0	5	15	25	40
Optimum asphalt content/%	6.50	6.70	7.60	8.10	9.20

Table 5. Performance test results of mixtures with different ceramsite proportions.

Ceramsite mixed ratio/%	0	5	15	25	40	Technical requirement
Splitting strength/MPa	5.82	7.67	7.19	6.26	6.08	-
Bending strength/MPa	28.25	31.56	29.34	28.68	26.55	-
Bending strain/$\mu\varepsilon$	4390	4186	4258	4072	3993	≥3000
dynamic stability (Cycles/mm)	15287	16845	17384	17962	19088	≥6000
Freeze-thaw splitting strength/MPa	4.48	7.21	6.87	6.09	5.21	-
TSR/%	92.55	92.87	91.58	93.46	92.16	≥80

pressure was vertical and loading speed is 50mm/min in test. The automatic Marshall instrument with displacement sensor was used.

3.2.2 Bending test
Characteristics of tension-bending resistance and elasticity of asphalt mixture were evaluated by bending test of asphalt mixture beam. The experiment was carried out according to *Bending Test of Asphalt Mixture (T0715-2011)* from *Test Methods of Asphalt and Asphalt Mixtures for Highway Engineering (JTG E20-2011)*. Test temperature was -10°C and loading speed was 50mm/min.

3.2.3 Rutting test
High temperature stability was evaluated by high temperature rutting test. The experiment was carried out according to *Rutting Test of Asphalt Mixture (T0719-2011)* from *Test Methods of Asphalt and Asphalt Mixtures for Highway Engineering (JTG E20-2011)*. The rutting test specimens were made by wheel-grind method. Mixtures with optimum asphalt content of GLEAC and ELEAC were prepared. Structure of actual deck pavement was simulated with rutting plate specimens. And specimens were solidified at 120°C. Test temperature was 70°C. Tire pressure was 0.7MPa.

3.2.4 Freeze-thaw splitting test
Water stability was evaluated by freeze-thaw splitting test. The experiment was carried out according to *Freeze-thaw Splitting Test of Asphalt Mixture (T0729-2011)* from *Test Methods of Asphalt and Asphalt Mixtures for Highway Engineering (JTG E20-2011)*. Marshall specimens, prepared with the same method, were solidified completely in test. Test specimens were immersed in water for 20 minutes at 25°C. Next, they were immersed in water with 0.09MPa and vacuumed for 15min. Specimens without freeze thawing were tested after being immersed in water for 2h.

3.2.5 Results and analysis
Performance test results of epoxy asphalt concrete with different ceramsite mixed ratios are shown in Table 5. As can be seen from Table 5, the maximum tensile stress and tensile strain of pavement surface are 0.919 MPa and 835.4 $\mu\varepsilon$ repectively. Compared with the most unfavorable response of large flexibility steel deck pavement in Table 2, splitting tensile strength of lightweight epoxy asphalt concrete with different ceramsite mixed ratios can meet the requirements. Secondly, temperature stability and water stability of lightweight epoxy asphalt concrete with different ceramsite mixed ratios can meet the requirements on epoxy asphalt concrete form *specifications for design and construction of pavement on highway steel deck bridge* (JTG/T3364-02-2019).

In addition, compared with performance test results of epoxy asphalt concrete with different ceramsite mixed ratios, it is found that performances of mixture improve after adding ceramsite. Splitting strength, tension-bending strength, and dynamic stability show obvious rules that splitting strength and tension-bending strength decrease slightly, yet dynamic stability increased significantly with the increase of ceramsite mixed ratio. Bending strain and TSR seldom change. It can be indicated that rounded shale ceramsite with size of 5-10mm is mainly used as lightweight aggregate in this paper. When proper amount of ceramsite replaces basalt aggregate, the strength and deformation resistance of asphalt concrete are improved to a certain extent. However, with the increase of ceramsite addition, gradation is influenced and high temperature stability increases while others decreases. In this paper, the amount of aggregate above 4.75mm is less than 40%. Therefore, ceramsite addition is set to 40% at most because of great influence for gradation with addition more than 40%.

The experiment results indicate that lightweight epoxy asphalt concrete with 40% ceramsite addition can still meet the performance requirements of large flexibility SBDP. For reducing self-weight, lightweight epoxy asphalt concrete with 40% (mass percentage) ceramsite addition is recommended as material of large flexibility SBDP. Lightweight epoxy asphalt concrete with 40% (mass percentage) ceramsite addition is recommended as material of large flexibility SBDP. The density of normal epoxy asphalt concrete is about 2.5 g/cm^3. The density of ceramsite asphalt concrete is about 2 g/cm^3. Through caclulation, nearly 1/4 of the mass can be saved compared to the normal design.

4 CONCLUSION

Feasibility for application of lightweight asphalt concrete in large flexibility steel bridge was studied. Initially, performance requirement of large flexibility steel bridge was analyzed by multiscale mechanical simulation. Subsequently, lightweight asphalt concrete was designed and performance tests were conducted. The conclusions are drawn as follows.

1) Mechanical response of Taizhou Yangtze River Highway Bridge under the most unfavorable condition is obtained through multiscale model. The maximum calculation results of stress and strain is about 17% more accurate than that in simplified model.
2) Lightweight asphalt concrete is designed and corresponding tests of mechanical performance and stability are carried out. Splitting tensile strength of lightweight epoxy asphalt concrete with different ceramsite mixed ratios can meet the requirements. Temperature stability and water stability can meet the requirements of specifications for epoxy asphalt concrete.
3) Change regularity of performance of lightweight epoxy asphalt concrete with different ceramsite additions are analyzed. Lightweight epoxy asphalt concrete with 40% ceramsite addition is recommended as material of large flexibility SBDP.

ACKNOWLEDGEMENTS

The authors gratefully appreciate the financial support for this research from the National Key R&D Program of China (Nos. 2018YFB1600100 and 2018YFB1600105).

REFERENCES

Chen, L. L., et al. 2020. Determination of Allowable Rutting Depth Based on Driving Safety Analysis[J]. *Journal of Transportation Engineering, Part B: Pavements*, 2020, 146(2).
Chen, L. L., et al. 2020. Feasibility evaluation of a long-life asphalt pavement for steel bridge deck. *Advances in Civil Engineering*, 2020(2020):1–8.
Chen L.L., Qian Z.D., Wang J.Y. 2016. Multiscale numerical modeling of steel bridge deck pavements considering vehicle-pavement interaction. *International Journal of Geomechanics* 16(1): 1–8.
EPPS, J. A., Gallaway, B. M., Brown, M. R. 1974. Synthetic Aggregate Seal Coat Research.
Huang, Wei. 2006 *Theory and method of deck paving design for long-span bridge*. Beijing: China Architecture & Building Press.
Liu, Q.H., Li, C.B., Zhang, C.M. et al. 2007. Experimental study on asphalt ceramsite concrete prepared with high strength shale ceramsite. *Journal of Guangzhou University(Natural Science Edition)* 6(3).

Liu X.Z., Zhao B.Q., Li J. 2018. Analysis on water stability performance of ceramic asphalt mixture. *Journal of China & Foreign Highway* 240(02): 304–306.

Liu Y., Qian Z.D., Zhang L., et al. 2012. Mechanical properties of epoxy asphalt mixture pavement with lightweight aggregate applied on bascule bridge. *Journal of Southeast University(English Edition)* 28(03): 321–326.

Mallick R.B., Hooper F.P., O'BRIEN S., et al. 2004. Evaluation of use of synthetic lightweight aggregate in hot-mix asphalt. In: *Transportation Research Board National Research Center*.

Qian Z.D., Chen L.L., Jiang C.L., et al. 2011. Performance evaluation of a lightweight epoxy asphalt mixture for steel bridge deck pavements. *Construction and building material*: 3117–3122.

Qian Z.D., Liu Y., Liu C.B., et al. 2016. Design and skid resistance evaluation of skeleton-dense epoxy asphalt mixture for steel bridge deck pavement. *Construction and Building Materials*: 114.

Song J.C., Zhou Y. 2019. Comparison on finite element methods of orthotropic steel bridge deck pavement. *Journal of China & Foreign Highway* 39(01): 82–86.

Su, Q.T., Wu, Chong., Dong Bing. 2005. Analysis of flat steel-box-girder of cable-stayed bridge by finite mixed element method. *Journal of Tongji University(Natural Science)* (06): 742–746.

Texas Transportation Institute. 1981. *Field Manual on Design and Construction of Seal Coats*.

Wang, F.Z., Hu, S.G., Ding, Q.J., et.al. 2005. Research on high performance lighweight concrete and its application in old-bridge deck rebuilding. *Journal of Highway and Transportation Research and Development*: 86-88+100.

Zhong J.F., Chen, X.F., Qian, L. 2010. Research on durability of lightweight concrete bridge deck pavement. *Journal of Highway and Transportation Research and Development(Application Technology)* 6 (05): 185–188.

Functional Pavements – Chen et al (eds)
© 2021 Taylor & Francis Group, London, ISBN 978-0-367-72610-2

Development of a multiscale simulation of asphalt pavements based on an FE-RVE approach

Z. Qian, L. Chen & J. Hu
Key Laboratory of Safety and Risk Management on Transport Infrastructures, Ministry of Transport, P.R. China
Intelligent Transportation System Research Center, Southeast University, Nanjing, P.R. China

G. Lu, P. Liu, H. Liao & M. Oeser
Institute of Highway Engineering, RWTH Aachen University, Aachen, Germany

ABSTRACT: The computational cost of pure microscale models at the macroscale is extremely high and the advantages of the microscale simulation are generally restricted to some specific regions of the macroscale structure. As a result, the Finite Element Representative Volume Element (FE-RVE) approach is proposed to do the multiscale analysis, which provides opportunities to more comprehensively describe the mechanical response of asphalt pavement, e.g., the microscale models enable parametric studies on the influence of such as aggregate morphology on the macroscopic response of asphalt pavements. In this study, a general workflow is put forward to do the FE-RVE based simulation. Afterward, a pavement structure model is set up as an example with microstructural RVE models. The example proves the algorithm to be sufficiently effective and reliable.

1 INTRODUCTION

The macroscale finite element (FE) model of asphalt pavements is often assumed to be homogeneous, whereas at the microscale the intrinsic heterogeneous structure is important to be considered, because the corresponding phenomenological constitutive model which is only a homogenized approximation of the microstructure is not able to describe the actual physical phenomena (Wollny et al. 2020). The multiscale modeling concepts have been applied to asphalt materials and pavements recently (Cucalon et al. 2016, Kim et al. 2013, Teixeira et al. 2014). An important assumption in the multiscale approach is separation of scales, which is obtained if the structural dimensions are significantly larger than the dimensions of material inhomogeneity (Nemat-Nasser & Hori 2013). In order to achieve this, a representative volume element (RVE) should be defined and the averaging should be performed for it. The RVE is a sample which is typical for the whole composite on average. It should contain sufficient inclusions for the overall moduli which are to be effectively independent of the surface values of traction and displacement (Hill 1963).

In this study, the Finite Element Representative Volume Element (FE-RVE) approach is briefly introduced to be used for the multiscale analysis. A general workflow is put forward to do the FE-RVE simulation for the asphalt pavement and mixtures. A pavement structure model is then created as a case study combined with microstructural RVE models. Some results are listed and described to prove the proposed algorithm sufficiently effective and reliable.

2 FE-RVE BASED APPROACH FOR MATERIAL HOMOGENIZATION

The Figure 1 shows the flowchart of the FE-RVE based computation approach. Firstly the geometry and material property of each component in the RVE model should be defined and

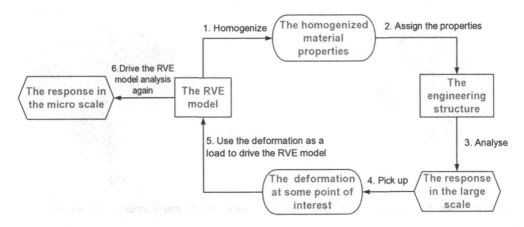

Figure 1. The flowchart of the FE-RVE based computation approach.

then the materials properties of the RVE model are homogenized. After that these homogenized materials properties can be assigned to engineering structure and the responses of the engineering structure in the large scale will be analyzed. Then the deformation at some points of interest in the engineering structure can be used to drive the RVE analysis. By doing that the response of the RVE model in the micro scale can be predicted. More details about homogenization of the material properties of composites can refer to Ji et al. (2017).

3 EXEMPLAR MULTISCALE SIMULATION OF PAVEMENT STRUCTURE

3.1 Definition of the FE model

An example is given for the multiscale simulation of pavement structure using the proposed algorithm. This pavement structure consists of four layers, such as asphalt surface course (SMA layer), asphalt base course (AC layer), subbase and sub-grade. The SMA and AC layers are considered as heterogeneous, which includes coarse aggregates and asphalt mortar. Their microstructure was derived from X-ray Computer Tomography (X-ray CT) techniques and Digital Image Processing (DIP). The details about the reconstruction of the microstructure can refer to the previous research (Kollmann et al. 2019a, b). Both of the length and width of the pavement are 6000 mm. The other definition about the geometry and material parameters are listed in Table 1. These material parameters are used to represent the material properties in the pavement structure with the temperature of 20°C at the pavement surface. The side surfaces and the bottom surface of the pavement are totally fixed. Besides, all the interactions between the adjacent layers are tied. A static circular load with diameter of 300 mm and constant pressure of 0.7 MPa is applied on the centre of the pavement top surface to simulate a tire load from a standard axle.

In order to get accurate results, the mesh size of the pavement and RVE models and the diameter of the RVE model should be determined. Particularly, different mesh sizes and diameters were defined to the model and their computational results were compared. When the

Table 1. Geometry and material parameters of the pavement structure.

Layer	Thickness (mm)	Component	E (MPa)	μ
SMA layer	40	Coarse aggregates	55000	0.2
		Asphalt mortar	3254	0.267
AC layer	140	Coarse aggregates	55000	0.2
		Asphalt mortar	3316	0.3
Subbase	450	-	120	0.49
Sub-grade	2000	-	45	0.49

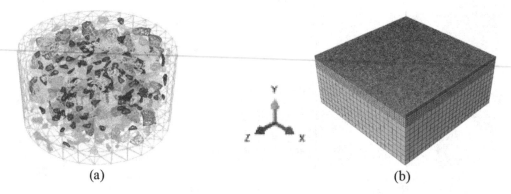

(a)   (b)

Figure 2.   The FE models. (a) RVE Model at the microscale; (b) Pavement model at the macroscale.

computational results started to be converged, the corresponding values of the mesh sizes and the diameter were adopted in the final FE model. Based on the comprehensive comparisons, proper parameters were determined and the final FE models are shown in Figure 2.

3.2 *Computational results*

After considering the proper mesh size of the FE models and the RVE volume, the homogenized orthotropic material parameters of SMA and AC layers are generated using the proposed algorithm, which are listed in Table 2. One can see that the E moduli in the vertical direction (E_2) of the SMA and AC are much larger than those in the horizontal directions (E_1 and E_3), which is reasonable and reliable, because the compaction makes the asphalt mixtures more condensed in the vertical direction during the pavement construction.

Applying these parameters into the pavement structure, the mechanical responses of the pavement structure can be computed. And then the computational mechanical responses are applied as the boundary conditions of the RVE model to generate the mechanical responses of the RVE model at the microscale. The distribution maximum principal stress in the macroscale pavement structure and the microscale RVE model (AC) is shown in Figure 3.

Table 2.   The homogenized elastic propertied of SMA and AC.

	E_1	E_2	E_3				G_{12}	G_{13}	G_{23}
Type	(MPa)	(MPa)	(MPa)	Nu_{12}	Nu_{13}	Nu_{23}	(MPa)	(MPa)	(MPa)
SMA	1085.92	3650.12	1065.64	0.07	0.08	0.23	585.47	435.31	571.68
AC	4554.74	14712.23	3080.30	0.08	0.10	0.27	2405.20	1564.99	1760.85

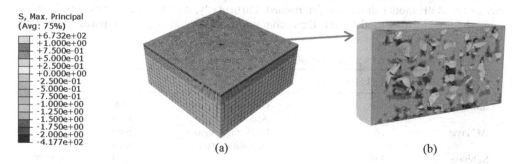

S, Max. Principal
(Avg: 75%)
+6.732e+02
+1.000e+00
+7.500e-01
+5.000e-01
+2.500e-01
+0.000e+00
-2.500e-01
-5.000e-01
-7.500e-01
-1.000e+00
-1.250e+00
-1.500e+00
-1.750e+00
-2.000e+00
-4.177e+02

(a)   (b)

Figure 3.   The computational results. (a) Response of the pavement; (b) Response of the RVE model.

The large principal stress occurs at the area under the load in the pavement model, as shown in Figure 3 (a). While the greatest concentrations of strong negative stresses occur in vertical direction near coarse aggregates, so one can imagine that the coarser aggregates have a greater influence on the load-bearing capacity of the asphalt mixture, as shown in Figure 3 (b).

4 CONCLUSIONS AND OUTLOOK

In this study, the homogenization of the heterogeneous material property of the asphalt mixtures is proposed by using the FE-RVE approach. A case study proves the algorithm to be sufficiently effective and reliable.

As an initial study, there are still several matters in the FE modelling which should be improved in the future. For example, a more realistic interface definition in the RVE model between the aggregates and the asphalt mortar would be considered (Liu et al. 2019). Moreover, a tire-pavement interaction model with different saturation levels of the subbase would be applied in the macroscale computation (Liu et al. 2018, Lu et al. 2020).

ACKNOWLEDGEMENTS

The financial support from the Fundamental Research Funds for the Central Universities (Grant Project No. 2242020k30050) and German Research Foundation (Grant Project No. FOR 2089/2, OE 514/1-2) is gratefully acknowledged.

REFERENCES

Cucalon, L.G.; Rahmani, E.; Little, D.N. & Allen, D.H. 2016. A multiscale model for predicting the visco-elastic properties of asphalt concrete. *Mechanics of Time-Dependent Materials*, Vol. 20 (3): 325–342.

Hill, R. 1963. Elastic properties of reinforced solids: some theoretical principles. *Journal of Mechanics and Physics of Solids*, Vol. 11 (5): 357–372.

Ji, H., Mclendon, R., Hurtado, J. A., Oancea, V., & Bi, J. 2017. Multi-scale Material Modeling with the Mean-Field Homogenization Method. *In NEFEMS World Congress*.

Kim, Y.R.; Souza, F.V. & Teixeira, J. 2013. A two-way coupled multiscale model for predicting damage-associated performance of asphaltic roadways. *Computational Mechanics*, Vol. 51 (2): 187–201.

Kollmann, J., Liu, P., Lu, G., Wang, D., Oeser, M., & Leischner, S. 2019a. Investigation of the microstructural fracture behaviour of asphalt mixtures using the finite element method. *Construction and Building Materials*, 227, 117078.

Kollmann, J., Lu, G., Liu, P., Xing, Q., Wang, D., Oeser, M., & Leischner, S. 2019b. Parameter optimisation of a 2D finite element model to investigate the microstructural fracture behaviour of asphalt mixtures. *Theoretical and Applied Fracture Mechanics*, 103, 102319.

Liu, P., Chen, J., Lu, G., Wang, D., Oeser, M., & Leischner, S. 2019. Numerical simulation of crack propagation in flexible asphalt pavements based on cohesive zone model developed from asphalt mixtures. *Materials*, 12(8), 1278.

Liu, P., Ravee, V., Wang, D., & Oeser, M. 2018. Study of the influence of pavement unevenness on the mechanical response of asphalt pavement by means of the finite element method. *Journal of Traffic and Transportation Engineering (English Edition)*, 5(3), 169–180.

Lu, G., Liu, P., Törzs, T., Wang, D., Oeser, M., & Grabe, J. 2020. Numerical analysis for the influence of saturation on the base course of permeable pavement with a novel polyurethane binder. *Construction and Building Materials*, 240, 117930.

Nemat-Nasser, S. & Hori, M. 2013. Micromechanics: overall properties of heterogeneous materials, Vol. 37, Amsterdam, Elsevier.

Teixeira, J.; Kim, Y.R.; Souza, F.; Allen, D. & Little, D. 2014. Multiscale model for asphalt mixtures subjected to cracking and viscoelastic deformation. *Transportation Research Record: Journal of the Transportation Research Board*, Vol. 2447: 136–145.

Wollny, I., Hartung, F., Kaliske, M., Liu, P., Oeser, M., Wang, D., Canon Falla, G., Leischner, S. & Wellner, F. 2020. Coupling of microstructural and macrostructural computational approaches for asphalt pavements under rolling tire load. *Computer-Aided Civil and Infrastructure Engineering*.

Functional Pavements – Chen et al (eds)
© 2021 Taylor & Francis Group, London, ISBN 978-0-367-72610-2

Meso-mechanical assessment of composite materials based on the finite element method

Cong Du, Pengfei Liu & Markus Oeser
Institute of Highway Engineering, RWTH Aachen University, Aachen, Germany

ABSTRACT: To deeply investigate the effects of the geometric features of inclusions on the mechanical performance of composite materials, this study established the heterogeneous composite models by using the random generation approach, and employed the finite element simulation to obtain the overall stiffness of the models. The model sizes of 5 cm and 10 cm were employed to investigate the size effect on the stiffness. In addition, the inclusion volume content and interactions are analyzed, and the Eshelby theory was used to take into account of the inclusion shapes and locations. As a result, the inclusion volume content and interactions are consistent with the stiffness; however, the Eshelby theory can only be used when the heterogeneous models are relatively small or simple.

1 INTRODUCTION

Due to the random and complex inclusions, it is difficult to precisely calculate and predict the mechanical responses of composite materials, e.g., concrete, asphalt mixture, etc. Particularly, asphalt mixture serves as the major material of pavement in highways and urban road, in which both the loading capacity and flexibility are satisfied (Kasu, Manupati & Muppireddy 2020, Azarhoosh & Koohmishi 2020). Such material poses a higher requirement on the selection and mixing process in order to guarantee higher performance. Therefore, a reasonable cross-scale relationship, which bridges the lower-scale structure and higher-scale response of a composite, can help engineers to improve the performance-oriented composite material design.

To this end, the microstructure-based finite element method (FEM) allows researchers to separately model the inclusions and matrix using the vertices' coordinates of inclusions. In addition, the homogenization can be employed on a representative volume element (RVE) to calculate the macro mechanical response of composite.

Currently, the random generation algorithm can establish mesoscale structures with randomly distributed polygonal inclusions (Castillo, Caro, Darabi & Masad 2017), which significantly improves the efficiency of mesoscale simulation.

However, the uncertainty and random of the inclusions make it difficult to obtain an accurate mechanical response of composite materials. Hence, different mechanical behaviors were calculated from different mesoscale structures, as well as different directions of a single structure. Therefore, the homogenized mechanical performance cannot afford a deeply insight into the roles played by various inclusions, and hence the selection of raw materials and the approach of mixing process for composite are still empirical-based.

To address this issue, the present study aims to investigate the geometric features of inclusions in the randomly generated FE meso-structures, and bridge a link to the homogenized mechanical response. Within this study, two sizes (5cm and 10cm) were employed to generate the mesoscale models. For each size, three different models were built to reduce errors of the simulation. The DIP technology, meso-mechanical method and statistics were utilized to analyze the distributions, shapes and interactions of inclusions.

2 THEORETICAL DEVELOPMENT

To reduce the difference between FE model and actual specimen, the random generation algorithm is supposed to account the inclusion shape, size, and spatial distribution within the composite. In this study, the generation algorithm allows the number of inclusion edges vary from 4 to 10. After the shape and size of inclusion has been determined, the random structure can be generated by "take-and-place" approach (Wang, Wang & Chen 2014). According to the gradation curve of the inclusions, the largest particle was generated and placed within the model boundary. Afterwards, this process was repeated with the checking step used to prevent the next particle from overlapping with the former one. The abovementioned "take-and-place" process was terminated until the total area of the placed particles follow the gradation curve. The randomly generated mesoscale models with two model sizes (5cm and 10cm) and three parallel samples for each size are shown in Figure 1. In the models, the bright gray area refers to the asphalt mortar; the dark gray areas refer to the coarse aggregates; the white areas represent the air voids.

The homogenization method can be used to construct the connection between the mesoscale and macroscale behavior when the heterogeneous medium meets statistical homogeneity. Thus, the macroscale mechanical issue can be solved by applying the volume averaging scheme on the results of the mesoscale model, defined as (Allen 2001),

$$f^{macro}\left(x_i^{macro}, t\right) = \frac{1}{V}\int_V f^{meso}\left(x_i^{meso}, t\right)dV \tag{1}$$

where f^{macro} and f^{meso} are variable functions at macroscale and mesoscale, respectively; x_i and t are the spatial coordinates and time; V is the volume of the mesoscale model.

The Eshelby tensor was used to describe the effect of inclusions' shapes, sizes and locations on the macro mechanical performance of mesoscale models. According to the Eshelby's theory (Trotta, Marmo & Rosati 2017), for an infinite, elastic, homogeneous and isotropic space having an inclusion Ω in which a uniform eigenstrain ε^* is defined, and the strain ε of matrix can be expressed as,

$$\varepsilon_{ij}(x) = S_{ijmn}(x)\varepsilon_{mn}^* \tag{2}$$

where the S_{ijmn} is the Eshelby tensor, which is closely related to the shapes and stiffness of inclusions. In particular, the Eshelby tensor is constant if and only if the inclusion has an elliptical shape.

In this study, the random generated mesoscale models were firstly simulated using the ABAQUS FE software, and the simulation results were subsequently homogenized to obtain the macroscale mechanical property. Afterwards, three efforts were conducted towards the geometric feature of the inclusions in mesoscale models, which were spatial distribution feature, meso-mechanics and interactions, respectively.

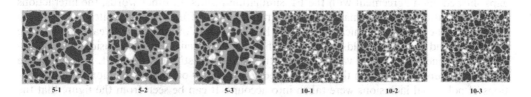

5-1   5-2   5-3   10-1   10-2   10-3

Figure 1. Image of the randomly generated mesoscale structure.

3 SIMULATION

To study the effect of the inclusions on the macro mechanical performance of composite, the static compression loading condition was simulated on the randomly established models. To figure out the roles played by inclusions, the Young's modulus of inclusions is much larger than that of matrix, which respectively were 1000MPa for matrix and 55000MPa for inclusion.

In addition, in order to take the anisotropy of the model into account, compression loadings were separately applied on the vertical direction, and the level of loading was $100\mu\varepsilon$.

In this case, the images of the mesoscale structures were processed using the DIP technology. Afterwards, the volume proportion of inclusion parts was counted and exhibited by Gaussian distribution. Within this effort, the influence of the spatial distribution of inclusions on the overall mechanical performance of composite can be demonstrated and quantitatively presented.

According to the Eshelby theory, the Eshelby tensors of polygonal inclusions in the mesoscale models were calculated. Subsequently, a uniform eigenstrain ε^* was adopted in each inclusion, and the resulting strain field ε can be obtained. Further, the average value of the strains caused by inclusions was calculated to represent the effects of shapes, locations and orientations of inclusions on the macroscale mechanics of composite.

In general, the skeleton of inclusions plays the dominate role in bearing the loadings. However, for some special materials, such as dense-grade asphalt mixture (AC), in which the aggregate inclusions are suspended in the asphalt matrix. Such materials perform both stiffness and flexibility in service. Hence, the interactions between inclusions can also be realized via the "stress transfer" process through the matrix. Within this case, the narrower gaps (smaller than 1.18mm) between inclusions were recognized by using the DIP technology.

4 RESULTS AND DISCUSSION

Figure 2a presents the FE mechanical simulation results and the volume proportions of inclusions. According to this figure, one can see that despite the gradations of inclusions for the two sizes were identical, models with a size of 5 cm exhibited larger uncertainty than 10cm ones. Due to the limitation of the random generation algorithm, all the volume proportions of inclusions in the mesoscale models are smaller than the gradation curve. However, the models with 5 cm size have less inclusion content than that with 10cm size. Correspondingly, the homogenized stiffness of the 5 cm models were smaller than that of the 10cm models. Therefore, it can be concluded that the stiffness of the composite is in positive correlation with the volume content of inclusions, and larger size of the model can bring higher accuracy of the simulation. However, for the different directions of a single model, the simulation results were different, which can be ascribed to the effects of the inclusion distribution.

As mentioned above, the gaps narrow than 1.18mm between inclusions were counted. Figure 2b illustrates the number of gaps per square millimeter in the models together with the stiffness of the models to characterize their correlation. The number of the gaps represents the interactions between inclusions within composites. The figure indicates that the number of gaps showed good agreement with the FE simulation results. In other words, the interactions between inclusions play dominate role in contributing the stiffness of composites.

Based on the meso-mechanical theory, the Eshelby tenors of inclusions in each model were calculated, and the strain fields, caused by the uniform eigenstrain of inclusions was subsequently obtained. The mean values of the strains and the stiffness from FE simulation are shown in Figure 2c. Based on the Eshelby theory, the effects of shapes, locations and orientations of polygonal inclusions were taken into account. It can be seen from the figure that the strains caused by inclusions showed good correlations with the mechanical performance amongst different models especially in small models (size=5 cm), which indicates that the Eshelby tensor shows powerful abilities in demonstrating the mechanical performance of the multiple inclusion problems when the heterogeneous models are relatively small or simple.

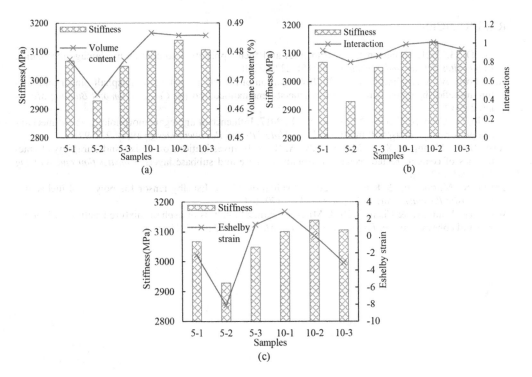

Figure 2.   Image of the FE simulation result: (a) Inclusion content; (b) interactions; (c) Eshelby strain.

5   SUMMARY AND CONCLUSION

This study investigates the mechanical performance of the composite materials with consideration of their geometric features. The heterogeneous models of asphalt mixtures were built using the random generation approach, and the model sizes of 5 and 10 cm were respectively employed. For the macroscale mechanics, the finite element simulation and homogenization calculation was used to determine the overall stiffness of the models. For the mesoscale geometric feature, the Eshelby strain and interacts of the inclusions were investigated. The results are as the following,

The overall stiffness of larger models (size=10cm) is higher than that in the smaller models (size=5cm), and difference of stiffness in larger models is lower. Therefore, it is recommended that the model size of 10cm can be more effectively characterize the overall mechanical behavior of composite materials.

The volume content of the inclusions and interactions between inclusions are consistent with the stiffness of the composites.

The Eshelby tensor results are consistent with the stiffness results in the small models, which indicates that the Eshelby tensile can be used to predict the mechanical responses of small or simple composites.

ACKNOWLEDGMENT

This research is funded by German Research Foundation (Grant Project No. FOR 2089/2, OE 514/1-2) and Excellence Strategy of the German Federal and State Governments (Grant Project No. StUpPD373-20). The authors gratefully acknowledge their financial support.

REFERENCES

Allen, D. H. 2001. Homogenization principles and their application to continuum damage mechanics. *Composites Science and Technology* 61(15): 2223–2230.

Azarhoosh, A., & Koohmishi, M. 2020. Investigation of the rutting potential of asphalt binder and mixture modified by styrene-ethylene/propylene-styrene nanocomposite. *Construction and Building Materials* 255(10).

Castillo, D., Caro, S., Darabi, M., & Masad, E. 2017. Influence of aggregate morphology on the mechanical performance of asphalt mixtures. *Road Materials and Pavement Design* 19(4): 972–991.

Kasu, S. R., Manupati, K., & Muppireddy, A. R. 2020. Investigations on design and durability characteristics of cement treated reclaimed asphalt for base and subbase layers. *Construction and Building Materials* 252(11).

Trotta, S., Marmo, F., & Rosati, L. 2017. Evaluation of the Eshelby tensor for polygonal inclusions. *Composite Pavement Part B-Engineering* 115: 170–181.

Wang, H., Wang, J., & Chen, J. 2014. Micromechanical analysis of asphalt mixture fracture with adhesive and cohesive failure. *Engineering Fracture Mechanics* 132: 104–119.

Functional Pavements – Chen et al (eds)
© 2021 Taylor & Francis Group, London, ISBN 978-0-367-72610-2

Creep analysis of viscoelastic beam by semi-analytical finite element method

Kairen Shen
Research assistant, School of transportation, Southeast University, Nanjing, China

Xianhua Chen*
Professor, Ph.D., School of transportation, Southeast University, Nanjing, China

ABSTRACT: The semi-analytical finite element method (SAFEM) has an advantage in computational efficiency since it only requires meshing in the cross-section while substituting meshing with interpolations of Fourier series along the third dimension. This study aims to develop this method for viscoelastic analysis further. More specifically, use the Generalized Maxwell model to describe the stress-strain relation, and the state variables approach to achieve the viscoelastic iteration. Then a program based on this method for simulating the BBR tests was developed, and its accuracy was verified by comparing with the test results of different asphalt binders. Besides, the comparison with the 3D finite element model in ABAQUS indicated the efficiency of SAFEM. Altogether, the modeling approach has some value for the widespread application of SAFEM in structural modeling.

1 INTRODUCTION

Although direct utilization of the three-dimensional finite element method (3D FEM) currently permits the modeling of all structures irrespective of their complexity, the computing costs involved are often considerable. In some constructions, the geometry and material properties remain unchanged along one direction, where the loading exhibits variation, however. Existing researches indicate that SAFEM is suitable for modeling these structures to enhance computing efficiency (Liu et al., 2017). This method only requires 2D meshing in the cross-section by incorporating the Fourier series along the third dimension, of which one term is corresponding to a 2D plate strain problem. Therefore, instead of solving an equation system with a large number of degrees of freedom, a smaller system is solved many times. This procedure provides a significant reduction in computational time.

Zienkiewicz summarized SAFEM firstly and verified its accuracy by analyzing the elastic prismatic bar (Zeinkiewicz & Too, 1972). Liu further introduced SAFEM into the elastic analysis of flexible pavement to evaluate the bearing capacity (Liu et al., 2018a). Besides, he assembled the Burgers model into a pavement SAFE model for the viscoelastic analysis (Liu et al., 2018b). The main inherent characteristic of viscoelastic materials is that their mechanical behavior is time-dependent. The response not only depends on the instantaneous loading conditions but also the full loading history. The equation system under each analysis step is solved based on the discrete-time iteration. With the increase of the step, the calculation time saved also accumulates. However, the Burgers model is usually used to describe the stress-strain relation of fluid materials due to the Kelvin model components. Therefore, it is necessary to implement the Generalized Maxwell model (GMM) into SAFEM to reflect the mechanical behavior of viscoelastic solid materials.

---

* Corresponding author

The objective of this study is to develop SAFEM for viscoelastic analysis further. The GMM was used to represent the stress-strain relation. Besides, the equation system was established by the work-energy principle. The state variables approach was used to derive the iteration of discrete-time analysis steps, of which time integrals were solved by the Newton-Cotes formula. Then a MATLAB code was written to achieve the above procedure, and a beam modeling program based on this code was developed. To verify the validity, the results of this program were compared with the BBR test of two kinds of asphalt binders under different temperatures, and 3D FEM in ABAQUS.

2 METHODOLOGY

Because Liu described the SAFEM for elastic analysis in detail (Liu et al., 2017), this section mainly introduces the implementation of GMM and the corresponding numerical calculation.

2.1 General description of SAFEM

Let (x, y, z) be the coordinates describing the domain, and z is the longitudinal direction, which is limited to lie between zero and a. For simplifying computation, it is reasonable to omit the displacements in the x-y plane at the front and rear edges, where only consider the motion in the z coordinate. Then the Fourier series only require odd expansion or even expansion. The displacement $\mathbf{u}$ defined by shape functions $\mathbf{N}$ can be written as follows:

$$\mathbf{u} = \begin{Bmatrix} u \\ v \\ w \end{Bmatrix} = \sum_{l=1}^{L} \sum_{i=1}^{n} \begin{Bmatrix} N_{si}(x,y) \sin \frac{l\pi z}{a} u_i^l \\ N_{si}(x,y) \sin \frac{l\pi z}{a} v_i^l \\ N_{ci}(x,y) \cos \frac{l\pi z}{a} w_i^l \end{Bmatrix} \tag{1}$$

where i is the node number, l is the Fourier series number, and u^l is the corresponding displacement. The shape functions $\mathbf{N}_s$ and $\mathbf{N}_c$ in the x-y domain are the same as the plane strain model.

2.2 Stress integral of viscoelastic materials

The Boltzmann superposition principle can transform the stress-strain relation into a hereditary integral as follows:

$$\sigma(t) = \int_0^t E(t-\tau) \frac{\partial \varepsilon}{\partial \tau} d\tau \tag{2}$$

where t is the current time, and τ is a time parameter starting from the beginning of the loading. It is necessary to express the $E(t)$ into a mathematical function for solving the hereditary integral. In GMM, the function can be written as follows:

$$E(t) = E_\infty + \sum_{i=1}^{n} E_i e^{-t/T_i} \,\&\, T_i = \eta_i / E_i \tag{3}$$

GMM has a real insight mhotivated by mechanical analogs depicted in Figure 1, where E_i and η_i are spring and dashpot, respectively.

Then the 3D stress tensor divided into spherical and deviatoric components can be written as follows:

$$\sigma_{ij}(t) = \delta_{ij}\sigma_0(t) + S_{ij}(t) = \delta_{ij} \int_0^t 3K(t-\tau) \frac{\partial \varepsilon_0}{\partial \tau} d\tau + \int_0^t 2G(t-\tau) \frac{\partial e_{ij}}{\partial \tau} d\tau \tag{4}$$

218

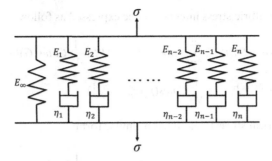

Figure 1. Generalized Maxwell model.

where K and G are the bulk and shear moduli respectively, ε_0 is the spherical stress tensor, and e_{ij} is the deviatoric stress tensor. Appling the GMM into the bulk and shear moduli, the stress can be written as follows:

$$\sigma_{ij}(t) = \delta_{ij}[3K_\infty\varepsilon_0(t) + \sum_{i=1}^{n} 3K_i \int_0^t e^{-(t-\tau)/T_i} \frac{\partial \varepsilon_0(t)}{\partial \tau} d\tau] + 2G_\infty e_{ij}(t) + \sum_{i=1}^{n} 2G_i \int_0^t e^{-(t-\tau)/T_i} \frac{\partial e_{ij}(t)}{\partial \tau} d\tau$$

(5)

2.3   Numerical solution of the stress integral

Since the integral time is not short, FEM usually uses the state variables approach to reduce the computer memory demand. It expresses the viscoelastic stress at time $t+\Delta t$ as a function of the stress and strain at time t. Then, there is no need to store the whole stress history. Assume the time increment Δt as t/n, and express the deviatoric stress as follows:

$$\Delta S_{ij} = S_{ij}(t+\Delta t) - S_{ij}(t) = 2 \sum_{i=1}^{n} G_i(e^{-\Delta t/T_i} - 1) \int_0^t e^{-(t-\tau)/T_i} \frac{\partial e_{ij}}{\partial \tau} d\tau + \int_t^{t+\Delta t} 2G(t + \Delta t - \tau) \frac{\partial e_{ij}}{\partial \tau} d\tau$$

(6)

Assuming P_i^n is the first integral in the equation, one can obtain the following relationship:

$$P_i^n = \int_0^{t-\Delta t} e^{-(\Delta t+t-\Delta t-\tau)/T_i} \frac{\partial e_{ij}}{\partial \tau} d\tau + \int_{t-\Delta t}^t e^{-(t-\tau)/T_i} \frac{\partial e_{ij}}{\partial \tau} d\tau = e^{-\Delta t/T_i} P_i^{n-1} + \int_{t-\Delta t}^t e^{-(t-\tau)/T_i} \frac{\partial e_{ij}}{\partial \tau} d\tau$$

(7)

Due to Δt usually being small, the strain can be treated as linear in the period. Then use the Newton-Cotes formula (Kalogiratou and Simos, 2003) to solve the integral as follows:

$$\int_{t-\Delta t}^t e^{-(t-\tau)/T_i} \frac{\partial e_{ij}}{\partial \tau} d\tau = \frac{1}{90} Cotes(f)(e_{ij}^n - e_{ij}^{n-1})$$

$$\int_t^{t+\Delta t} 2G(t + \Delta t - \tau) \frac{\partial e_{ij}}{\partial \tau} d\tau = \frac{1}{45} Cotes(G)(e_{ij}^{n+1} - e_{ij}^n)$$

(8)

$$Cotes(f) = (7 + 32e^{-0.25\Delta t/T_i} + 12e^{-0.5\Delta t/T_i} + 32e^{-0.75\Delta t/T_i} + 7e^{-\Delta t/T_i})$$

$$Cotes(G) = [90G_\infty + \sum_{i=1}^{n} G_i(7 + 32e^{-0.25\Delta t/T_i} + 12e^{-0.5\Delta t/T_i} + 32e^{-0.75\Delta t/T_i} + 7e^{-\Delta t/T_i})]$$

219

As a result, the deviatoric stress integral can be expressed as follow:

$$S_{ij}(t+\Delta t)=S_{ij}(t)+2\sum_{i=1}^{n}G_i(e^{-\Delta t/T_i}-1)P_i^n+\frac{1}{45}Cotes(G)(e_{ij}^{n+1}-e_{ij}^n)$$

$$P_i^n=e^{-\Delta t/T_i}P_i^{n-1}+\frac{1}{90}Cotes(f)(e_{ij}^n-e_{ij}^{n-1})$$

(9)

Similarly, the spherical stress integral has a similar form:

$$\sigma_0(t+\Delta t)=\sigma_0(t)+3\sum_{i=1}^{n}K_i(e^{-\Delta t/T_i}-1)Q_i^n+\frac{1}{30}Cotes(K)(\varepsilon_0^{n+1}-\varepsilon_0^n)$$

$$Q_i^n=e^{-\Delta t/T_i}Q_i^{n-1}+\frac{1}{90}Cotes(f)(\varepsilon_0^n-\varepsilon_0^{n-1})$$

(10)

2.4   Equilibrium equation in the iteration of SAFEM

Based on the virtual work-energy principle, the work done by the load is equal to the increment of strain energy during each time increment Δt. One can obtain the equation as follows:

$$\iiint_V \left\{ \int_t^{t+\Delta t}[d\varepsilon(t)^T]\sigma(t)\right\}dV = \int_t^{t+\Delta t}[dU(t)^T]F(t)$$

(11)

Then use the trapezoidal integral approximation as follows:

$$\iiint_V \{[\varepsilon(t+\Delta t)^T-\varepsilon(t)^T][\sigma(t+\Delta t)+\sigma(t)]/2\}dV = [U(t+\Delta t)^T-U(t)^T][F(t+\Delta t)+F(t)]/2$$

(12)

The process of viscoelastic analysis always has initial calculation and iterative calculation. The material is assumed as elastic when the initial moment, so the equation can be expressed as follows:

$$\iiint_V (\sum_{l=1}^{L}[B^l]^T)[3K(0)C_1+2G(0)C_2](\sum_{m=1}^{L}[B^l])dV\sum_{l=1}^{L}\{U(0)\}^l-\sum_{l=1}^{L}\{F(0)\}^l=0$$

(13)

where the $K(0)$ and $G(0)$ is initial bulk moduli and shear moduli. Then the initial displacement at each Fourier series $\{U(0)\}^l$ can be solved.

In the iteration, the equation at each analytical step can be written as follows:

$$\iiint_V (\sum_{l=1}^{L}[B^l]^T)[\frac{1}{30}Cotes(K)C_1+\frac{1}{45}Cotes(G)C_2](\sum_{m=1}^{L}[B^l])dV\sum_{l=1}^{L}[\{U(n+1)\Delta t\}^l+\{U(n)\Delta t\}^l]+$$

$$\iiint_V (\sum_{l=1}^{L}[B^l]^T)[3\sum_{i=1}^{n}K_i(e^{-\Delta t/T_i}-1)Q_{il}^n+2\sum_{i=1}^{n}G_i(e^{-\Delta t/T_i}-1)P_{il}^n](\sum_{m=1}^{L}[B^l])dV$$

$$-\sum_{l=1}^{L}[\{F((n+1)\Delta t)\}^l+\{F((n)\Delta t)\}^l]=0$$

(14)

220

The iteration can solve the displacement in each Fourier series term, and the Fourier inversion can obtain the actual displacement $\{U(n\Delta t)\}$.

3 ANALYSIS AND VERIFICATION

3.1 Materials and test method

BBR is a useful instrument to calibrate the flexural-creep property of asphalt binders. Load the specimen with a constant concentrated force for 240s, and the force and the mid-point deflection are monitored versus time. This study conducted the BBR test of one neat and one styrene-butadiene-styrene modified asphalt binder at four different temperatures (-12°C, -18°C, -24°C and -30°C).

3.2 The GGM parameters

Asphalt binders can be treated as linear viscoelastic materials at low temperatures. By the Euler-Bernoulli beam theory, creep compliance $D(t)$ can be written as follows (Pingaro and Venini, 2016):

$$D(t) = \frac{\varepsilon(t)}{\sigma} = \frac{4b\delta(t)h^3}{PL^3} \tag{15}$$

where b is the beam width, $\delta(t)$ is the beam mid-span deflection, h is the beam thickness, P is the applied constant load, and L is the distance between the beam supports. Furthermore, based on the convolution integral principle, the creep compliance $D(t)$ and the relaxation modulus $E(t)$ has the following relationship:

$$\int_0^t D(t) \times E(t - \tau)d\tau = 1 \tag{16}$$

By the inverse Laplace transform, the relation between $D(t)$ and $E(t)$ can be written as follows (Hajikarimi et al., 2018):

$$D(t) \times E(t) = \frac{1}{\Gamma(1 + \alpha) \times \Gamma(1 - \alpha)} = \frac{\sin(\alpha\pi)}{\alpha\pi}$$
$$\Gamma(t) = \int_0^\infty x^{t-1}e^{-x}dx \tag{17}$$

where Γ is the gamma function, and α is the slop of $\lg[D(t)]$-$\lg(t)$.

Use the GMM to represent the $E(t)$, and the Prony series in Table 1 were obtained after fitting. Besides, the Poisson ratio was assumed as 0.35.

Table 1. The GMM parameters of all asphalt binders.

Specimens			Neat asphalt binder				SBS-modified asphalt binder			
T_i(s)		E(MPa)	-12°C	-18°C	-24°C	-30°C	-12°C	-18°C	-24°C	-30°C
		E_∞	60.3	257.8	768.7	1392.2	29.6	90.5	247.0	531.4
T_1	100	E_1	140.0	310.8	515.5	520.8	55.5	117.2	218.6	286.5
T_2	10	E_2	205.4	281.7	324.4	267.1	78.0	133.6	190.7	208.4
T_3	1	E_3	185.8	162.5	184.4	88.1	82.9	125.6	151.7	139.3

3.3 Program development and verification

A program was developed in MATLAB to achieve the above SAFEM for viscoelastic analysis. Figure 2 presents the interface. After loading the test data and inputting the GMM parameters, it will show the time-history curves of the mid-span displacement.

Further, the comparison of two kinds of asphalt binders at different temperatures was shown in Figure 3. It can be found the test results have a good agreement with the simulation.

Since the displacement at 60s is vital for analyzing the creep property, Table 2 shows the difference between the results. It can be found the error of all results is less than 5%.

To verify the efficiency of SAFEM further, the SAFEM was compared with a 3D FEM in ABAQUS. These models have the same cross-section meshing. In the Z coordinate direction,

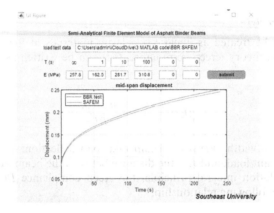

Figure 2.   The program interface.

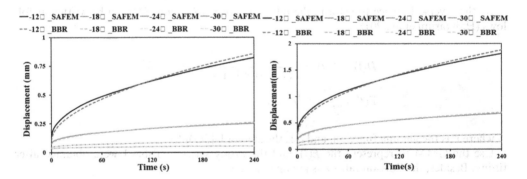

Figure 3.   The mid-span displacements of (a) Neat asphalt binder and (b) SBS-modified.

Table 2.   The difference between the test results and the program.

Displacement (mm)	Neat asphalt binder				SBS-modified asphalt binder			
	-12°C	-18°C	-24°C	-30°C	-12°C	-18°C	-24°C	-30°C
Test results	0.465	0.172	0.074	0.047	1.092	0.457	0.205	0.113
BBR-SAFEM	0.487	0.175	0.074	0.047	1.121	0.462	0.205	0.113
Error (%)	4.52	1.71	0	0	2.59	1.08	0	0

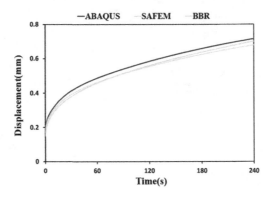

Figure 4. Comparison of mid-span displacements between three results.

SAFEM has 10 Fourier series terms, while the 3D FEM has 40 grids. The calculation results were shown in Figure 4. Besides, the computing time of ABAQUS was about 860s, while the SAFEM about 50s.

4   CONCLUSION

This study has proposed the SAFEM for viscoelastic analysis in detail, and a program for simulating the BBR test has been developed in MATLAB. By comparing the test results with the program, the accuracy of this method was verified. Besides, the efficiency of SAFEM was indicated by comparing it with 3D FEM in ABAQUS. More importantly, the implementation of GGM and the iteration approach have some reference value for the further application of SAFEM in structural engineering.

REFERENCES

Hajikarimi, P., Moghadas Nejad, F., Mohammadi Aghdam, M., 2018. J. Transp. Eng. Part B Pavements 144, 04018010.
Kalogiratou, Z., Simos, T.E., 2003. J. Comput. Appl. Math. 158, 75–82.
Liu, P., Wang, D., Hu, J., Oeser, M., 2017. J. Test. Eval. 45, 20150456.
Liu, P., Wang, D., Otto, F., Hu, J., Oeser, M., 2018a. Int. J. Pavement Eng. 19, 479–488.
Liu, P., Xing, Q., Wang, D., Oeser, M., 2018b. Adv. Civ. Eng. 2018, 1–15.
Pingaro, M., Venini, P., 2016. Comput. Struct. 168, 46–55.
Zeinkiewicz, O., Too, J., 1972. Proc. Inst. Civ. Eng. 53, 147–172.

Pavement construction and maintenance

Study on control index of steel deck pavement for medium-small span bridges in seasonal frozen area

L.D. Zhang, Y.T. Zhang & Y.F. Cao
Jilin Provincial High Class Highway Construction Bureau, Changchun, Jilin, China

J.Y. Zong* & K.Q. Zhang
JSTI GROUP & National Engineering Laboratory for Advanced Road Materials, Nanjing, Jiangsu, China

ABSTRACT: In order to study the control index of steel deck pavement for medium-small span bridges in seasonal frozen area, mechanical characteristics of pavement under different working conditions (different longitudinal and transverse load arrangement) were simulated by using ANSYS software. The influence of load (overload), automotive braking force and temperatures on steel deck pavement mechanical responses was studied. The results demonstrate that overload has great effect on pavement performance, and the shear strength of interlayer bonding is the key control index. The shear stress increases gradually with the increase of the braking force, and the shear strength of interface should be considered seriously as well. The pavement surface tensile stress should be taken as the control index is due to the pavement surface tensile stress is larger at low temperature conditions (-40°C~0°C), and the pavement surface vertical displacement should be taken as the control index is due to the pavement surface vertical displacement is larger at high temperature conditions (40°C~60°C).

1 INTRODUCTION

In recent years, with the rapid development of China's transportation infrastructure construction and the rapid development of bridge construction, steel bridges have shown many advantages, such as having a low dead weight, excellent seismic performance, fast construction, novel bridge types and the ability to meet the requirements of sustainable development. Steel bridges have been widely used for bridges that are already built or under construction in China. Domestic scholars have also conducted extensive and in-depth research on steel bridge deck pavement and have achieved abundant results (Huang et al. 2002; Qing et al. 2015; Wang et al. 2015; Chen et al. 2017; Luo et al. 2019; Tang et al. 2018). However, at present, domestic research on steel deck pavement mostly focuses on long-span bridges in warm or hot climate areas. Research on the steel deck pavement of medium- and small-span bridges in seasonally frozen areas has not received sufficient attention, and relevant research results are rarely reported. Therefore, this paper studies the control index of steel deck pavement for medium-small span bridges in seasonal freezing zone.

2 ESTABLISHMENT MODEL

2.1 Basic requirements

(1) It is assumed that the pavement materials and steel box girder bridge panels are completely elastic, uniform, continuous and isotropic linear elastic materials.
(2) All the interfaces between the upper and lower layers of the pavement and the pavement and steel bridge deck are in full contact.
(3) The shear studs on steel bridge deck are not considered separately.

2.2 Model

A finite element calculation and analysis model (Figure 1) is established in five U-shaped stiffeners in the transverse direction and three spans in the longitudinal direction of a steel bridge deck pavement, which solid unit SOLID 45 is used for simulation of the pavement layers and shell unit SHELL63 is used for the simulation of steel bridge panel, U-shaped stiffeners and diaphragms. The finite element model sizes and parameters are detailed in Table 1 and 2 respectively.

2.3 Load and boundary conditions

The two wheels are simplified as a single wheel load, i.e., the two wheels on each side are converted into a single wheel weight of 70 kN and considering an impact coefficient which refers to the increase coefficient of the vertical dynamic effect of the moving vehicle load on the bridge structure of 30%, the total weight of the single wheel is 91 kN. The calculation loads and boundary conditions are shown in Table 3 and 4 respectively.

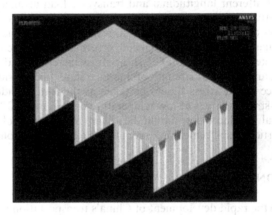

Figure 1. Finite element model.

Table 1. Finite element model sizes.

thickness /mm	U-shaped stiffener /mm					Diaphragm plate /mm		Upper pavement thickness /mm	Lower pavement thickness /mm
	Upper	Lower	Height	Spacing	Thickness	Thickness	Spacing		
20	300	170	280	450	8	14	1500	100	80

Table 2. Finite element model parameters.

Structural position	Elastic modulus /MPa	Poisson's ratio
Steel bridge deck plate	$2.1*10^5$	0.3
Pavement upper layer	$1.5*10^3$	0.25
Pavement lower layer	$3.5*10^4$	0.25

Table 3. Calculation loads.

Load type	Load level	Vertical load /MPa	Longitudinal bridge width range /mm	Transverse bridge width range /mm
Single wheel load	Highway Class I	0.758	200	600

Table 4. Boundary conditions.

Positions	Boundary conditions
Diaphragm/Longitudinal clapboard	Bottom complete restraint
The edge of transverse bridge	No transverse horizontal displacement
The edge of longitudinal bridge	No longitudinal horizontal displacement

3 INFLUENCE FACTOR ANALYSIS

3.1 Loading conditions

Based on the current overloading situation in China, this paper sets the load conditions as a normal load, an overload of 50%, an overload of 75%, an overload of 100% and an overload of 120%, respectively, to analyze the shear stress changes at different interfaces of the pavement under the corresponding loads.

As shown in Table 5, the shear stress between the pavement layers and between the pavement layer and steel bridge deck have a good linear correlation with the traffic load. In the case of overload, the adhesive layer in the pavement structure will face a severe test. Therefore, in this paving scheme, the shear strength of the bonding layer between the paving layers is the key control index. If the material design of the bonding layer is unreasonable, shear slip damage easily occurs between the pavement layers.

3.2 Vehicle braking force

When a car travels on a medium-and small-span steel bridge with a large longitudinal slope, such as a ramp, it is not only affected by the vertical stress perpendicular to the pavement but also by the horizontal stress parallel to the pavement due to the braking of the car, especially emergency braking under heavy loading, which will produce a large shear stress on the pavement (Liu et al. 2016; Liu et al. 2017). In this paper, the friction coefficients are set to $f = 0$, $f = 0.25$, $f = 0.5$, and $f=0.75$ to simulate different braking conditions.

From Table 6, it can be seen that with the increase in the braking force, each shear stress index gradually increases, with the increase in the shear stress on the pavement surface and between the pavement layers being the most obvious. Therefore, in the design of steel bridge

Table 5. Stress values of pavement under different load conditions.

Load	Maximum lateral tensile stress on asphalt pavement/MPa	Maximum longitudinal tensile stress on asphalt pavement/MPa	Maximum shear stress between pavement layers/MPa	Maximum shear stress between pavement and steel bridge deck/MPa
Regular load	0.5406	0.6797	0.2358	0.1280
Overload 50%	0.8110	1.02	0.3546	0.1925
Overload 75%	0.9461	1.19	0.4140	0.2247
Overload 100%	1.08	1.36	0.4734	0.2569
Overload 120%	1.19	1.50	0.5210	0.2827

Table 6. Shear stress values of pavements with different friction coefficients.

The friction coefficient	Maximum shear stress of pavement surface /MPa	Maximum shear stress between pavement layers /MPa	Maximum shear stress between pavement and steel bridge deck /MPa
0	0.0910	0.1097	0.2151
0.25	0.1934	0.1956	0.2259
0.5	0.3134	0.2974	0.2752
0.75	0.4333	0.3992	0.3244

deck pavement, the requirements for emergency braking of heavy-duty vehicles on the shear resistance of the pavement layer should be fully considered to prolong the service life of the steel bridge deck pavement.

3.3 Temperatures

Due to the lack of ventilation measures inside the box girder and slow heat dissipation speed, the temperature of the deck of the box girder bridge is 25°C ~ 35°C higher than the air temperature, and the extreme high temperature is close to 60°C in the hot summer season. In the low temperature season, the extreme service temperature of steel bridge deck pavement can be lower than -30°C. According to the climate characteristics of seasonally frozen areas in our country, this study sets the temperature range of the pavement layer to -40°C~60°C as shown in Table 7. Referring to the bridge deck pavement temperature calculation model which proposed by Ji Xiaoping and Zheng Nanxiang of Chang' an University (Ji et al. 2014) and the asphalt mixture modulus of resilience prediction model at different temperatures which proposed by Jia Lu of Tongji University (Jia et al.2008), the elastic modulus of asphalt mixture at different levels is calculated and determined which combination with the material modulus of resilience at 20°C.

From the above test results in Table 8, it is found that under low temperatures(-40°C~0°C), the elastic modulus of the asphalt pavement layer is larger, and the vertical displacement of the pavement layer surface is smaller, but the tensile stress of the pavement layer surface is

Table 7. Elastic modulus of asphalt pavements under different temperatures (Yin et al. 2018).

Temperature of pavement surface	-40 °C	-20 °C	0 °C	20 °C	40 °C	60 °C
Elastic modulus/MPa	56000	14000	4000	1500	600	300

Table 8. Results of the maximum stress, strain and vertical displacement of pavements under different temperatures.

Temperatures	D_{max-v} [a] /μm	$Stress_{max-t}$ [b] /MPa		$Shear_{max-p-s}$ [c] /MPa	$Shear_{max-b-p}$ [d] /MPa	$Shear_{max-b-d}$ [e] /MPa
		T	L			
-40°C	46.9	1.76	2.31	0.2807	0.2846	0.2580
-20°C	71.4	1.1348	1.4301	0.2415	0.2490	0.2626
0°C	104	0.7369	0.9113	0.2242	0.2256	0.2435
20°C	139	0.5407	0.6799	0.2180	0.2170	0.2377
40°C	204	0.4448	0.5905	0.2153	0.2130	0.2346
60°C	304	0.4084	0.5604	0.2143	0.2116	0.2335

[a] D_{max-v} : maximum vertical displacement of pavement surface;
[b] $Stress_{max-t}$: maximum tensile stress of pavement surface;
[c] $Shear_{max-p-s}$: maximum shear stress of upper pavement surface
[d] $Shear_{max-b-p}$: maximum shear stress between pavements
[e] $Shear_{max-b-d}$: maximum shear stress between pavement and bridge deck

larger at this time. Therefore, under low temperature conditions, pavement material design should take the tensile stress on the pavement surface as the control index to prevent crack defects caused by the insufficient low temperature crack resistance of the pavement. In contrast, under high temperature conditions (40°C~60°C), the elastic modulus of asphalt pavement is small, and the vertical displacement of the pavement is large. Therefore, under this temperature condition, the vertical displacement of the pavement surface should be taken as the control index, and the deformation defects of the pavement materials should be prevented.

4 CONCLUSION

(1) The results show that the overload condition has a significant effect on the pavement performance, and the shear stress between the pavement layers increases linearly with increasing vehicle load.
(2) With the increase in the braking force, each shear stress index gradually increases, of which the shear stress increases most obviously on the pavement surface and between the pavement layers.
(3) Under low temperature conditions (-40°C~0°C), the tensile stress on the surface of the pavement is relatively large, so the pavement material design should take the tensile stress on the surface of the pavement as the control index, while under high temperature conditions (40°C~60°C), the vertical displacement of the pavement is relatively large, and under this temperature condition, the vertical displacement of the pavement surface should be taken as the control index.

REFERENCES

Chen XH, Huang W, Qian ZD, Zhang L. 2017. "Design principle of deck pavements for long-span steel bridges with heavy-duty traffic in China." *Road Materials and Pavement Design* 18: 226–239 DOI: 10.1080/14680629.2017.1329877.

Huang W, Zhang X C, Hu G W.2002. "New advance of theory and design on pavement for long-span steel bridge." *Journal of Southeast University (Natural Science Edition)* 32(3): 480–487 (in Chinese).

Ji X P, Zheng N X. 2014. "Prediction model and characteristics of temperature field on bridge deck asphalt pavement." *Journal of Chang' an University (Natural Science Edition)* 34(3): 60–65 (in Chinese).

Jia L, Sun L J. 2008. "A Numerical Temperature Prediction Model for Asphalt Concrete Pavement." Journal of tongji university (natural science) 35(8):1039–1043 (in Chinese).

Liu X, Feng DC, Tang H, Zhou CJ, Li JJ.2016. "Investigation on the Bending Fatigue and Shear Failure in Steel Bridge Deck Pavement Systems." *Journal of Testing and Evaluation* 44(2): 895–906 DOI: 10.1520/JTE20150236.

Liu Y, Zhang HP, Liu YM, Deng Y, Jiang N, Lu NW.2017. "Fatigue reliability assessment for orthotropic steel deck details under traffic flow and temperature loading." *Engineering Failure Analysis* 71: 179–194 DOI: 10.1016/j.engfailanal.2016.11.007.

Luo S, Liu ZM, Yang X, Lu Q, Yin J.2019. "Construction Technology of Warm and Hot Mix Epoxy Asphalt Paving for Long-Span Steel Bridge" Journal of Construction Engineering and Management 145 (12): 12–25 DOI: 10.1061/(ASCE)CO.1943-7862.0001716.

Qing L, John B.2015. "Alternate uses of epoxy asphalt on bridge decks and roadways." *Construction and Building Materials* 78:18–25 DOI: 10.1016/j.conbuildmat.2014.12.125.

Tang T, Zha XD, Xiao QM, Chen YQ.2018. "Laboratory Characterization and Field Validation of ROADMESH-Reinforced Asphalt Pavement in China" International Journal of Civil Engineering 16(3): 299–313 DOI: 10.1007/s40999-016-0128-9.

Wang HC, Li GF.2015. "Study of factors influencing gussasphalt mixture performance." *Construction and Building Materials* 101: 193–200 DOI: 10.1016/j.conbuildmat.2015.10.082.

Yin YM, Huang WK, Lv J B, Ma X, Yan JH.2018. "Unified Construction of Dynamic Rheological Master Curve of Asphalts and Asphalt Mixtures."International Journal of Civil Engineering 16(9A): 1057–1067 DOI: 10.1007/s40999-017-0256-x.

Functional Pavements – Chen et al (eds)
© 2021 Taylor & Francis Group, London, ISBN 978-0-367-72610-2

Evaluation of highway intelligent site construction based on entropy-TOPSIS method

Jianping Huang
JSTI Group, Nanjing, Jiangsu, China

Jizuo Zeng
Shandong Luqiao Group Co. ltd, Jinan, Shandong, China

Chunying Wu
JSTI Group, Nanjing, Jiangsu, China

Kun Zhang
Rizhao Transportation Development Group, Rizhao, Shandong, China

ABSTRACT: In response to the rise of highway intelligent site construction, this paper uses the TOPSIS-based evaluation method to comprehensively evaluate the implementation quality of intelligent site, including engineering, personnel, safety, materials, quality, progress and environmental information data, so as to determine the integrity of the intelligent implementation of the project together with quality performance, in the form of ranking the pros and cons of ongoing or completed intelligent site projects, which plays a vital role in regulating the development and promotion of highway intelligent site construction.

1 INTRODUCTION

In recent years, the development of intelligent site construction is booming. Many enterprises have increased their efforts in exploration and practice, and have launched comprehensive intelligent site systems. However, at this stage, there is still a lack of unified and clear evaluation method for intelligent site construction. This paper determines the performance of ongoing or completed intelligent site projects by collecting relevant data, experts scoring, TOPSIS-based method to judge its pros and cons (Zou, 2006).

2 EVALUATION PRINCIPLE OF INTELLIGENT SITE CONSTRUCTION

Technique for Order Preference by Similarity to Ideal Solution (TOPSIS) is a commonly used comprehensive evaluation method, which can make full use of the original data information, and the result can accurately reflect the gap between the various evaluation schemes(Gu, 2015). In this paper, 11 steps are taken to evaluate the construction level of different intelligent sites.

Step 1: making the form used for expert scoring, together with instructions.

Step 2: Data standardization. According to the principle of combining qualitative and quantitative, an evaluation matrix of multiple objects on multiple indicators R is obtained:

Step 3: Define H and d,

$$H_i = -k \sum_{j=1}^{n} f_{ij} \ln f_{ij} \quad i = 1, 2, \ldots, m \tag{1}$$

Where,

$$f_{ij} = r_{ij} / \sum_{j=1}^{n} r_{ij} \tag{2}$$

$$k = \frac{1}{\ln n} \tag{3}$$

And, $d_i = 1 - H_i \tag{4}$

Step 4: Determine the weight of evaluation indicator.

$$\omega_i = d_i / \sum_{i=1}^{m} d_i i = 1, 2, \ldots, m \tag{5}$$

Step 5: Construct the decision matrix R and normalize it, as follows:

$$r_{ij} = d_{ij} / \sqrt{\sum_{j=1}^{m} d_{ij}^2} (1 \leq i \leq m, 1 \leq j \leq n) \tag{6}$$

Step 6: Calculate the weighted matrix:

$$X = [\omega_j r_{ij}]_{m \times n} \tag{7}$$

Step 7: Determine the ideal solution and negative ideal solution, as follows:

$$F^* = \left(\left(\max_i \omega_j r_{ij} | j \in J \right) or \left(\min_i \omega_j r_{ij} | j \in J^0 \right) \right)^T \tag{8}$$

$$F^0 = \left(\left(\min_i \omega_j r_{ij} | j \in J \right) or \left(\max_i \omega_j r_{ij} | j \in J^0 \right) \right)^T \tag{9}$$

Step 8: Calculate the distance from each solution to the ideal solution:

$$L_2^* = \sqrt{\sum_{j=1}^{n} \left(\omega_j r_{ij} - f_j^* \right)^2}, 1 \leq i \leq m \tag{10}$$

Step 9: Calculate the distance from each solution to the negative ideal solution:

$$L_2^0 = \sqrt{\sum_{j=1}^{m} \left(\omega_j r_{ij} - f_j^0 \right)^2}, 1 \leq i \leq m \tag{11}$$

Step 10: Calculate the relative proximity of each solution to the ideal solution:

$$l_i = \frac{L_2^0(i)}{[L_2^*(i) + L_2^0(i)]}, 1 \leq i \leq m \tag{12}$$

Step 11: According to the relative proximity l_i, the pros and cons of each solution are ranked.

233

3 EVALUATION EXAMPLE FOR INTELLIGENT SITE CONSTRUCTION

3.1 Data collection

In this paper, 7 completed highway intelligent sites in Jiangsu province, are comprehensively compared and ranked, they are A_1, A_2, A_3, A_4, A_5, A_6, A_7, respectively. Experts are invited to the sites to inspect the quality performance.

And, there are 7 indicators for each intelligent site, they are engineering information data z_1, personnel information data z_2, safety information data z_3, material information data z_4, quality information data z_5, progress information data z_6, and environmental information data z_7. Based on the scoring criteria, they give scores to all indicators and fill them in the corresponding score form. Then take the average value of each expert's score and summarize them, we obtain the scores of each indicator of 7 intelligent sites, as shown in Table 1.

3.2 Evaluating process

3.2.1 Standardization of indicators
The value of the indicators in Table 1 is standardized, with the results shown in Table 2.

3.2.2 Calculating the weights of the indicators
Using formulas (1)(2)(3)(4)(5), the weights of the indicators are obtained, with the results shown in Table 3.

3.2.3 Comprehensive calculation using TOPSIS
Combined with the weights already determined above, a weighted normalization matrix can be constructed. According to formulas (6)(7)(8)(9), the ideal solution F^* and the negative ideal solution F^0 of each indicator are determined. The specific values are shown in Table 4.

Therefore, using formulas (10) and (11) to calculate the distance between the weighted value of each indicator of the above 7intelligent site projects and the ideal solution and the negative ideal solution, the distance L_2^* (i) is obtained from the ideal solution, while the distance L_2^0 (i) from the negative ideal solution, and finally calculate the relative proximity of the indicator l_i according to the formula (12) and rank the smart construction site items according to l_i. The specific values are shown in Table 5.

As can be seen from Table 5, by calculating the minimum and maximum distance between each project and the ideal solution as well as the negative ideal solution, the close value is obtained. According to the size of the close value, a comprehensive evaluation ranking is made, as A4>A1>A2>A7>A6>A5>A3, as shown in Figure 1.

Table 1. The average indicator scores of intelligent sites.

indicator \ Project	A_1	A_2	A_3	A_4	A_5	A_6	A_7
z_1	83	92	88	75	96	80	84
z_2	91	82	89	92	81	87	80
z_3	77	89	95	83	67	75	86
z_4	80	66	76	67	70	86	74
z_5	85	74	70	83	85	82	89
z_6	77	82	73	67	91	84	80
z_7	92	84	73	87	79	88	75

Table 2. The results after standardizing the value of the 7 indicators.

Project indicator	A_1	A_2	A_3	A_4	A_5	A_6	A_7
z_1	0.381	0.809	0.619	0.000	1.000	0.238	0.429
z_2	0.917	0.167	0.750	1.000	0.083	0.583	0.000
z_3	0.357	0.786	1.000	0.571	0.000	0.286	0.679
z_4	0.700	0.000	0.500	0.050	0.200	1.000	0.400
z_5	0.789	0.211	0.000	0.684	0.789	0.612	1.000
z_6	0.417	0.625	0.250	0.000	1.000	0.708	0.542
z_7	1.000	0.579	0.000	0.737	0.316	0.789	0.105

Table 3. The results after standardizing the value of the indicators.

Project indicator	$\Sigma f_{ij}\ln f_{ij}$	$-k$	H_i	d_i	ω_i
z_1	-1.689	-0.514	0.868	0.132	0.126
z_2	-1.571	-0.514	0.808	0.192	0.184
z_3	-1.710	-0.514	0.879	0.121	0.116
z_4	-1.551	-0.514	0.797	0.203	0.194
z_5	-1.716	-0.514	0.882	0.118	0.113
z_6	-1.711	-0.514	0.879	0.121	0.115
z_7	-1.637	-0.514	0.841	0.159	0.152

Table 4. The weighted normalization value and ideal solution of the indicators.

Project indicator	A_1	A_2	A_3	A_4	A_5	A_6	A_7	F^*	F^0
z_1	0.048	0.149	0.072	0.000	0.113	0.027	0.065	0.149	0
z_2	0.116	0.031	0.087	0.194	0.009	0.067	0.000	0.194	0
z_3	0.045	0.145	0.116	0.111	0.000	0.033	0.103	0.145	0
z_4	0.088	0.000	0.058	0.010	0.023	0.115	0.061	0.115	0
z_5	0.099	0.039	0.000	0.133	0.089	0.070	0.152	0.152	0
z_6	0.053	0.115	0.029	0.000	0.113	0.081	0.082	0.115	0
z_7	0.126	0.107	0.000	0.143	0.036	0.091	0.016	0.143	0

Table 5. The weighted normalization value and ideal solution of the indicators.

Project indicator	$L_2^*(i)$	$L_2^0(i)$	l_i	Rank
A_1	0.184	0.232	0.558	2
A_2	0.232	0.265	0.533	3
A_3	0.269	0.174	0.393	7
A_4	0.219	0.297	0.576	1
A_5	0.284	0.188	0.398	6
A_6	0.233	0.198	0.459	5
A_7	0.258	0.221	0.461	4

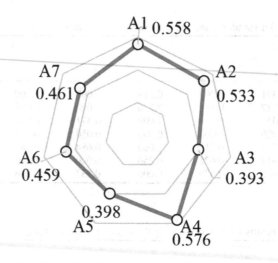

Figure 1.   The ranking spider chart of intelligent sites.

4   CONCLUSION

This paper establishes evaluation standards for intelligent site construction by using TOPSIS evaluation method based on weight to conduct comprehensive analysis of intelligent site projects. The ranking of them can be more comprehensively ranked according to the size of the close value, so as to determine the integrity of the intelligent implementation of the project together with quality performance. However, due to the primary data coming from experts scoring, the evaluation result may have some subjective effect. Therefore, the following research will concentrate on improving the result assessment as objective and fair as possible.

ACKNOWLEDGMENT

Funding: This research is funded by Natural Science Foundation of Jiangsu Province, grant number BK20170156.

REFERENCES

Zou, Z.H. & Yi, Y. & Sun, J.N.2006. Entropy method for determination of weight of evaluating indicators in fuzzy synthetic evaluation for water quality assessment. *Journal of Environmental Sciences* 18(05): 1020–1023.
Gu, Z.Y. & Qiu, J.A.2015. Evaluation model for NBA coaches based on entropy-TOPSIS method. *Journal of Nanjing Sport Institute*(Natural Science) 14(5): 152–157.

Functional Pavements – Chen et al (eds)
© 2021 Taylor & Francis Group, London, ISBN 978-0-367-72610-2

Application of intelligent control technology in asphalt pavement construction for N-M expressway

Peng Zhang
JSTI Group, Nanjing, Jiangsu, China

Xiaoning Chen & Shilei Xu
Nanjing Highway Development Center, Jiangsu, China

Chunying Wu
JSTI Group, Nanjing, Jiangsu, China

ABSTRACT: In Nanjing-Maanshan (N-M) expressway project, intelligent control technology was developed and used in asphalt pavement construction, through which, mixing, transportation, paving and compacting data, were collected and transmitted separately to data center, so as to make statistical analysis and intelligent feedback. Due to the realization of real-time collecting construction quality data, rapid evaluation for the key indicators such as asphalt aggregate ratio, transporting time, paving temperature and compacting times etc. were done. Results showed that developing the intelligent control technology for asphalt pavement achieved the whole construction process quality inspection and data monitoring, so as to effectively avoid the unstable factors affecting construction quality, guaranteeing the certified rate for expressway construction.

1 INTRODUCTION

With the continuous increase of expressway mileage in Jiangsu provinces, the demand for quality and service level rises (Liu, 2014). As we all know, asphalt pavement construction is an irreversible process. From asphalt mixture mixing and transportation, to pavement paving and compacting, problems in any link of pavement construction will lead to loss of the final quality(Wang, 2017). Compared to the traditional expressway pavement construction management methods such as "pre-check", "post-test", "sampling test", "limited feedback", there is no effective countermeasures to quality problems that are fait accompli, and the test results are also lacking in comprehensiveness (Wei, 2017).

Against the shortcomings of traditional methods, in N-M expressway project, this paper develops intelligent control technology(ICT), which has the functions of real-time, continuous data collection, transmission, statistical analysis and information feedback on the whole process of asphalt mixture mixing, transportation, paving and compacting. ICT improves "pre-inspection and post-test" into construction process control; improves random sampling of inspection into overall inspection of pavement quality; improves limited feedback during construction into real-time feedback(Ma, 2018).

2 WORKING PRINCIPLE OF ICT

Based on internet of things and cloud calculation, asphalt pavement intelligent construction combines data checking, transferring, analysis and feedback as a whole, to realize the real-time control on the overall process of asphalt pavement mixing, transporting, paving and

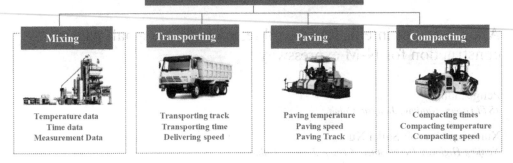

Figure 1. Working principle of ICT.

compacting, as shown in Figure 1. ICT is mainly used for guiding asphalt pavement construction real time with warning when construction data collected beyond the required limitation. In this case, pavement quality is guaranteed immediately, so as to reduce the cost for reconstruction later when checked out by labor. Considering the situation that data package is lost or missed when transferring, the abnormal data will be treated through median filter algorithm.

2.1 ICT for asphalt mixture mixing

In this technology, temperature sensors and data collection units are deployed in mix plant to get the grading data, asphalt content, admixture quality, time data and temperature data. These data are real-time transmitted to data center. After calculation and analysis, ICT makes intelligent feedbacks. The goal is to guarantee the mix proportion within the allowed range from the lab.

2.2 ICT for asphalt mixture transporting

By use of RFID, the loading time, mixture information, driver information and vehicle information are obtained; during delivering, GPS is used to get vehicle position and driving speed; when offloading the mixture, by use of RFID installed on pavers, the offloading time and other information are obtained. The goal is to guarantee the mixture temperature after arrived the site.

2.3 ICT for asphalt pavement paving

Infrared temperature sensors are used to get the temperature data of the whole paving section. The distance sensors are used to get the speed of pavers. Paving temperature and speed are the key indicators to be collected and analyzed, so as to reduce mixture segregation real time.

2.4 ICT for asphalt pavement compacting

By using the mobile station and base station, we can make rollers position to centimeter level; compacting temperature can be obtained by infrared sensors. Compacting times, compacting temperature and compacting speed are the collected indicators at construction site, which will be shown in LED and PAD real time, so as to guide the operator. And also, the information can be real-time sent to data center to issue and query.

3 APPLICATION OF ICT IN N-M EXPRESSWAY PROJECT

3.1 *ICT for asphalt mixture mixing*

3.1.1 *Real-time monitoring on production conditions*
In this paper, the grading of Sup-19 high-performance asphalt mixture is matched and collected. And the collected indicators include the measurement data of each hot silo, mineral powder, asphalt aggregate ratio, mixing cycle, etc., as shown in Table 1.

3.1.2 *Fluctuation analysis on production data*
The fluctuation indicators include the measurement data of each silo, the key sieve passing rate, the measurement data of the mineral powder, the asphalt aggregate ratio, the mixing cycle, and the mixture discharged temperature. Figure 2 is the hot silo (0~4mm) fluctuation figure.

It can be seen from the figure that the dosage of the hot silo (0~4mm) is extremely stable, which has always been near the target value, and has not exceeded the set upper and lower limits, so that the production quality is more reliable.

3.2 *ICT for asphalt mixture transporting*

The asphalt mixing plant is about 30 kilometers away from the construction field of N-M expressway, and the driving time of the mixed material is about 1 hour. Through ICT, the vehicle information, mixing information, transportation information and paving information are displayed in a unified manner to achieve traceability of asphalt pavement quality problems. In this project, the average transport time is 57.8 min, the overtime percentage is 9.1%, and the overall duration of the transport of the mixture is well controlled.

Table 1. The real-time monitoring of material in N-M expressway mixing plant.

Hot silo	Actual weight (t)	Actual percentage (%)	Production ratio (%)	deviation (%)
1# 0-4mm	378.12	35	35	0
2# 4-6mm	150.47	13.9	14	0.1
3# 6-11mm	303.15	28.1	28	0.1
4# 11-22mm	204.18	18.9	19	0.1
Mineral powder	43.24	4	4	0
Asphalt	45.39	4.2	4.4	0.2

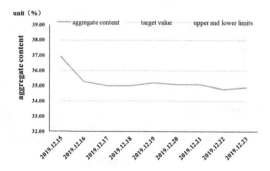

Figure 2. Hot silo (0~4mm) fluctuation figure.

3.3 ICT for asphalt pavement paving

In the construction process of N-M expressway, the mode of joint operation of 1 paver and 3 rollers is adopted. Figure 3-4 are data analysis of paving temperature and paving speed, respectively.

It can be seen from Figure 3, that the paving temperature curve fluctuates greatly, but the temperature control is better, almost all are distributed within the upper and lower limits, indicating that the paving temperature control meets the requirements of the construction specifications, but the uniformity of the paving temperature need to be further strengthened.

And it can be seen from Figure 4 that the distribution of paving speed collected in real time is relatively uniform, with an average paving speed of 2.98m/min, which meets the requirements.

3.4 ICT for asphalt pavement compacting

In the compacting process, a double-drums roller (initial compact), 1 rubber-tyre roller (re-compact), and 1 double-drums roller (final compact) combined operation mode is adopted.

Figures 5-6 show the initial compacting temperature following chainages, and data analysis of compacting speed, respectively.

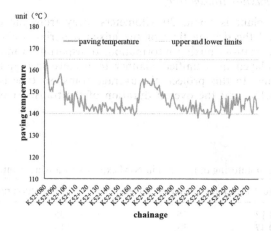

Figure 3. Data analysis of paving temperature.

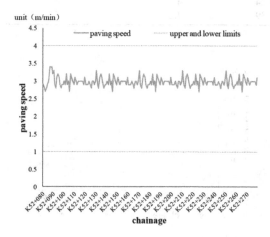

Figure 4. Data analysis of paving speed.

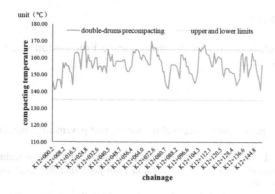

Figure 5. Initial compacting temperature following chainages.

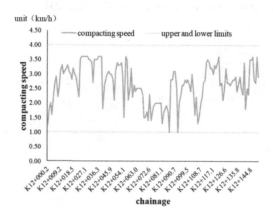

Figure 6. Data analysis of compacting speed.

It can be seen from Figure 5 that the average initial compacting temperature is 156.08°C, and the curve fluctuates significantly. There are even several chainages where temperature is higher than the set upper limit, but most of them meet the construction temperature requirements.

It can be seen from Figure 6 that the average value of the compacting speed is 2.66km/h, and the curve fluctuation is very obvious. Although the speed values are distributed within the set upper and lower limits, speed stability adjustments are urgent to make as soon as possible, to ensure the compaction quality of the asphalt pavement.

4 CONCLUSION

The application of ICT in N-M expressway project, realizes the dynamic control of the construction process, provide an analysis and early warning mechanism, timely analyze quality problems, find quality fluctuations, and ensure the achievement of project quality goals by real time monitoring and collecting the key data information. ICT for the entire process of the asphalt pavement can reduce the rework, improve the construction quality and efficiency, save the initial construction cost and the corresponding maintenance cost in the later period, and has direct economic and social benefits. However, due to the good internet condition, in this paper, abnormal data process is not discussed. The system reliability and robust analysis will be discussed in future work.

ACKNOWLEDGMENT

Funding: This research is funded by Natural Science Foundation of Jiangsu Province, grant number BK20170156.

REFERENCES

Liu, X. 2014. Intelligent quality control technology for asphalt pavement compaction. *Road Machinery & Construction Mechanization* 31(10):46–50.
Wei, W.J. & Zhang, L. L. & Han, C.2017. Research on intelligent control technology in asphalt pavement construction process. *Journal of Highway and Transportation Research and Development* (application edition) 13(09):83–85.
Wang, X. & Tang, J. Y. & Xing, Y.Z.2017. Application of intelligent monitoring system in asphalt pavement construction based on internet of things. *Technology of Highway and Transport* 33(03):23–28.
Ma, D.L. 2018. Application of intelligent monitoring system in asphalt pavement construction quality control based on internet of things. *Value engineering* 37(14):232–234.

Functional Pavements – Chen et al (eds)
© 2021 Taylor & Francis Group, London, ISBN 978-0-367-72610-2

Research on EAC maintenance strategy based on mould and comprehensive evaluation

M. Zhang, K. Zhong & M. Sun
Research Institute of Highway Ministry of Transport, Beijing, China
Key Laboratory of Transport Industry of Road Structure and Material, Beijing, China

Z. Qian
Southeast University, Beijing, China

H. Wang
Delft University of Technology, Delft, The Netherlands

T. Gan
Chongqing Jiaotong University, Chongqing, China

ABSTRACT: In order to study the maintenance strategy of Epoxy Asphalt Concrete (EAC) in low temperature environment, the similarity of four anti-cracking performance indexes (plane strain fracture toughness, bending stiffness modulus, ultimate tensile stress and ultimate elongation linear strain) of EAC were analyzed, and the mathematical model for investigating the relationship between the number of freeze-thaw cycles and the comprehensive evaluation index of the crack resistance of epoxy asphalt mixture (EAM) was established in this paper. At first, the weight of each index is determined by one-way ANOVA. Then the anti-cracking performance of EAC under freezing-thawing condition is evaluated by fuzzy comprehensive evaluation method. The results show that the pavement maintenance bureau needs to carry out the maintenance of epoxy asphalt pavement on steel bridge deck every 30 days (1 month) for ensuring the crack resistance of EAC in winter with low temperature.

1 INTRODUCTION

Under the condition of low temperature in winter, EAC is easily affected by the freezing-thawing cycle, it exacerbates the attenuation of anti-crack performance of pavement, which will lead to cracks (Islam et al. 2014). Without timely maintenance of cracks, it will easily lead to delamination of pavement layer, corrosion of steel deck and the safety problem of bridge structure. However, the existing research was mainly focused on the construction, workability and mechanical properties of Epoxy Asphalt Mixture (Ji et al. 2017). At present, the damage caused by cracks in EAC in low temperature environment has not been solved, so it is of great significance to study the maintenance measures of EAC with cracks.

The analytic hierarchy process (AHP) was adopted to establish the AHP model of the influence degree of the crack resistance of EAM in this paper. And the bending stiffness modulus, the fracture energy, the plane strain fracture toughness (K_{IC}), the ductile fracture toughness (J_{IC}), the ultimate tensile stress and the ultimate elongation linear strain were taken as the index of the plan layer and the comprehensive evaluation index of the anti-cracking performance of the EAM is put forward. At last, the anti-cracking performance of the epoxy EAM under different freeze-thaw cycles was comprehensively evaluated by means of mold and comprehensive evaluation method, the comprehensive evaluation results and maintenance strategy for epoxy asphalt pavement with cracks are provided.

2 SIMILARITY ANALYSIS OF EAC CRACK RESISTANCE EVALUATION

According to the theory of fracture mechanics, the three parameters of fracture energy, plane strain fracture toughness (K_{IC}) and ductile fracture toughness (J_{IC}) are similar (Portillo 2013). Therefore, the plane strain fracture toughness (K_{IC}) and the remaining 3 sub-item evaluation index were retained to carry on the next step similarity analysis, finally the mold and the cluster method was used to eliminates the remaining similarity index for guaranteeing the rationality of anti-crack performance evaluation. After the original data was standardized and the dimension was eliminated, the range variation data was obtained (Zhang 2019). The final result is shown in Table 1.

After the original data were standardized, the correlation coefficient method was used to establish the fuzzy similarity matrix between the four sub-evaluation indexes of the cracking resistance of EAM:

$$\underline{R} = \begin{bmatrix} 1 & 0.72 & 0.72 & 0.88 \\ 0.72 & 1 & 1 & 0.64 \\ 0.72 & 1 & 1 & 0.65 \\ 0.88 & 0.64 & 0.65 & 1 \end{bmatrix}$$

The corresponding parameters of row/column 1~4 are bending stiffness modulus (u_1), plane strain fracture toughness K_{IC} (u_2), ultimate tensile stress (u_3) and ultimate elongation linear strain (u_4). Meanwhile, the dynamic cluster graph of 4 sub-evaluation indexes of EAC anti-cracking performance was obtained, as shown in Figure 1.

In order to make the classification more clear, F test can be used to eliminate the unsuitable classes, determine the optimal threshold of the mold and clustering analysis. The final calculation is shown in Table 2.

According to the principle of statistical variance analysis, when F > F_α (r-1, n-r), the difference between classes is significant, which shows that the classification is relatively reasonable. Therefore, it can be seen from Table 2 that the classification is relatively reasonable when λ=1. The fracture energy, ductile fracture toughness (J_{IC}) and ultimate tensile stress in the original index layer B can be eliminated after optimization by F test, and a hierarchy model for evaluating the crack resistance of EAM can be obtained, as shown in Figure 2.

Table 1. Data after dimensional elimination of similarity index of EAC.

Itemized evaluation index	Number of freeze-thaw cycles						
	0	5	10	15	20	25	30
Bending stiffness modulus	1.00	0.77	0.68	0.64	0.56	0.30	0.00
Plane strain fracture toughness (K_{IC})	1.00	0.70	0.35	0.00	0.35	0.34	0.66
Ultimate tensile stress	1.00	0.69	0.35	0.00	0.29	0.38	0.69
Ultimate elongation linear strain	0.15	0.09	0.00	0.04	0.21	0.55	1.00

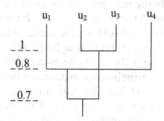

Figure 1. The dynamic clustering diagram of each sub-item evaluation index.

Table 2. F test table for fuzzy clustering of each sub-item evaluation index.

Number of categories	λ	F	$F_{0.1}(r-1, n-r)$
3	1	66.73	49.5
2	0.8	0.17	8.53

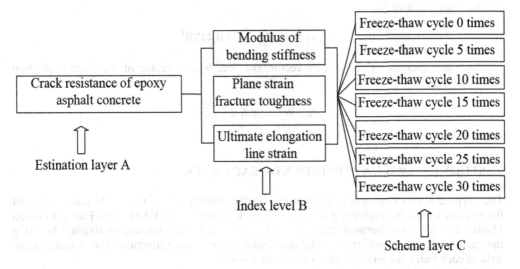

Figure 2. Hierarchical structure model for crack resistance evaluation of EAC after optimization.

The influence of freeze-thaw cycles on the crack resistance of EAM was analyzed by One Factor Analysis of Variance (ANOVA) (Nadir et al. 2014), and the results as shown in Figure 3 were obtained.

It is assumed that the weight relationship of the above three sub-evaluation indicators is inversely proportional to the level of significance of the impact of the project level on the indicators. Therefore, the following weight relations can be obtained initially:

$$A_0 = [0.54 \ 0.09 \ 0.37]$$

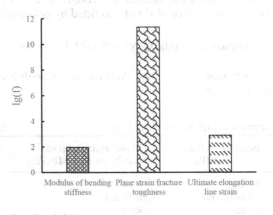

Figure 3. *Lg (f)* values of each sub-index.

At the same time, the following three evaluation indicators can be established to determine the sub-matrix:

$$A = \begin{bmatrix} 1 & 6 & 2 \\ 1/6 & 1 & 1/3 \\ 1/2 & 3 & 1 \end{bmatrix}$$

C=0 can be obtained from the formula, and the discernibility matrix is consistent. Then we get the eigenvector of A_1:

$$W_0 = [0.8847\ 0.1474\ 0.4423]^T$$

After normalization of the feature vector, the final weight vector of the three evaluation indexes is:

$$A = W^T = [0.6\ 0.1\ 0.3]$$

3 STUDY ON EVALUATION INDEX OF EAC CRACK RESISTANCE

The purpose of this section is to analyze the crack resistance of cracked EAM under different freeze-thaw cycles by evaluating indexes of crack resistance of EAM. The Fuzzy Relation Matrix $R = R=(r_{ij})_{3 \times 7}$ between the factor set and the judgment set was established by using the partial large-scale semi-trapezoidal distribution membership function. The evaluation criteria of each index are established, as shown in Table 3.

The following Matrix can be obtained:

$$R = \begin{bmatrix} 1 & 0.823 & 0.742 & 0.703 & 0.631 & 0.393 & 0.111 \\ 1.205 & 1 & 0.542 & 0.187 & 0.545 & 0.536 & 0.859 \\ 0.181 & 0.101 & 0 & 0.013 & 0.274 & 0.797 & 1 \end{bmatrix}$$

According to the final weight vector and evaluation matrix of the three evaluation indexes known above, the weighted average model of the fuzzy comprehensive evaluation model can be used to obtain the following evaluation results:

$$B = AR = [0.7748\ 0.6241\ 0.4994\ 0.4444\ 0.5153\ 0.5285\ 0.4525]$$

Finally, the relationship between the number of freeze-thaw cycles and the comprehensive evaluation index of the crack resistance of EAM was fitted by polynomial method:

$$EACACI = -0.00005\ N^3 + 0.0031\ N^2 - 0.0539\ N + 0.7895, \quad R^2 = 0.9517 \qquad (1)$$

When EACACI=Epoxy Asphalt Concrete Anti-fracture Capability Index; and N = the number of freeze-thaw cycles (times).

Table 3. Evaluation criteria for crack resistance of EAC.

Classification	Bending stiffness modulus/MPa	Plane strain fracture toughness K_{IC}/(MPa$\sqrt{m}$)	Ultimate elongation linear strain/‰
★★★	>12000	>0.4	>4.5
★★	4000-12000	0.20-0.4	2.5-4.5
★	<4000	<0.2	<2.5

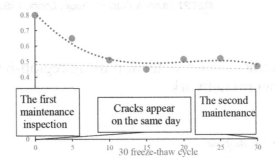

Figure 4.   Curing period.

Considered the frequency of epoxy asphalt pavement maintenance will increase manpower, equipment and other economic costs (Wang et al. 2015, Özgan et al. 2013), the pavement maintenance unit should maintain the steel bridge asphalt pavement every 30 days (1 month) to ensure the anti-cracking performance of epoxy concrete in winter with low temperature (Jian et al. 2011).The results are shown in Figure 4.

4   CONCLUSION

(1) With the increase of the number of freeze-thaw cycles, the comprehensive evaluation index of anti-cracking performance of EAM shows a decreasing trend. When the number of freeze-thaw cycles is less than 10 times, the decay rate is increases gradually; However, when the number of freeze-thaw cycles is more than 10 times and less than 30 times, the decay rate decreases gradually. The comprehensive evaluation index of anti-cracking property of EAM decreases by 37%.
(2) In order to ensure the anti-cracking performance of EAM in winter temperature condition, the pavement maintenance bureau needs to maintain the epoxy asphalt pavement of steel bridge deck every 30 days (1 month).

REFERENCES

Islam, M.R. & Tarefder, R.A.2014. Effects of large freeze thaw cycles on stiffness and tensile strength of asphalt concrete. *Journal of Cold Regions Engineering* 30(1): 06014006.
Jian, C. & Liu, J.J.& Wu, L. & Yang J. 2011. Studies review of the technology for snow and ice control for winter road maintenance. *ICTE2011-Proceedings of the 3rd International Conference on Transportation Engineering*: 3245–3254.
Ji, J. & Liu, L.H. & Suo, Z. & Zhang, Y.J. 2017. Influence of epoxy asphalt concrete anti-fatigue layer on structure of perpetual asphalt concrete pavement with flexible base. *Journal of Traffic and Transportation Engineering* 8: 1–8.
Nadir, O. F. E. & Othman, I. & Ahmed, H. O. 2014. A Novel Feature Selection Based on One-Way ANOVA F-Test for E-Mail Spam Classification. Research Journal of Applied Sciences. *Engineering and Technology* 7(3): 625–638.
Özgan, E. & Serin, S. 2013. Investigation of certain engineering characteristics of asphalt concrete exposed to freeze thaw cycles. *Cold Regions Science and Technology* 85: 131–136.
Portillo, O. & Cebon, D. 2013. Experimental and Numerical Investigation of Fracture Mechanics of Bitumen Beams. *Engineering fracture mechanics* 97(1): 281–296.
Wang, W.X. & Zhao, Q. & G, R.J. 2015. A Hybrid Approach Based on Grey Correlation Analysis and Fuzzy Comprehensive Judgment for Evaluating Service Quality of Passenger Train. *Advance in Mechanical Engineering* 7(1).
Zhang, M. & Qian, Z.D. & Huang, Q.B. 2019. Test and Evaluation for Effects of Freeze Thaw Cycles on Fracture Performance of Epoxy Asphalt ConcreteComposite Structure. *Journal of Testing and Evaluation* 47(1): 556–572.

Functional Pavements – Chen et al (eds)
© 2021 Taylor & Francis Group, London, ISBN 978-0-367-72610-2

Automated pavement crack detection using region-based convolutional neural network

Jingwei Liu
Monash University, Clayton, Australia

Xu Yang*
Chang'an University, Xi'an, China
Monash University, Clayton, Australia

Vincent C.S. Lee
Monash University, Clayton, Australia

ABSTRACT: Cracking is one of the main defects of pavement deterioration, which is caused by the environment and pressure. Automated pavement crack detection is being developed to obtain the number, classification and location of pavement cracks in images through object detection, which is a challenging task because of the irregular pattern and the noises on pavement. However, most of the previous methods are more related to object segmentation, which segments crack pixels from pavement background. As a result, an automated pavement crack detection method was proposed based on region-based convolutional neural network (R-CNN) like you only look once version three (YOLOv3) in this paper, whose network architecture was modified to fit the special features of pavement crack detection. A dataset of pavement crack images was built, which was used to train and test the proposed method. Results show that precision, recall and F1 score of proposed method are 91.95%, 89.31% and 90.61% respectively, which are higher than other state-of-the-art pavement crack detection methods.

1 INTRODUCTION

Pavement cracking is one of the most important objects of road maintenance, which is caused by climate change and tensile stress due to traffic load (Dinegdae and Birgisson, 2016). Bigger distresses like spalling, raveling, potholes and pavement structure failure could be developed from pavement cracks, if the cracks were not repaired timely (Jassal, 1999). As a result, pavement service life can be extended by regular maintenance, in which pavement crack detection is the essential step of crack repair (Pellecuer et al., 2014). Automated pavement crack detection is a challenging task because of the irregular pattern of pavement cracks and the low contrast between the cracks and surrounding pavement, in addition to other factors like lighting conditions, shadows, asphalt oil stain, water, snow, ice, pavement markings, aggregate outlines, soil, sands, branches and other debris that cause noise in images.

In the recent years, many automated pavement crack detection methods have been developed using deep learning. Shi et al. proposed a crack detection framework CrackForest based on random structured forests (Shi et al., 2016). Fan et al. proposed a supervised method based on multi-label problem (Fan et al., 2018). Li et al. proposed an unsupervised multi-scale fusion crack detection method (Li et al., 2018). Zou et al. proposed an end to end trainable deep convolutional neural network DeepCrack (Zou et al., 2018). However, most of the

---

* Corresponding author

previous methods are more related to object segmentation that segments crack pixels from pavement background, while the objective of pavement crack detection is to obtain the number, classification and location of pavement cracks in images. In the definition of deep learning, object detection is defined as the combination of object classification and localization, while object segmentation is defined as the extraction of object from background. As a result, previous pavement crack segmentation methods could not be used for crack detection, while there are few methods for actual pavement crack detection.

An automated pavement crack detection method based on region-based convolutional neural network (R-CNN) like you only look once version three (YOLOv3) was proposed in this paper, whose network architecture and was modified to fit the special features of pavement crack detection. A dataset of pavement crack images was built, which was used to train and test the proposed method. In the end, test results were analyzed to evaluate the performance of proposed method on pavement crack detection.

2 METHODOLOGY

2.1 Backbone of proposed method

An automated pavement crack detection method was developed based on YOLOv3, which is an efficient and accurate method of object detection. Because YOLOv3 was not developed for civil engineering, the network architecture cannot meet the requirement of pavement crack detection. Considering the special features of pavement cracks, such as irregular pattern, clustering of alligator and block cracks, tiny width, small area compared to the pavement image and low contrast between cracks and pavement, the network structure of YOLOv3 method was modified for automated pavement crack detection. The backbone of proposed network architecture was still Darknet-53, which consists of 53 convolutional layers, 5 residual layers, an average-pooling layer, a fully connected layer and a softmax layer. Darknet was selected as the backbone of proposed method to detect the pavement cracks more efficiently and accurately, whose efficiency and accuracy on object detection have been proved in previous researches (Redmon and Farhadi, 2018).

2.2 Network architecture

The network architecture of proposed method was not developed into very complex, because there were only four classes of object for detection. Four network architectures with different YOLO layer group structures were proposed for pavement crack detection, including three YOLO layers, three spatial-full YOLO layers, three simplified YOLO layers and five YOLO layers. Network architecture with three YOLO layers was the original YOLOv3 architecture, while network architectures with three spatial-full YOLO layers, three simplified YOLO layers and five YOLO layers were modified from original YOLOv3 architecture.

The modification was mainly conducted on the YOLO layer group structure of network architecture, in which the original three YOLO layers were replaced with three spatial-full YOLO layers, three simplified YOLO layers or five YOLO layers. Spatial-full YOLO layer was made from original YOLO layer with max-pooling layer as spatial-partial blocks, which was developed to detect small objects. Simplified YOLO layer was made from original YOLO layer without cyclical convolutional layers, which was developed to simplify the structure and focus on detecting simple objects.

3 EXPERIMENT

Proposed method was programmed in C++ language. The parallel computing framework used was CUDA 10.1 with cuDNN 7.5.1, and the visual library used was OpenCV 2.4.2. The training and test experiments were conducted by using supercomputer centre in Monash University, and the specification of the computer is: Nvidia Tesla K80 24GB GPU, Intel® Xeon® Processor E5-2680 v4 CPU, 64 GB RAM and Linux OS.

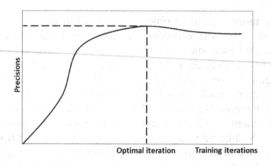

Figure 1. Overfitting in deep learning.

Thousands of pavement crack pictures were taken from roads, including cracks of different shapes and scales. Data augmentation method was used on the images, including rotation and flipping. The images were then annotated to make ground truths for deep learning.

Overfitting is an unavoidable problem in training procedure of deep learning, which means that proposed method trains more iteration than the ultimately optimal iteration, as shown in Figure 1. As a result, a function was developed and added into the training algorithm, in which the best iteration could be automatically selected while training and the best training result weights could be automatically output. The codes were developed based on the validation while training, in which the training result of different iterations were validated and the iteration with the best validation result was output as the optimal iteration.

Precision (Pr), recall (Re) and F1 score (F1) were applied as the evaluation metrics of the proposed method, which are widely used evaluation metrics in deep learning. Precision is the ratio of true positive (TP) among the detected values consisting of true positive and false positive (FP), as shown in Eq. (1). Recall is the ratio of true positive over the observed values consisting of true positive and false negative (FN), as shown in Eq. (2). F1 score is the harmonic mean of precision and recall, as shown in Eq. (3). The Pr, Re and F1 scores of all validation images were averaged values.

$$Pr = \frac{TP}{TP + FP} \tag{1}$$

$$Re = \frac{TP}{TP + FN} \tag{2}$$

$$F1 = \left(\frac{Pr^{-1} + Re^{-1}}{2} \right) = \frac{2 \times Pr \times Re}{Pr + Re} \tag{3}$$

where, TP is the correctly detected cracks, FP is the non-cracks wrongly detected as cracks and FN is the cracks wrongly detected as not cracks.

4 RESULTS AND DISCUSSION

The proposed method was trained and tested with four different network architectures on the pavement crack dataset, in order to find the most suitable network architecture for pavement crack detection, and results are shown in Figure 2. Most of the pavement cracks in images were detected, while there were still some mistakes in detection. The accuracy of crack detection was significantly affected by the noises in images, like water, shadow and lighting condition. Clustered block crack was confused with alligator crack in some circumstances, and the cracks were even divided into several parts. The images were randomly selected in the

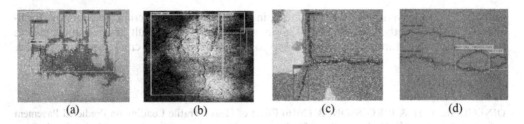

Figure 2.   Test result samples of different network architectures.

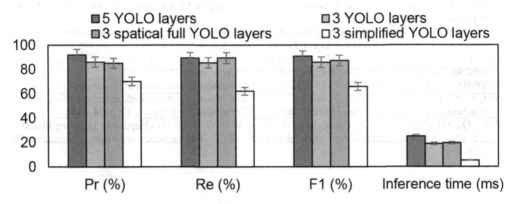

Figure 3.   Test results of network architectures.

detection results of methods with different network architectures, which was hard to be pre-cisely compared. As a result, the histogram of detection results was drawn, including preci-sion, recall, F1 score and inference time of different network architectures.

As shown in Figure 3, detection speed of network architectures with 5 YOLO layers, 3 YOLO layers and 3 spatial-full YOLO layers was close to each other, while the accuracy of network architecture with 5 YOLO layers was higher than other two structures in a small margin, which could be caused by the deeper network structure and more YOLO layer pro-cessing. Recall of network architecture with 3 spatial-full YOLO layers was higher than other architectures while the precision was lower, which means FN could be effectively reduced by focusing on small objects with the cost that FP would be higher. Moreover, performance of network architecture with 3 simplified YOLO layers was much worse than the other three net-work architectures, although with the fastest speed, which could be caused by the simplified structure. The precision, recall, and F1 score of proposed method with five YOLO layers were 91.95%, 89.31% and 90.61% respectively, which were highest among the different network architectures.

5   CONCLUSION

This paper studied the automated pavement crack detection method based on RCNN like YOLOv3. Different network architectures were developed and evaluated to improve the per-formance of proposed method. Based on the results and discussion presented, the main find-ings and conclusions can be summarized as follows:

Pavement cracks can be effectively detected by proposed method with precision, recall and F1 score at 91.95%, 89.31% and 90.61% respectively. Four different network architectures of proposed method was studied, including 3 YOLO layers, 3 spatial-full YOLO layers, 3 simplified YOLO layers and 5 YOLO layers, in which the pavement crack detection performance of proposed method with 5 YOLO layers was the best. As a result, it is recommended to use network architecture with 5 YOLO layers for proposed pavement crack detection method.

REFERENCES

DINEGDAE, Y. H. & BIRGISSON, B. (2016) Effect of Heavy Traffic Loading on Predicted Pavement Fatigue Life. *8th RILEM International Conference on Mechanisms of Cracking and Debonding in Pavements*. Dordrecht, Springer, Netherlands.

FAN, Z., WU, Y., LU, J. & LI, W. (2018) Automatic pavement crack detection based on structured prediction with the convolutional neural network. *arXiv preprint arXiv:1802.02208*.

JASSAL, K. (1999) *Development of potholes from cracks in flexible pavements*, ProQuest Dissertations Publishing.

LI, H., SONG, D., LIU, Y. & LI, B. (2018) Automatic pavement crack detection by multi-scale image fusion. *IEEE Transactions on Intelligent Transportation Systems*, 1–12.

PELLECUER, L., ASSAF, G. & ST-JACQUES, M. (2014) Life cycle environmental benefits of pavement surface maintenance. *Canadian Journal of Civil Engineering*, 41, 695.

REDMON, J. & FARHADI, A. (2018) YOLOv3: An Incremental Improvement. *arXiv e-prints*.

SHI, Y., CUI, L., QI, Z., MENG, F. & CHEN, Z. (2016) Automatic road crack detection using random structured forests. *IEEE Transactions on Intelligent Transportation Systems*, 17, 3434–3445.

ZOU, Q., ZHANG, Z., LI, Q., QI, X., WANG, Q. & WANG, S. (2018) Deepcrack: Learning hierarchical convolutional features for crack detection. *IEEE Transactions on Image Processing*, 28, 1498–1512.

Functional Pavements – Chen et al (eds)
© 2021 Taylor & Francis Group, London, ISBN 978-0-367-72610-2

Study on evaluation method of energy efficiency and carbon emission intensity of highway maintenance

Cheng-yong Chen
Shandong Hi-speed Group Co., Ltd., Jinan, Shandong, China

Zhen-qing Liu
Beijing New Road Times Transportation Technology Co., Ltd., Beijing, China

Chao Hu, Hong-feng Li, Hao Yu & Jian-ping Xia
Shandong Hi-speed Group Co., Ltd., Jinan, Shandong, China

Yintao Guo
Beijing New Road Times Transportation Technology Co., Ltd., Beijing, China

Feng-lei Zhang
Southeast University, Nanjing, Jiangsu, China

ABSTRACT: In this paper, we have collected 14 kinds of data for maintenance projects in 6 sections of 4 cities, mainly including plant cold recycling of asphalt pavement, cold in-place recycling of asphalt surface course, asphalt surface course + asphalt pavement base or full-depth recycling of pavement base, plant cold recycling of pavement base material, plant hot recycling and warm recycling of asphalt surface course, cold in-place recycling of asphalt surface course. Based on the analysis of the energy efficiency data of the above research objects, LCA (Life Cycle Assessment) indirect model is established to evaluate the carbon emission intensity of the recycling technology of asphalt pavement. The evaluation results show that in terms of input-output energy efficiency, the plant hot recycling scheme is the highest, the hot mix modified asphalt and base asphalt are next, the plant warm recycling and plant cold recycling are similar and the cold in-place recycling is the worst.

1 INTRODUCTION

Highway is an important component of the infrastructure construction system, which plays an important role in the economic development, social progress and the improvement of people's life quality (Anastasopoulos et al., 2010). For the concept of green cycle and low-carbon, scholars have carried out positive thinking and exploration (Zhang, 2013, Yang, 2017). Some countries have carried out research and practice in green building and green road evaluation (Zuo and Zhao, 2014). Therefore, the LCA model and data quality evaluation method of road maintenance based on the field data are constructed in this study. Through the data analysis of several road maintenance sections, the energy consumption and carbon emission of the asphalt pavement maintenance scheme are compared, and the goal of energy conservation and emission reduction is set up. Finally, the energy efficiency of the asphalt pavement maintenance scheme is comprehensively evaluated.

2 FIELD DATA COLLECTION

This paper collects all kinds of data of 14 maintenance projects in 6 road sections of 4 cities (NJ, ZJ, YZ and XZ), and establishes a database of energy consumption and carbon emission

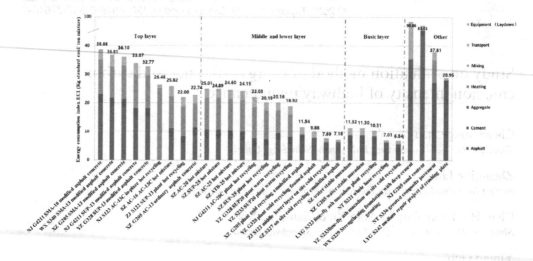

Figure 1.   Comparison of energy consumption index (ECI) of various maintenance technologies.

of highway maintenance technology in this province. Among them, the calculation of energy consumption includes four main links: raw material, mixture production, mixture paving and material transportation, and each link has its own calculation index value, energy consumption proportion and carbon emission value, as shown in Figure 1.

3   MODEL BUILDING

The construction and maintenance activities will affect the operation of the road traffic, which will lead to congestion, detour or queuing. The disturbed traffic flow will generate additional fuel consumption and air pollutants compared with the normal operation state. Quantifying the difference between the two scenarios, i.e. congestion state minus normal operation state, is the problem to be solved by congestion module, i.e. indirect calculation model. The calculation formula is as follows:

$$Y_{total} = \sum_{i=1}^{i=n} VMT_{queue,i} \times Y_{queue,i} + VMT_{det\,our,i} \times Y_{det\,our,i} + VMT_{workzone,i} \times Y_{workzone,i} - VMT_{normal,i} \times Y_{normal,i} \quad (1)$$

$$Y = BaseFuel \times \xi_{IRI} \times \xi_G \quad (2)$$

$$VMT = Q \times T \times L \quad (3)$$

Where, Y – is energy consumption and carbon emission index;
VMT – is total mileage of a vehicle, the subscript represents different traffic operation scenarios, such as normal operation, queuing, deceleration crossing construction belt, detour, etc;
i – is different vehicle types;
n – is number of vehicle types;
$BaseFuel$ – is reference fuel consumption coefficient of different vehicle types;
ξ_{IRI} – is roughness;
ξ_G – is slope correction factor;
Q – is number of i vehicle types in traffic flow;
T – is construction duration;
L – is length of construction section;

4 EVALUATION OF ENERGY CONSUMPTION AND CARBON EMISSION INTENSITY OF HIGHWAY MAINTENANCE

At present, the single index mode is often used in the environmental impact of highway construction and maintenance projects, such as carbon emission per unit mass energy consumption, carbon emission per unit investment energy consumption, etc. Although these designed indexes can reflect the environmental impact of different maintenance schemes to a certain extent, however, more similar and even more concerned indexes, such as cost, road performance, etc., are ignored. And different indicators often conflict with each other, for example, the project with superior performance has higher cost and more significant environmental impact. Therefore, we comprehensively evaluate the energy efficiency of different maintenance schemes and propose quantitative indicators.

4.1 Evaluating indicator

The following evaluation indexes are proposed for the maintenance scheme, including project cost, energy consumption, carbon emission, raw material consumption and performance. Different indicators are evaluated qualitatively and quantitatively in different ways, as follows:

Project cost (¥); objective equation: the lower the better;
Energy consumption (kgce); target equation: the lower the better;
Carbon emission (kg-CO_2), target equation: the lower the better;
Raw material consumption (kg); Objective equation: the less the better;
Performance: five-point evaluation, (very good, good, general, poor, very poor), objective equation: the higher the evaluation, the better;

According to the indicators and evaluation principles determined above, collect relevant information of different schemes, and the summary is shown in Table 1. Due to the limited cases of cement pavement maintenance, this study only evaluates the asphalt pavement maintenance scheme.

4.2 Evaluation results

Firstly, the initial weight of each index is defined as: project cost: road performance: energy consumption: aggregate consumption: carbon emission = 35:40:5:10:10. Using the established methodology, the calculation results are shown in Figure 2.

Table 1. Information on different maintenance schemes.

Schemes	Engineering cost (¥)[1]	Energy consumption (kgce)	Carbon emission (kg-CO_2)	Raw material consumption (t)[2]	Performance
Hot mix base asphalt	241.2	24.51	60.27	0.931	very good
Hot mix modified asphalt	286.8	35.41	86.96	0.931	good
Plant hot recycling	209.3	21.13	52.66	0.719	very good
Plant warm recycling	199.7	19.55	48.04	0.719	general
Plant cold recycling	191.9	10.81	26.65	0.195	very poor
Cold in-place recycling	174.6	7.11	18.34	0.195	poor

Note: [1] only includes direct engineering cost, and the price is estimated by the author;
[2] raw material consumption refers to the quality of new aggregate used.

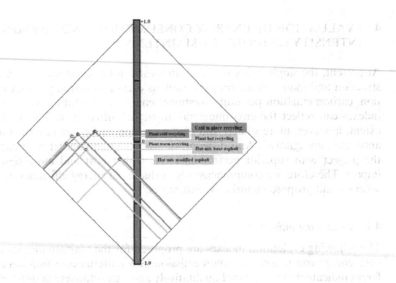

Figure 2.   Diamond chart of local comparison of different maintenance schemes.

In Figure 3, if one scheme *a* is completely wrapped by another scheme *b*, scheme *a* is better than scheme *b*; if the two schemes overlap but do not wrap each other, it means that there are trade-off indicators for the two schemes, such as high cost, superior performance and low cost, performance degradation. It can be clearly observed that in terms of local comparison, the cold in place regeneration scheme covers the matrix hot mixing modified hot mixing scheme.

Figure 3 shows a more detailed breakdown of the scores for each option. A negative number indicates that the indicator is at a disadvantage compared to other options and vice versa. The amplitude of plant warm recycling and plant hot recycling is the smallest, which shows that it is a moderate maintenance scheme, which has no obvious advantages or disadvantages, and its absolute index ϕ is close to zero. The advantages and disadvantages of hot mix modified asphalt and cold in-place recycling are obvious. The previous analysis is based on the fixed weight proportion.

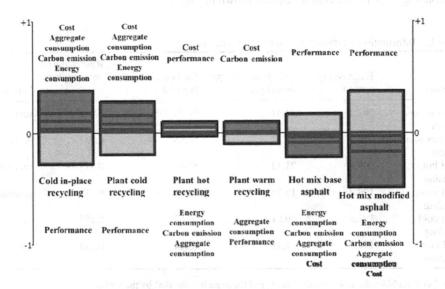

Figure 3.   Comparison histogram of different maintenance schemes.

5 CONCLUSION

Through the research of this paper, the main conclusions are as follows:

(1) The LCA model and the reliability evaluation methodology of the model results can be used to evaluate the reliability of the calculation results of the environmental impact model under the conditions of data quality, model hypothesis and parameter variation.
(2) Data envelopment analysis shows that in terms of the input-output energy efficiency level, plant hot recycling scheme is the highest and cold in-place recycling scheme is the worst.

REFERENCES

ANASTASOPOULOS, P. C., FLORAX, R. J. G. M., LABI, S. & KARLAFTIS, M. G. (2010) Contracting in highway maintenance and rehabilitation: Are spatial effects important? *Transportation Research Part A-policy and Practice*, 44, 136–146.
YANG, X. (2017) Studyof Low-carbon Technology Decision-making System of Green Road based on LCA Theory. *Shanghai Highways*.
ZHANG, Y. F. (2013) Research on Low-Carbon Architectural Development Based on Green Life Cycle. *Applied Mechanics and Materials*, 443, 263–267.
ZUO, J. & ZHAO, Z. Y. (2014) Green building research-current status and future agenda: A review. *Renewable & Sustainable Energy Reviews*, 30, 271–281.

The significance of situation awareness theory for freeway risk analysis

C. Yuan & Q. Dong*
Department of Transportation, Southeast University, Nanjing, Jiangsu, China

X.Q. Chen
Department of Civil Engineering, College of Science, Nanjing University of Science and Technology, Nanjing, Jiangsu, China

L.Y. Wang
China Communications Construction Company Second Highway Consultants Co. Ltd, Wuhan, Hubei, China

ABSTRACT: The primary objective of this research is to clarify its guiding significance for freeway risk analysis based on the comprehension of situational awareness (SA) theory, and to provide a theoretical basis for determining freeway risk factors and building a risk warning system in the future. The theory of situational awareness is originated in the military field and later has been widely applied in the field of network security. Situation awareness theory includes three stages: perception of environmental factors, understanding of the current state, and prediction of the future state, which will be applied to deeply analyze freeway risk problems and promote active traffic management.

1 INTRODUCTION

The development of freeways characterized by high speeds and high capacity has brought time and space changes to people, however, these characteristics determines that once traffic problem occurs, its social impact and loss of life and property are immeasurable. In recent years, extensive researches have been conducted on real-time risks of freeways, which provides an important theoretical basis for freeway risk warning. Real-time monitoring of freeway operating status has received much research attention, however, the research results on perception, assessment of freeway risk status and early warning are not consistent (Roy et al., 2018). In addition, researchers have carried out freeway risk monitoring and early warning system design and the differences in their research results mainly come from the definition of potential risk monitoring objects and the manifestations of risk warning system (Bachmann et al., 2013, Fang and Guo, 2013, Zengqiang et al., 2008). Therefore, it is necessary to conduct in-depth research on freeway risks to determine potential risk factors and detect emergencies in advance.

Situation awareness theory (SA) includes three stages: perception of environmental factors, understanding of the current state, and prediction of the future state. The main purpose of this research is to study the significance of situational awareness

---

* Corresponding author

theory for freeway risk analysis, and to provide guidance for determining potential risk factors of freeways.

2 COMPREHENSION OF SITUATION AWARENESS THEORY

Situation awareness theory is originated in the military field for effective command and decision-making in war. During the 1980s and 1990s, the theory of situation awareness was initially formed and received widespread attention in the late 1990s. In the early stage of the formation of situation awareness theory, many scholars have slightly different definitions (Table 1). The most concise definition of situation awareness was proposed by Endsley, which refers to the acquisition and understanding of environmental factors and the prediction of the future under certain time and space (Ensley, 1995). Therefore, the understanding of situational awareness is generally divided into three independent levels:

Level 1-Perception of elements in the environment: This is the identification of key factors and potential threats for a higher level of sampling in subsequent modelling and analysis.

Level 2-Understanding of the current situation: Combining perceived elements and measuring the status of current incident, then making decisions and taking actions.

Level 3-Projection of the future status: Predicting the future state based on the current situation, aiming to understand and determine the development trend of the incident.

In the late 1990s, Tim Bass first proposed the concept of situation awareness in the field of network security (Bass, 1999). Since then, Network Situation Awareness has attracted great attention and has been widely used in the field of network and information security (JAKALAN, 2013, Liu et al., 2019, Zhao and Liu, 2018), aiming to strengthen cybersecurity early warning and monitoring by building a situational awareness system to deal with increasingly complex and severe cyberspace security issues.

Table 1. Definitions of situation awareness (Council, 1998).

Author	Definitions
Endsley (1987, 1995)	Perception of the elements in the environment within a volume of time and space Comprehension of their meaning Projection of their status in the future
Stiffler (1988)	The ability to envision the current and near-term disposition of both friendly and enemy forces
Sarter and Woods (1991, 1995)	Just a label for a variety of cognitive processing activities that are critical to dynamic, event-driven, and multitask fields of practice
Dominguez (1994)	Continuous extraction of environmental information, integration of this knowledge to form a coherent mental picture, and the use of that picture in directing further perception and anticipating future events
Flach (1995)	Perceive the information Interpret the meaning with respect to task goals Anticipate consequences to respond appropriately

3 SIGNIFICANCE OF SA FOR FREEWAY RISK ANALYSIS

3.1 *Current study on freeway risk*

Crashes are the main factor threatening freeway safety, therefore, early research mainly considered the relationship between traffic information or weather characteristics and crashes, such as traffic volume, average speed, average occupancy, and weather types. Songchitruksa and Balke (2006) predicts the possibility of in-lane incidents based on real-time loop detection data. Weather and environmental data (such as lighting conditions) were also applied to predict the types of events and the research results can help to improve freeway risk perception (Songchitruksa and Balke, 2006). Abdel-Aty et al. (2007) applied traffic data to identify and predict collisions, simultaneously, the use of ITS strategies was proposed to avoid potential traffic accidents (Abdel-Aty et al., 2007). The latest research often considers more indicator factors to perceive freeway risks in real time, such as vehicle factors and driver factors. In addition, the sub-indices of each indicator are divided more clearly to achieve crash risk prediction and provide accurate basis for active traffic management (Wen et al., 2019, Zhai et al., 2020).

However, freeway emergencies are not limited to traffic accidents, but also include various abnormal events such as inclement weather and abnormal vehicle behavior. Liu et al. (2010) developed a freeway early warning information system, including wireless communication equipment along the freeway, car electronic assistants, and management center, which can realize accident alarm, speed alarm, early warning and quick response (Liu et al., 2010). However, the system relies on higher measurement, control and communication technology, and the cost is very high. Duan et al. (2013) proposed a traffic safety information service early warning system based on vehicular network big data, including data collection layer, communication layer, business processing layer and information presentation layer, which can realize the judgment of vehicle speeding, low speed parking, etc. and release early warning information in time (Duan et al., 2013). Ding (2017) designed a feature-based large-scale freeway way traffic flow monitoring system, which consists of three subsystems including data storage, service support, and central processing, as well as 11 functional modules (Ding, 2017). The core of the system is to apply cellular data to traffic state perception and realize the monitoring and early warning of freeway traffic flow.

3.2 *Risk analysis under the guidance of SA*

3.2.1 *Level 1 of SA – Perception and identification of the risk factors*
In view of freeway safety issues, situational awareness theory requires that the potential risk factors that may threaten freeway safety be identified and comprehensively analyzed. Furthermore, many scholars have conducted analyses and preventive countermeasures research on traffic accidents to determine the rules and causes of freeway emergencies, and have reached an agreement that freeway emergencies are mainly caused by drivers, vehicles, roads and environmental factors (Sha, 2006, Zhang, 2008, Sun, 2018).

3.2.2 *Level 2 of SA – Determination of key indicators and risk evaluation model construction*
The second stage of situation awareness theory requires the combination of perception elements to realize the understanding of the current state. Therefore, for freeway safety issues, risk factors need to be abstracted and classified into specific evaluation indicators according to the analysis in Section 1 and establish evaluation index system of freeway risk. Finally, a risk evaluation model should be established to evaluate the real-time risk of freeway.

3.2.3 *Level 3 of SA - Prediction and early-warning of freeway risk*
Furthermore, the third level of situational awareness theory is to predict the future situation based on the current situation. Therefore, in addition to completing the current freeway risk assessment, it is also required to predict the trend of real-time risk in future research, so as to realize more accurate risk warning and rapid decision-making. First, the key indicators are predicted based on the collected raw data. This process can be achieved through various

machine learning models, such as simple linear models or complex neural network models. Then, the prediction results of key indicators are substituted into freeway risk evaluation model so as to realize the prediction of freeway risk, and provide strong support for real-time risk warning and rapid decision-making.

Finally, based on the situational awareness theory, the evaluation index system and freeway risk early warning system are established, therefore, it can not only achieve the risk assessment and warning of the current freeway status, but also predict the trend of real-time risk, which is of great significance for active traffic management.

This research aims to study the significance of situational awareness theory for freeway risk analysis, and provide guidance for determining potential risk factors of freeways. Through the understanding of the three levels of situational awareness theory, the real-time monitoring of freeway risk factors and the active traffic management system can be divided into four steps: 1. Determine the key risk factors of the freeway; 2. Establish a freeway risk evaluation index system and risk assessment model; 3. Freeway risk classification and early warning level grading; 4: Freeway risk early warning system design.

4 CONCLUSIONS

Based on the comprehension of the concept of situation theory, this research has determined the significance of situation awareness theory for freeway risk analysis, which can provide guidance for determining potential risk factors of freeway. Furthermore, this study elaborates how to apply SA to freeway risk analysis, which is mainly divided into three stages: perception, understanding and prediction. First, in the perception stage, it is required to analyze the causes and possible risk factors of freeway emergencies. Second, in the understanding stage, it is necessary to build a freeway risk evaluation index system and risk evaluation model to evaluate freeway risk. Third, in the prediction stage, the predicted values of key indicators are obtained through model prediction to realize real-time risk prediction and make quick decisions which needs further research in the future.

REFERENCES

ABDEL-ATY, M., PANDE, A., LEE, C., GAYAH, V. & SANTOS, C. D. (2007) Crash risk assessment using intelligent transportation systems data and real-time intervention strategies to improve safety on freeways. *Journal of Intelligent Transportation Systems*, 11, 107–120.

BACHMANN, C., ROORDA, M. J., ABDULHAI, B. & MOSHIRI, B. (2013) Fusing a bluetooth traffic monitoring system with loop detector data for improved freeway traffic speed estimation. *Journal of Intelligent Transportation Systems*, 17, 152–164.

BASS, T. (1999) Multisensor data fusion for next generation distributed intrusion detection systems. *Proceedings of the IRIS National Symposium on Sensor and Data Fusion.* Citeseer.

COUNCIL, N. R. (1998) *Modeling human and organizational behavior: Application to military simulations*, National Academies Press.

DING, F. (2017) A Feature Based Traffic Monitoring System for Large Scale Freeway Using a Big Data Resource. University of Wisconsin–Madison.

DUAN, Z., KANG, J., TANG, L., FAN, N., CHENG, H. & LU, M. (2013) Early warning of highway traffic safety information service system. *Information Technology Journal*, 12, 3849–3854.

ENSLEY, M. (1995) Toward a theory of situation awareness in dynamic systems. *Human factors*, 37, 85–104.

FANG, Y. & GUO, Z. (2013) Risk Warning System for Complex Operating Environment Sections of Expressway. *ICTIS 2013: Improving Multimodal Transportation Systems-Information, Safety, and Integration.*

JAKALAN, A. (2013) Network security situational awareness. *The International Journal of Computer Science and Communication Security.*

LIU, X., CAO, L. & HUANG, X. (2010) Highway early warning information system. *2010 2nd International Conference on Information Engineering and Computer Science.* IEEE.

LIU, X., YU, J., LV, W., YU, D., WANG, Y. & WU, Y. (2019) Network security situation: From awareness to awareness-control. *Journal of Network and Computer Applications*, 139, 15–30.

ROY, A., HOSSAIN, M. & MUROMACHI, Y. (2018) Enhancing the prediction performance of real-time crash prediction models: a cell transmission-dynamic Bayesian network approach. *Transportation research record*, 2672, 58–68.

SHA, A. (2006) Research on the freeway traffic accidents and countermeasures. Southeast University.

SONGCHITRUKSA, P. & BALKE, K. N. (2006) Assessing weather, environment, and loop data for real-time freeway incident prediction. *Transportation research record*, 1959, 105–113.

SUN, W. (2018) Study on the freeway traffic accident analysis and countermeasures based on data mining. Jilin University.

WEN, H., ZHANG, X., ZENG, Q. & SZE, N. (2019) Bayesian spatial-temporal model for the main and interaction effects of roadway and weather characteristics on freeway crash incidence. *Accident Analysis & Prevention*, 132, 105249.

ZENGQIANG, M., GUOSHENG, G., WANMIN, S. & YAN, Y. (2008) Wireless monitoring system of vehicle overspeed on freeway based on GPRS. *2008 27th Chinese Control Conference*. IEEE.

ZHAI, B., LU, J., WANG, Y. & WU, B. (2020) Real-time prediction of crash risk on freeways under fog conditions. *International Journal of Transportation Science and Technology*.

ZHANG, C. (2008) The cause and forewarning of freeway transportation accident. Southwest Jiaotong University.

ZHAO, D. & LIU, J. (2018) Study on network security situation awareness based on particle swarm optimization algorithm. *Computers & Industrial Engineering*, 125, 764–775.

Functional Pavements – Chen et al (eds)
© 2021 Taylor & Francis Group, London, ISBN 978-0-367-72610-2

Study of the optimum compaction time for PU mixture based on impact penetration equipment

Y. Xu, M. Duan, Y. Li, J. Ji & S. Xu
Beijing University of Civil Engineering and Architecture, Beijing, China

ABSTRACT: Based on the self-developed impact penetration equipment, a method was proposed to determine the optimal compaction time of a polyurethane (PU) mixture using a penetration depth index. In this study, the variation curves of penetration depth with penetration times under four types of contact energy index and three different penetration strut diameters were compared based on the principle of better differentiation of penetration depth difference of mixture in the range of the best compaction time. The results showed that the optimum penetration strut diameter and number of penetration cycles are 20 mm and 40, respectively. Thus, the method of determining the best compaction time of the PU mixture based on the penetration depth was verified. In the outdoor construction environment with a temperature of 30 °C and humidity of 40%, the optimal time range determined by the penetration test was 41–47 mm, and the real value of the optimal time was 44 mm, which proves that the method proposed in this study is reasonable and effective.

1 INTRODUCTION

Polyurethane (PU) adhesives have various advantages including high bonding strength, good toughness, and strong applicability, and they can be cured at room temperature to achieve sufficient strength (Yang et al. 2017). These adhesives can be divided into two types: one-component and two-component. The one-component PU adhesive produces adhesive force through a curing reaction with moisture in the air at a certain temperature, and the curing reaction speed can be adjusted using the amount of catalyst. Therefore, it is relatively simple to use and has a wider application range (Gao, 2013). The one-component PU mixture is a new potential road pavement material, which could completely replace asphalt.

Similar to the conventional asphalt mixture, the PU mixture also needs to be fully compacted to ensure a long life and good performance of the pavement. However, owing to the different strength formation mechanisms between the PU and asphalt mixtures, the viscosity–temperature curve cannot be used to determine the optimum compaction time for the PU mixture. At present, the relationship between the compaction under different curing temperatures and the performance of PU mixtures after compaction and curing is often established through laboratory tests to determine the optimum compaction time. The state of the PU mixture with various curing temperatures in laboratories can be expressed by the contact energy index (CEI) (Dessouky et al. 2004), which is based on the shear gyratory compactor (SGC). However, construction sites are generally not equipped with SGC. Therefore, it is difficult to determine the best compaction of the PU mixture with CEI at the construction site. Presently, the compaction time of the PU mixture is still estimated based on the field experience.

The purpose of this study was to develop a simple and rapid mechanism for estimating the degree of curing of a PU mixture in order to determine the best compaction state in the field. To that end, this study aimed to develop an impact penetration equipment to characterize the degree of curing of the mixture from the penetration depth or duration for the PU mixture. Moreover, an attempt was made for quickly and accurately estimating the optimum compaction time for the PU mixture on site by establishing a relationship between the best compaction time determined by laboratory tests and the degree of curing of the PU mixture on site.

2 MATERIALS AND METHODS

2.1 Materials

2.1.1 PU adhesives
The test results for the performance of one-component PU adhesive selected in this study are shown in Table 1.

2.1.2 Aggregates and mineral powder
The coarse aggregate used in the test was basalt, the fine aggregate was limestone-manufactured sand, and fine mineral powder of basalt was used as the filler. The technical indices of coarse and fine aggregate and mineral powder meet the requirements of "Technical Specification for Construction of Highway Asphalt Pavement" (JTG F40-2004). Furthermore, AC-13 gradation was selected, and according to previous engineering experience, the rubber aggregate ratio of the one-component PU mixture was 6.5%.

2.2 Experimental equipment and test methods

2.2.1 Design of experimental equipment
A set of impact penetration test equipment was designed. A cylindrical specimen mold with a diameter of 100 mm at the bottom of the equipment was used to hold the mixture specimen with a size of Φ 100 mm $\times$ 68 mm. The upper and lower plates were fixed on three immovable rods to fix the horizontal position of the drop hammer and the penetration strut. Further, the flathead penetration strut with a certain diameter was used, and a drop hammer with a weight of 4 kg was dropped along the penetration strut from a certain height to impact the specimen in the cylinder to simulate the compaction of a mixture by a roller. The overall structure of the equipment is shown in Figure 1.

In the experiment, the falling height and impact velocity of the drop hammer were 300 mm and 2425 mm/s, respectively. When the drop hammer collided with the 10 kg penetration strut, the impact velocity of the penetration strut became 1385 mm/s under the condition of perfectly elastic collision momentum conservation, and 693 mm/s under the condition of inelastic collision momentum conservation. The impact velocity of the penetration strut on the specimen was 693–1385 mm/s. In the process of road rolling, the instantaneous shear rate between the mixture particles is approximately equal to the forward speed of the roller. When the rolling speed is 3–5 km/h, the instantaneous shear rate of the mixture is approximately 833–1389 mm/s, with an average of 1111 mm/s. It can be seen that the instantaneous shear velocity of the mixture when the falling hammer falls from a height of 300 mm is similar to that under on-site rolling. Therefore, it is appropriate to select a drop hammer height of 300 mm.

Table 1. PU-adhesive properties.

Index	Unit	Test result	Test method
Density	g/cm^3	1.1	GB/T 4472-2011
Viscosity (23 °C)	mPa·s	3000	JB/T 9357-1999
Tack-free time (23 °C)	h	1.5	GB/T 16777-2008
Curing time (23 °C)	h	12	GB/T 16777-2008
Tensile strength (23 °C)	MPa	10	GB/T 16777-2008
Elongation at break	%	323	GB/T 16777-2008

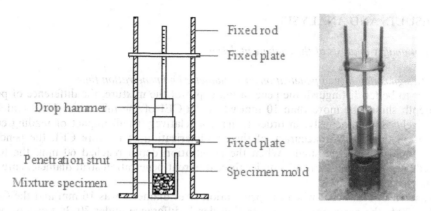

Figure 1.   Impact penetration test equipment.

2.2.2 *Design rationality of penetration strut diameter and penetration time*

With an increase in the diameter, the penetration resistance of the penetration strut increased, and increased penetration cycles were required to reach the same penetration depth. Therefore, it is necessary to determine the appropriate penetration strut diameter and penetration times in order to observe significantly visible change in the degree of curing of the mixture shown by the evaluation index. Tan (2002) found that the size of the specimen penetration struts is an important factor that affects the results of the impact penetration test. When the ratio of diameter of the penetration strut to that of the mixture specimen was less than 1:4, the maximum shear stress under the penetration strut was close to the shear stress in the actual situation. The diameter of the Marshall specimen used in this study was 100 mm, and the corresponding penetration strut diameter should be less than 25 mm. Therefore, three penetration strut diameters of 16 mm, 20 mm, and 24 mm were selected for the study. According to the CEI of AC-13, one-component PU mixture measured under different conditions, 463.8, 662.5, 757.2, and 1130.5, were selected as representative CEIs. The variation curves of penetration depth with penetration times under four CEIs were drawn by the same penetration method to determine the reasonable penetration strut diameter and penetration times. The penetration test was carried out manually, and the drop hammer dropped 10 times per minute. The CEI of the PU mixture is the area under the compaction curve with a compactness ratio of 89% to 97%. The calculation method is shown in Figure 2.

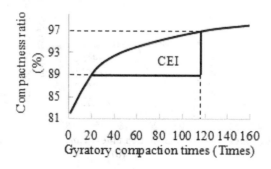

Figure 2.   Calculation method of CEI of PU mixture.

3 RESULTS AND ANALYSIS

3.1 *Rationality analysis of the equipment design*

3.1.1 *Design rationality of penetration strut diameter and penetration time*
In order to better distinguish the penetration depth of the mixture, the difference of penetration depth should be more than 10 mm when the CEI of the mixture is 463.8 and 1130.5,
respectively. Simultaneously, in order to have a relatively small impact of reading error of
penetration times on the accuracy of the characterization of mixture CEI, the penetration
times should not be too small. When the penetration depth reached 60 mm, the test was
stopped. The experimental results under three different penetration strut diameters are shown
in Figures 3a, b, and c.

As shown in Figure 3a, when the penetration strut diameter was 16 mm and the CEI was
463.8–1130.5, the corresponding penetration depth difference under 30–36 times of penetration was greater than 10 mm. However, when the CEI is small, the curing degree of the PU
mixture is low and the mixture is soft. In the process of impact penetration into a PU mixture,
a 16 mm penetration strut has an evident sinking phenomenon under the action of self-weight.
Owing to the small penetration resistance of the 16 mm penetration strut diameter and the
rapid change in penetration depth, the width of the engineering-observable penetration times
is insufficient. Therefore, it is not suitable to select a 16 mm diameter.

As shown in Figure 3c, when the penetration strut diameter is 24 mm, the penetration depth
does not change significantly with CEI owing to the large penetration resistance, and the penetration depth error accumulates significantly. When the CEI is 463.8–1130.5, the corresponding penetration depth difference in the entire penetration process is less than 10 mm, which
does not have a visible differentiation and is not suitable for engineering observation. Therefore, the selection of 24 mm as the penetration strut diameter is inappropriate.

As shown in Figure 3b, when the penetration strut diameter is 20 mm, and the CEI is
463.8–1130.5, the corresponding penetration depth difference under 30–48 times of penetration is greater than 10 mm, and the width of the engineering-observable penetration times
is appropriate, and the selection of the diameter of the 20 mm penetration strut is suitable.
Since the penetration depth difference of the specimen under 30–48 times of penetration
was greater than 10 mm, we established a relationship diagram between the penetration
times and the penetration depth difference (Figure 4). In order to not have too few penetration times, and the change in the penetration depth difference after 40 penetration cycles
was not evident, the penetration times were selected as 40.

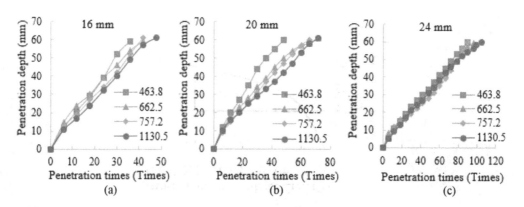

Figure 3.   Variation curves of penetration depth with penetration times.

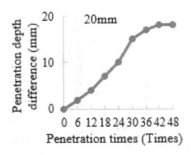

Figure 4. Relationship between penetration times and difference in penetration depth.

3.2 *Proposal and verification of the test method*

3.2.1 *Proposal of test method*

1) According to the laboratory test, the optimum compaction time for the PU mixture under different temperatures and humidity was determined, and the CEI of the mixture was obtained. The penetration test was carried out in optimum compaction state to determine the corresponding penetration depth "h."
2) The labor on the site of construction was divided into the construction and test group. The construction group conducted normal mixing and paving work. The experimental group and the construction group mix at the same time, take an appropriate amount of mixed mixture every 10–20 min to compact the Marshall specimen (Φ 100 mm × 68 mm), and then put the specimen into the specimen mold for 40 cycles of impact penetration test.
3) The formation time t and penetration depth h (t) of the specimen is recorded. When h (t) reaches the penetration depth h under the optimal compaction state, the corresponding time t is the optimum compaction time, and the construction team can perform the compaction operation.

3.2.2 *Verification of the test method*

According to the previous results reported by a research group (He, 2019), the relationship between the optimal compaction time of AC-13 one-component PU mixture and temperature, humidity, catalyst content, and standing time were established. When the splitting strength reaches the maximum, the CEI of AC-13 one-component PU mixture under the optimal compaction time was 600–800, and the corresponding depth for the 40th penetration was 41–47 mm.

Under the conditions of outdoor temperature T = 30 °C, humidity H = 40%, and catalyst dosage of 0.2%, AC-13 one-component PU mixture was mixed and placed in the outdoor environment for the curing reaction. During 40–120 min standing time, 6 Marshall specimens were formed every 20 min, and 3 of them were subjected to an impact penetration test to determine their penetration depth. The splitting strength of the other three specimens was determined after curing according to T0716-2011.

The depth for the 40th penetration at the peak splitting strength was compared with that under the optimal compaction time to verify the rationality of determining the optimal compaction time of a one-component PU mixture based on impact penetration equipment. The test data are shown in Figure 5.

It can be seen from Figure 5 that under this condition, the splitting strength of the specimen reached its peak value (3.02 MPa) when standing for 100 min, and this can be considered as the best compaction time. The corresponding penetration depth value is 44 mm, which is in the range of 41–47 mm. Therefore, the verification results prove that the method of determining the optimal compaction time with impact penetration equipment is reasonable and effective.

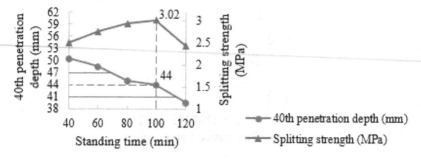

Figure 5. Results of outdoor validation experiment.

4 SUMMARY AND CONCLUSIONS

(1) A set of impact penetration equipment was developed. The hammer drop height was determined to be 300 mm based on the principle that the impact velocity of the penetration strut should be equivalent to the instantaneous shear rate of the mixture by the roller. In order to ensure that the penetration depth of the mixture has a significant division in the best compacting time range, it was determined that the diameter of the penetration strut was 20 mm by comparing the curves of penetration depth with penetration times under 4 types of CEI and 3 different penetration strut diameters, and the penetration cycles were 40.

(2) A method was proposed to determine the best compaction time of a one-component PU mixture in situ using penetration depth. The method was verified in an outdoor construction environment with a temperature of 30 °C and humidity of 40%. The penetration depth under the best compaction time determined by the penetration test was 41–47 mm, while the actual penetration depth under the optimal compaction time was 44 mm, which lies within the range. This proves that the method proposed in this study is reasonable and effective.

ACKNOWLEDGMENTS

This work was funded by the Science and Technology Project of the Beijing Municipal Education Commission (Grant No. SQKM201810016003), the Beijing Advanced Innovation Center for Future Urban Design (Grant No. UDC2019032624), and the research project of the National Natural Science Foundation of China (Grant No. 51978035).

REFERENCES

Dessouky, S. et al. 2004. Prediction of hot mix asphalt stability using the superpave gyratory compactor. *Journal of Materials in Civil Engineering* 16: 578–587.
Gao, J. 2013. Study on the synthesis, modification and performance of one-component moisture-cure polyurethane adhesive. *Guangdong University of Technology.*
He, J. 2019. Study on compaction characteristics of polyurethane mixture. *Beijing University of Civil Engineering and Architecture.*
Tan, Y. 2002. Research on shear properties of asphalt mixture based on stress field distribution of asphalt pavement. *Tongji University.*
Yang, Z. et al. 2017. Flexible and stretchable polyurethane/waterglass grouting material. *Construction and Building Materials* 138: 240–246.

Functional Pavements – Chen et al (eds)
© 2021 Taylor & Francis Group, London, ISBN 978-0-367-72610-2

Study on the long-life pavement scheme of the reconstruction and extension project of highway

Zhenhui Mo, Zhaohui Min, Yingqin Sun, Jun Yang & Jianwei Wang
School of Transportation Engineering, Southeast University, Nanjing, China

Lei Zhang*
ITS Research Center, Southeast University, Nanjing, China

ABSTRACT: Different structural schemes of epoxy asphalt long-life pavement were studied. Finite Element Method was employed to calculate the mechanical responses of the long-life pavement under the cases with different epoxy asphalt layer and thicknesses. The results showed that the epoxy asphalt mixture layer can reduce the vertical compressive strain of pavement and improve the anti-rutting performance as well. In most structural schemes, the maximum tensile strain of the asphalt layer maintains at around 80 micro strain. Considering the maximum horizontal tensile strain of the asphalt layer, the vertical compressive strain at the top of the subgrade, the anti-rutting performance of the pavement, the Scheme B is recommended for the epoxy asphalt long-life pavement structure, and the performance will be better if the total thickness of the asphalt pavement is greater than 22 cm.

1 INTRODUCTION

In recent years, due to the rapid growth of traffic volume on highway sections in Jiangsu Province, the service level of some road sections has fallen below level 3, and the master plan of Jiangsu highway network has clearly proposed the expansion of this highway on research [1]. As the semi-rigid base layer commonly used in China, is easy to produce shrinkage cracks and contraction cracks, and then extend to the asphalt pavement to form reflective cracks, the transportation department hopes to construct some long-life asphalt pavements with composite bases to deal with this problem [2]. Aiming at the key problems faced in the construction of long-life pavement and considering the excellent mechanical properties of epoxy asphalt materials, this paper proposed the structural schemes of epoxy asphalt long-life pavement based on the characteristics of semi-rigid bases of existing highway.

The concept of long-life pavement was first proposed in 2000 by Huddeston and others of the American Asphalt Pavement Alliance [3]. Harold [4] studied the thickness design method of hot-mixed asphalt layer for long-life asphalt pavement, and proposed two indexes to control the tensile strain at the bottom of the asphalt layer and the compressive strain of the top of subgrade. Other scholars proposed that the tensile strain should be limited to 60 to 100 micro strain based on the test results [5].

In this paper, different epoxy asphalt long-life pavement structural schemes were formed by placing the epoxy asphalt concrete at different level of the pavement structure. Finite Element models were established for mechanical simulation, and the multi-index results were studied to obtain the optimal structural scheme of epoxy asphalt long-life pavement.

---

* Corresponding author

2 PAVEMENT SIMULATION MODEL

2.1 *Pavement structural scheme*

As mentioned above, epoxy asphalt has both strong fatigue resistance and anti-rutting perform-ance, however the initial cost of construction is relatively high, therefore, long-life pavement structural schemes were proposed, of which the epoxy asphalt concrete was paved as different layers of the asphalt pavement. The structural schemes contain 5 categories, as shown in Figure 1. Scheme A means epoxy asphalt is used as the top layer, scheme B means epoxy asphalt is used as the middle layer, scheme C means epoxy asphalt is used as both the top layer and middle layer, and scheme D means epoxy asphalt is used as both the middle layer and bottom layer with add-itional 2cm stress absorbing layer (SAL) below, scheme X0 is the traditional plan for highway expansion (without epoxy asphalt layer), the vertical ordinate indicates the depth (cm), the abscissa indicates the structural scheme, and the numbering after the capital letter indicates the sub-category of different epoxy asphalt layer thickness in the same category; The underlying layer structure of all structural schemes is exactly the same, 38cm subbase (cement-stabilized aggregate), 20cm base (low content cement-stabilized aggregate), and the compacted subgrade.

2.2 *Finite element model*

In order to simulate the actual condition of the pavement structure and consider the symmetry of the pavement as well, a three-dimensional finite element half model of the road structure was established as shown in Figure 2. The length of the model is 100m, the width is 21m, and the height of the model is 8.74-8.84m (8m subgrade + 0.2m subbase + 0.38m base + 0.16-0.26m pavement), and the side slope is 1:1.5. The bottom boundary and edge boundary condi-tions of the model is fixed, the contact conditions between each layer is completely tied, and the element type is C3D8. Considering that there are many overloaded vehicles in this section, 130kN is selected as the typical axle load (the overload is 30% compared to the standard axle load of 100kN). In order to appropriately simplify the load pattern, a square area of 0.213×0.213m is used for loading, and the tire pressure is set at 0.91MPa.

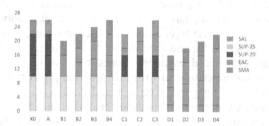

Figure 1.   Structural schemes of the pavement.

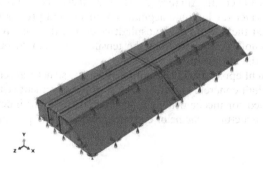

Figure 2.   Finite element model.

3 RESULT

3.1 *Mechanical response*

This paper mainly selects the maximum tensile strain of the asphalt mixture and the maximum vertical compressive strain at the top of subgrade as the key indexes of long-life pavement. The maximum tensile strain limit of the asphalt mixture is 100 micro strain, and the maximum vertical compressive strain limit at the top of subgrade is 200 microstrains. The maximum tensile strain of the asphalt mixture layer of each pavement structural scheme is shown in Figure 3(a). The bar graph represents the tensile strain, and the blue horizontal line represents the tensile strain limit value of 100 microstrains. The maximum horizontal tensile strain of the four schemes (X0, A, B, and C) are equivalent, fluctuating around 80 microstrains; the tensile strain in D is relatively large, but it gradually decreases as the thickness of the pavement increases. When the thickness of the asphalt pavement is 20 cm, the maximum tensile strain of the asphalt layer is 99 microstrains, therefore it can meet the requirements and has long lifespan under traffic load. It should be noted that for some schemes such ad B, the maximum tensile strain does not decrease linearly with the reduction of thickness.

The maximum vertical compressive strain at the top of the subgrade of each pavement structural scheme is shown in Figure 3(b), and the red horizontal line represents the limit value of the vertical compressive strain. It can be seen that the minimal compressive strain which appears in C3 is 156 microstrains, and the maximum compressive strain which appears in D1 is 208 microstrains. Except for D1, the vertical compressive strains of the other schemes all meet the requirement of no more than 200 microstrains. And for the same type of pavement, the vertical compressive strain of the top of subgrade decreases by about 5% when the thickness of the asphalt pavement increases by 2 cm. For scheme B, when the thickness increases, the vertical compressive strain deceases and become lower than X0.

Assume that the vertical compressive strain of asphalt concrete materials is significantly correlated with its high-temperature plastic deformation (non-linear proportionality), the vertical compressive strain distribution of the asphalt layer along the depth direction was calculated. The anti-rutting performance is preliminarily evaluated by observing the envelope area of vertical compressive strain curve with vertical axis.

It can be seen from Figure 4 that when epoxy asphalt mixture is used, the vertical compressive strain of the layer will be greatly reduced. For A and C, the vertical compressive strain of the top layer does not exceed 300 microstrains, which are much smaller than SMA of 533 microstrains. But for A and C, the compressive strain of middle and bottom layer will rapidly increase to be close to X0. For B and D, the compressive strain of the top layer is close to X0, but when epoxy asphalt is used for the middle layer, the vertical compressive strain decreases rapidly and does not exceed 300 microstrains. As the depth increases in B, when reaching the SUP layer, the compressive strain increases rapidly and then gradually decreases. In D, the vertical strain at the SAL is greatly increased, because it uses a material with low modulus and good elasticity. the vertical

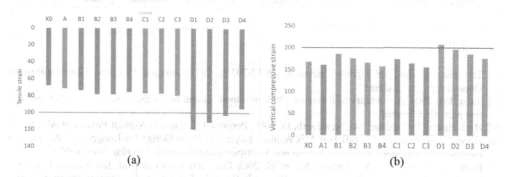

(a) (b)

Figure 3. Mechanical response of pavement structure (a) Maximum tensile strain of asphalt layer (b) Maximum vertical compressive strain at the top of subgrade.

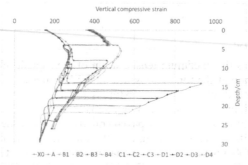

Figure 4. Vertical compressive strain along the depth curve.

compressive strain of the layer can basically be restored when unloaded, and the thickness of this layer is relatively thin, therefore the influence of this layer can be ignored. It is obvious that the envelope area of each curve (A, B, C, D) is smaller than X0. Further comparing the envelope area of curves B and D with curves A and C, it can be seen that the microstrain greater than 300 in B and D only appear in the top layer, while in A and C only the microstrain of the top layer is less than 300. The envelope area of the B, D curve is smaller, and its anti-rutting performance is better.

4 CONCLUSION

(1) The maximum horizontal tensile strains of the asphalt layer in Scheme A, B, C are comparable, basically fluctuating around 80 microstrains. However, most cases of scheme D exceed the limit of 100 microstrains.
(2) The maximum vertical compressive strain of the subgrade of each pavement structure is greater than 150 microstrains, and the vertical compressive strains of other asphalt pavement meet the requirement of no more than 200 microstrains, except Scheme D1. And for the same asphalt pavement type, every 2cm increment in asphalt pavement thickness will make the vertical compressive strain of the top of subgrade decrease by about 5%. When the thickness increases, B is relatively better than other schemes.
(3) In B and D, the vertical compressive strain greater than 300 only appears in the upper layer, while in A and C only the microstrain in the upper layer is less than 300. The envelope area of the B, D curve is smaller, and their anti-rutting performance is better.
(4) Considering the maximum horizontal tensile strain of the asphalt layer, the vertical compressive strain at the top of the subgrade, the anti-rutting performance of the pavement, the Scheme B is recommended for the epoxy asphalt long-life pavement structure, and the performance will be better if the total thickness of the asphalt pavement is greater than 22 cm.

REFERENCES

[1] The plan of Jiangsu highway network (2017-2035), 2018. Jiangsu Province Communications Department. (in Chinese)
[2] Shen, J.A. 2004. A summary of foreign asphalt pavement design methods. Beijing: *China Communication press* (in Chinese)
[3] Huddelston, J., Buncher, M., Newcomb, D. 2000. Perpetual Pavements Asphalt Pavement Alliance.
[4] Harold, L. & Quintus, V. 2001.Hot-Mix Asphalt Layer Thickness Design for Longer-Life Bituminous Pavements. *Perpetual Bituminous Pavements*. Transportation Research Circular Number503.
[5] Romanoschi, S., Gisi, A., Portillo, M., et al. 2008.The First Findings from the Kansas Perpetual Pavements Experiment. Transportation Research Record: *Journal of the Transportation Research Board*.

Pavement surface properties and vehicle interaction

Functional Pavements – Chen et al (eds)
© 2021 Taylor & Francis Group, London, ISBN 978-0-367-72610-2

Study on permeability performance decay law and clogging test of porous asphalt mixture

Haoran Zhu, Mingming Yu & Shaochan Duan*
JSTI Group Co. Ltd., Nanjing, China
National Engineering Laboratory for Advanced Road Materials, Nanjing, China

ABSTRACT: In order to solve the pores clogging of porous asphalt mixture, the self-designed constant water head permeameter was used in the clogging simulation test of PA-13 standard Marshall specimens with different porosities, and the decay law of the water permeability was studied. The results showed that the clogging process of porous asphalt mixture was divided into four stages: fast clogging, partial recovery, slow and stable clogging. The particulate matter was mainly blocked at a depth of 15-20 mm in the asphalt mixture.

Keywords: Road engineering, Pores asphalt pavement, Permeability coefficient

1   INTRODUCTION

Porous asphalt pavement enjoys the benefits of noise reduction, anti-skid, cooling and drainage. Clogged pores are the main reason restricting the popularization and application of porous asphalt pavement.

In recent years, road researchers have studied the drainage and anti-clogging properties of porous asphalt mixtures. Moriyoshi et al. (Moriyoshi et al., 2013) reported the permeability coefficient of porous asphalt pavement which gradually decreases after opening the traffic. Coleri et al. (Coleri et al., 2013) found that the decrease of the permeability coefficient of porous asphalt pavement had a clear correlation with the attenuation of the pore structure. Tan et al. (Tan et al., 2000) showed that the permeability coefficient of porous asphalt pavement had a quadratic function relationship with the mass of particles inside the specimen. Hamzah et al. (Othman, 2005) studied the anti-clogging performance of single-layer and double-layer drainage asphalt pavement by loading the suspension.

The current reports only indicated the influence of a single factor (decay or clogging of the pore structure) on the permeability of porous asphalt pavement. In this study, the effects of porosity, pore depth and other factors on clogging and water permeability were studied through clogging simulation test and CT scanning technology, and a four-stage model of clogging process of porous asphalt mixture was established.

2   EXPERIMENTAL

2.1   *Material*

The concentration of total suspended solids (TSS) in urban rainwater runoff is 52.5~1946.5 mg/L, and the particle size smaller than 0.075 mm accounted for about 50%. Therefore, a suspension simulation with a concentration of 500 mg/L was used for the clogging simulation test. The particle size range of the suspension is shown in Table 1.

---

* Corresponding author

Table 1. Information on particulate matter in suspension.

Size	< 0.075	0.075~0.15	0.15~0.30
Mass percent (%)	54.2	29.8	16

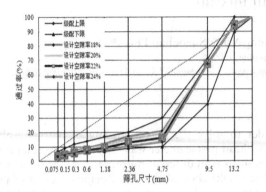

Figure 1. Synthetic gradation curves of all PA-13 specimens.

According to the standard JTG F40-2004, PA-13 standard Marshall specimens (10 cm in diameter and 4-7 cm in height) with air void of 18%, 20%, 22% and 24% were produced, among which the binder was high-viscosity modified asphalt. The synthetic gradation curves of all PA-13 specimens are shown in Figure 1.

2.2 Equipment and methods

In this study, the self-designed constant water head permeameter was designed by reference to domestic and foreign equipment which was used to determine the permeability coefficient of the standard Marshall specimens. The permeability coefficient was tested according to the standard "Technical Specification for Pervious Cement Concrete Pavement" (CJJ/T 135-2009).

An industrial CT scanner (Y. CY Precision S) was used to perform scanning tests before and after clogging of all PA-13 specimens.

3 RESULTS AND DISCUSSION

3.1 Clogging simulation test

It can be seen from Figure 2 that the normalized permeability coefficients (C) of porous asphalt mixtures with different porosities have similar changes. When the cumulative blocking suspension flow (Q) reached 100 L, C significantly decreased to about 53%. C increases slowly with the increase of Q between 100 and 120 L. When Q is between 120 and 160 L, C decreases slowly again to about 50%. And finally, C doesn't change with Q. Therefore, the clogging process is divided into a fast clogging phase, a partial recovery phase, a slow clogging phase, and a clogging stable phase. The fitted functional relationship is as follows.

$$C = 1.2769 - 0.165\ln Q \quad \text{Fast clogging } (0 \leq Q \leq 100) \tag{1}$$

$$C = 0.0017Q + 0.361 \quad \text{Partial recovery } (100 \leq Q \leq 120) \tag{2}$$

$$C = -0.0009Q + 0.6719 \quad \text{Slow clogging } (100 \leq Q \leq 160) \tag{3}$$

$$C = 0.5400 \quad \text{Clogging stable } (160 \leq Q \leq 200) \tag{4}$$

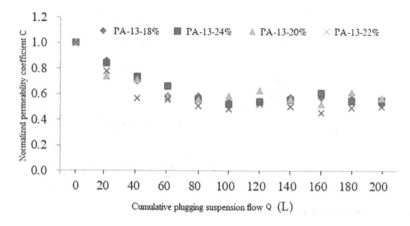

Figure 2.   Variation curve of the normalized permeability coefficient of all Marshall specimens with the cumulative plugging suspension flow.

3.2   *CT scanning test*

PA-13-20% was selected for CT scanning test before and after clogging, with a scanning interval of 0.1 mm. Matlab was used to analyze the two-dimensional image of the specimen after CT scanning, as shown in Figure 3. The regularity of clogging was studied by analyzing the air void at different depths of the specimen. The relationship between the air void and the connected air void of the PA-13-20% specimen before and after clogging and depth are shown in Figures 4 and 5.

It can be seen from Figures 4 and 5 that the scanning depth increases and the difference in air void decreases. In the depth range of 0-10 mm, the difference in air void before and after clogging is relatively large followed by small when the scanning depth is greater than 10 mm. This is attributed to the fact that large particles are clogged on the surface of the specimen, but small particles can penetrate into the inside of the specimen and can eventually reach a depth of 60 mm. Considering that the ends of the CT scan image are blurry, the upper and lower ends of all the specimens have been removed by 6.8 mm. It can be concluded that most of the particles are clogged in the depth range of 15-20 mm, and the bottom is almost not blocked. Therefore, after the road is cleaned by a high-pressure cleaning vehicle, there is almost no blockage on the surface, which is quite necessary to maintain the drainage function of the road for a long time.

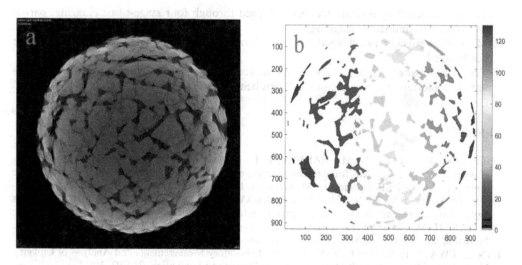

Figure 3.   Two-dimensional image of the specimen after Matlab processing.

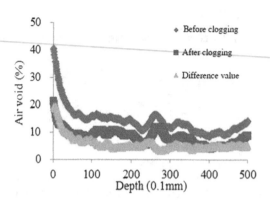

Figure 4.   Variation curve of air void of specimen with depth.

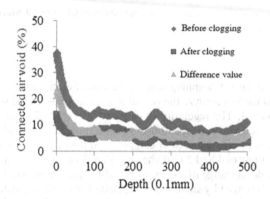

Figure 5.   Variation curve of connected air void of specimen with depth.

4   CONCLUSIONS

The standard Marshall specimens of porous asphalt mixture were subjected to clogging simulation and CT scanning test. This study can be summarized into the following conclusions.

(1) The porous leaching mixture has been plugged through four stages: fast clogging, partial recovery, slow and stable clogging.
(2) Most of the particles are clogged in the depth range of 15-20 mm, and the bottom is almost not blocked.
(3) The high-pressure cleaning truck cleans the road regularly, which can reduce road blockages and alleviate the problem of road gaps being blocked.

REFERENCES

COLERI, E., KAYHANIAN, M., HARVEY, J. T., YANG, K. & BOONE, J. M. (2013) Clogging evaluation of open graded friction course pavements tested under rainfall and heavy vehicle simulators. Journal of Environmental Management, 129, 164–172.
MORIYOSHI, A., JIN, T., NAKAI, T. & ISHIKAWA, H. (2013) Evaluation methods for porous asphalt pavement in service for fourteen years. Construction and Building Materials, 42, 190–195.
OTHMAN, H. M. (2005) Characterization of the clogging behaviour of double layer porous asphalt. Journal of the Eastern Asia Society for Transportation Studies, 6, 968–980.
TAN, S., FWA, T. F. & GUWE, V. Y. K. (2000) Laboratory Measurements and Analysis of Clogging Mechanism of Porous Asphalt Mixes. Journal of Testing and Evaluation, 28, 207–216.

Functional Pavements – Chen et al (eds)
© 2021 Taylor & Francis Group, London, ISBN 978-0-367-72610-2

The effect of aggregate type and polishing levels on long-term skid resistance

Song Li & Kaiyin Zhang
School of Transportation, Wuhan University of Technology, Wuhan, China

Rui Xiong
School of Materials Science and Engineering, Chang'an University, Xi'an, China

Jiahui Zhai
School of Materials Science and Engineering, Northwestern Polytechnical University, Xi'an, China

ABSTRACT: The aggregates used in surface layer must own sufficient skid resistance performance such as high polished stone value (PSV), low abrasion value and low crush value to ensure traffic safety. The skid resistance of the aggregate is affected by chemical composition/mineral composition. In this study, four type aggregates were selected. The main chemical compositions of aggregate were analyzed using X-ray diffraction XRD. The long-term skid resistance of different aggregates were investigated by British Pendulum Tester. The mineral composition, micro and macro texture of aggregates were linked to the skid resistance of aggregates. A new hardness ratio and multivariate analysis were introduced to understand the polishing mechanism of calcined bauxite aggregate. The results show that the attenuation of PSV was affected mainly by the attenuation rate of micro texture compared with macro texture.

1 INTRODUCTION

Flexible asphalt pavements are widely used because of their good road performance, running safety, and comfortable ability (Yin et al. 2017). However, polishing and the consequent decreases in skid resistance are the world-wide issues regarding road safety. The skid resistance of pavements refers to the friction developed during the tire/road contact process (Chen et al. 2016). There are several factors that would affect the surface texture of pavement (Masad et al. 2004). Normally, aggregate properties are regarded as one of essential factors to determine the frictional characteristics of asphalt pavements (Guan et al. 2018, Jin et al. 2018). Mineral aggregates, which have a significant effect on the skid resistance of asphalt pavement, consist more than 70% by weight in asphalt mixtures (Huang et al. 2007, Zhu et al. 2012). In order to ensure the safety of traffic, the pavement surface is generally required to use the aggregates with excellent anti-wear and anti- abrasion properties.

As is known to all, the properties of aggregates are affected by the mineral composition. A number of studies have been conducted to evaluate the effect of mineral composition on aggregate (YU et al. 2019, Li et al.2019). Properties of rocks are influenced by the following six factors: mineral composition, size, shape, porosity and spatial distribution of mineral grains. Scanning electron microscope (SEM) and X-Ray Diffraction (XRD) can be used to determine the mineral composition of rock materials.

The objective of this work is to analyze the long-term skid resistance of coarse aggregate. The four types of 88 #calcined bauxite, 75#calcined bauxite limestone aggregates, and basalt aggregate were selected. A new aggregate hardness ratio was introduced based on the mineralogical composition and the hardness of the individual minerals to understand the polishing mechanism of aggregate. This study is expected to offer practical guidance for correctly select aggregate used in the pavement surface layer.

2 MATERIALS

The four type aggregates in the experiments were 88# calcined bauxite, basalt, limestone and 75# calcined bauxite.

Table 1 summarizes some of the properties of these aggregates. It can be seen that 88# calcined bauxite had better resistance to polishing, abrasion, as compared with others. However, density and water absorption values are found to be higher for 88# calcined bauxite.

It is known that chemical components influence hardness, durability, soundness, and toughness properties of aggregates. To analyze the effect of chemical compositions on the properties of different aggregates. The chemical elements compositions of aggregates were determined by X-ray diffraction (XRD), and the results are presented in Table 2. XRD were carried out using an AXS D8 ADVANCE X-ray diffractometer (Bruker Corporation, New York, NY, USA).

From Table 2, it can be found that SiO_2 and Al_2O_3 are main chemical components of aggregates except limestone. Al_2O_3 is mainly present in corundum, and SiO_2 is mainly found in mullite in calcined bauxite; SiO_2 is present in quartz in basalt; CaO is mainly found in calcite in limestone. The Moh's hardness of corundum, mullite, quartz and calcite is 9,6,7,3, respectively. Combined with Table1, this maybe the reason that different aggregates show different properties.

Table1. Mechanical and physical properties of different aggregates.

Aggregate type	Cruse value (%)	Los Angeles Abrasion value (%)	Density (g/cm³)	Polishing Stone Value (BPN)	Water absorption (%)
88# calcined bauxite	7.7	10.6	3.328	55	1.463
75# calcined bauxite	18.6	13.8	3.034	45	5.189
Basalt	11.5	12.9	3.035	43	0.787
Limestone	22.6	20.6	2.554	38	1.950

Table 2. The chemical composition of different aggregates.

Aggregate type	SiO_2	TiO_2	Al_2O_3	Fe_2O_3	MgO	CaO	Na_2O	K_2O	P_2O_5	LOI
88# calcined bauxite	3.32	4.47	90.29	1.55	0.15	0.17	<0.01	0.17	0.24	0.01
75# calcined bauxite	19.59	3.29	75.51	1.12	0.16	0.3	<0.01	0.18	0.17	0.27
Basalt	49.51	3.61	14.00	12.62	0.16	4.97	8.67	1.92	1.4	0.05
Limestone	6.65	0.038	0.68	0.21	0.71	51.1	0.03	0.15	0.013	0.004

3 METHODS

The British Pendulum Tester (BPT)

The British Pendulum Tester has been universally applied to measure skid resistance of aggregates and asphalt mixtures respectively, according to EN1097-8 and NLT-175/98. The Pendulum Test evaluates low-speed friction (about 10 km/h).

The influence of macro and micro texture on the attenuation rate of polishing value

Long-term skid resistance of different aggregate was shown in Figure 1. When entering the polishing attenuation interval, the polishing values of the four aggregates begin to decrease significantly. According to the attenuation curve, it can be found that the rapid attenuation interval is 0-160,000 times. At the 160,000 polishing cycles, the 88# calcined bauxite aggregate, 75# calcined bauxite aggregate, and basalt aggregate all begin to enter the gradual attenuation stage of the abrasive value, while the limestone sample still continues the trend of a small decline after 160,000 times of polishing. The results indicate that 88# calcined bauxite shows the highest friction coefficient while limestone has the lowest level of friction.

Main mineral parameter of different aggregate is shown in Table 3. It can be found that 88# calcined bauxite own biggest Moh's hardness ration, and shows the biggest skid resistance. It is differential polishing principle, namely aggregates which have very different intrinsic properties is beneficial to both pavement and environment.

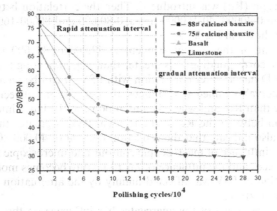

Figure 1.   PSV attenuation curve of coarse aggregate.

Table 3.   Main mineral parameter of different aggregates.

Aggregate type	Main mineral types	Mineral composition	Moh's hardness	Moh's hardness ration
88# calcined bauxite	Corundum	86.0	9	1.50
	Mullite	14.0	6	
75# calcined bauxite	Corundum	76.0	9	1.50
	Mullite	24.0	6	
Basalt	Pyroxene	35.8	6	1.09
	Plagioclase	40.7	6	
	Hornblende	23.5	5.5	
Limestone	Calcite	71.4	4	0.85
	Dolomite	28.6	3.5	

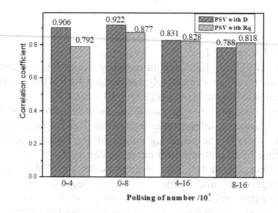

Figure 2.   The correlation between PSV and D, Rq.

Gradually speaking, the macro profile and the surface micro texture of coarse aggregate together determine the aggregate polishing value, and the interval of the rate of change of the decay rate of the abrasive value may be related to the change of the influence of macro profile and micro texture on the polishing value in the long-term polishing process. In this paper, the macro profile -fractal dimension (D) calculated by MATLAB software. Micro texture- the root mean square height (Rq) was introduced. Then the correlation between PSV the corresponding D and Rq in 0-40000, 40-160000 times, 0-80,000 and 80000-160000 interval are analyzed, the results are shown in Figure 2.

When the end point of the current segment is 40,000 or 80,000 times, the correlation between PSV and the corresponding D is greater than Rq, indicating that the first half of the attenuation segment D has a better correlation with PSV. It can be seen that when the initial point of the latter interval is valued 40,000 times, the correlation between PSV and the corresponding D is greater than Rq, and when the initial point of the latter interval is valued 80,000 times, the correlation between the Rq of the residual PSV is greater than D. The results of correlation analysis also indicate that the transition interval of 40,000-80,000 is the transition interval, in which the influence of macroscopic contour and microscopic texture on the polishing value is transformed, and the influence of micro texture becomes more significant. In conclusion, the attenuation of PSV was affected mainly by the attenuation rate of micro texture compared with macro texture.

As is known, the mineralogy of the aggregates has influence on the skid resistance of the asphalt mixtures. In order to quantify the aggregate's mineralogical composition with long-term skid resistance, the Aggregate Hardness Parameter, was defined by equation (1)- (3).

$$dmp = \sum dv_i * p_i \qquad (1)$$

$$Cd = |dv_i - dv_p| \qquad (2)$$

$$AHP = dmp + Cd \qquad (3)$$

Where: AHP means the Aggregate Hardness Parameter, dmp is the Average Hardness of the aggregates, Cd is the Contrast of Hardness of the aggregates, dv_i is the Moh's hardness of each mineral constituting the aggregates and p_i is the percentage by mass of each mineral constituting the aggregate. dv_p is the Moh's hardness of the most abundant mineral constituting the aggregate.

According to the slope of decay curves, the polishing attenuation boundary is 16000 polishing cycles. So, the PSV after 16000 polishing cycles was defined PSV_{end}. Figure 3 shows the relationship between PSV_{end} and AHP. The testing results indicate that polishing resistance is affected by the mineral structure or mineral compositions of aggregate and the aggregate

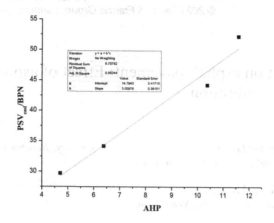

Figure 3. The relationship between PSV$_{end}$ and AHP.

hardness parameter gives a good indication of the ability of an aggregate to retain its micro texture and consequently its friction properties.

4 CONCLUSIONS

88# calcined bauxite owns the best long-term skid resistance compared with basalt, limestone and 75# calcined bauxite. The chemical composition of calcined bauxite aggregate including Al_2O_3 and SiO_2 has significant influence on the skid resistance of aggregates. This provides a new and interesting way for selecting suitable aggregate used in the pavement surface layer.

The attenuation of PSV was affected mainly by the attenuation rate of micro texture compared with macro texture.

The aggregate hardness parameter gives a good indication of the ability of an aggregate and its asphalt mixtures to retain long –term skid resistance.

REFERENCES

Chen, X.H., Dai, S.B., Guo, Y.Q., Yang, J. & Huang, X.M., 2016. Polishing of asphalt pavements: From macro- to micro-scale. *Journal of Testing and Evaluation*, 44 (2), 885–894.
Jin, C., Yang, X., You, Z. & Liu, K., 2018. Aggregate shape characterization using virtual measurement of three-dimensional solid models constructed from x-ray images of aggregates. *Journal of Materials in Civil Engineering*, 30 (3), 04018026.
Huang, Y., Bird, R.N. & Heidrich, O., 2007. A review of the use of recycled solid waste materials in asphalt pavements. *Resources, Conservation and Recycling*, 52 (1), 58–73.
Guan, B.W., Wu, J.Y., Xie, C., Fang, J.H., Zheng, H.L. & Chen, H.X., 2018. Influence of macrotexture and microtexture on the skid resistance of aggregates. *Advances in Materials Science and Engineering*, 2018, 9.
Masad, E., Little, D. & Sukhwani, R., 2004. Sensitivity of performance to aggregate shape measured using conventional and image analysis methods. *Road Materials and Pavement Design*, 5 (4), 477–498.
Yin, Y., Chen, H., Kuang, D., Song, L. & Wang, L., 2017. Effect of chemical composition of aggregate on interfacial adhesion property between aggregate and asphalt. *Construction and Building Materials*, 146, 231–237.
Zhu, J., Wu, S., Zhong, J. & Wang, D., 2012. Investigation of asphalt mixture containing demolition waste obtained from earthquake-damaged buildings. *Construction and Building Materials*, 29, 466–475.
D. Yu, R. Xiong, S. Li, P. Cong, A. Shah, Y. Jiang, Laboratory Evaluation of Critical Properties and Attributes of Calcined Bauxite and Steel Slag Aggregates for Pavement Friction Surfacing, Journal of Materials in Civil Engineering 31(8) (2019).
S. Li, P. Cong, D. Yu, R. Xiong, Y. Jiang, Laboratory and Field Evaluation of Single Layer and Double Layer High Friction Surface Treatments, Transportation Research Record: Journal of the Transportation Research Board 2673(2) (2019) 552–561.

Functional Pavements – Chen et al (eds)
© 2021 Taylor & Francis Group, London, ISBN 978-0-367-72610-2

Cycling comfort on asphalt pavement: Effect of asphalt pavement macro-texture on vibration

Jie Gao
School of Transportation and Logistics, East China Jiaotong University, Nanchang, China
School of Highway, Chang'an University, Xi'an, China

Aimin Sha, Liqun Hu & Di Yun
School of Highway, Chang'an University, Xi'an, China

ABSTRACT: The vibration experienced by cyclist greatly influences the cycling comfort on urban asphalt pavement. However, the relationship between pavement's surface characteristics and cycling vibration remains undiscovered. In this work, the cycling vibration on urban asphalt pavement was evaluated using a self-developed device, and the 3D digital data of pavement surface was captured via the 3D laser scanning technology. Furthermore, the macro-texture of pavement surface was described via 8 parameters, and the correlation between parameters and cycling vibration is analyzed. The results show that Arithmetical mean height and average peak spacing are two main factors influencing the cycling comfort. Results of this study should be interested by cyclists, bicycle manufacturers, transport planners and road authorities.

1 INTRODUCTION

Cycling is becoming a popular means of transport because of being enjoyable, environmentally-friendly and cost-effective (Zhang et al., 2015). At present, there are 77 companies in the Chinese market offering 23 million shared bicycles to more than 400 million registered users, 17 billion trips have been made using shared bicycles (Luo, 2018).

How pavement quality influences the cycling comfort? It is proved in previous study that the cycling vibration intensity is responsible for cyclists' comfort (Gao et al., 2018), but the essential question is: What qualities of the pavement determine the cycling vibration? Pavement macro-texture, a road surface characteristic with wavelengths from 0.5 mm to 50 mm, is reported to be crucial to cycling vibration (Chou et al., 2015). For instance, Li's group (Li et al., 2013a, Li et al., 2013b) first introduced the asphalt pavement macro-texture in terms of mean profile depth (MPD) to the investigation of cycling quality. Results showed that the MPD is significant with the cycling vibration. However, the characterization of pavement macrotexture should involve more parameters.

In this study, the pavements are 3D laser scanned to obtain their roughness surface, and height, hybrid, and feature information of pavement are described via 8 parameters. The correlation between macrotexture parameters and cycling comfort was established. Outcomes of this study will help road contractors to build high-comfort bike lane, for transport planners and road authorities to monitor cycling lane quality.

2 EXPERIMENTS

2.1 *Tested road sections*

A total of 11,500 m asphalt pavement was selected for field test, which involves 46 sections of 24 urban roads, as shown in. The tested sections located in a low traffic disturbance environment, and its surface characteristic has rich diversity. Furthermore, sections with severely damaged pavements, such as cracks or potholes, were not counted as tested sections; whilst pavements of minor defects were included. The slope of tested sections is in range of 0.5 % – 1.5 % while the curve radius is between 600 m – 1200 m.

2.2 Vibration measuring

A self-developed system Dynamic Cycling Comfort (DCC) Measure System was used for collecting cycling vibration signal, as shown in Figure 1. Three parts are involved, they are accelerometer, GPS coupled with a smart phone, and bicycle. A HOBO Pendant G Acceleration Data Logger was used, the three-channel 8-bit resolution vibration signal can be collected during cycling. GPS employed a VBOX Sport data logger. It records the time, position, and velocity during cycling. Meanwhile, a VBOX APP was installed on a smart phone, the smart phone functioned as a monitor that provided such info as velocity, trail, and distance to the cyclist during testing. In addition, a shared bicycle was used, which is very common in Chinese cites. The selected cyclist is a 28-years old heathy male (weight 83 Kg, height 177 cm) who has had over 6,000 km of cycling experience over the past 10 years.

The quantitative assessment of human exposure to vibration level (a_{wv}) was conducted in accordance with international standard ISO 2631 (ISO-2631, 1997) using the vibration signals on x (a_{wx}), y (a_{wy}), and z (a_{wz}) axis for the duration of the measurement T. The used frequency weighting curve was W_k. The a_{wx}, a_{wy}, and a_{wz} can be calculated from Eq. 1 while the whole-body vibration level a_{wv} was obtained from Eq. 2.

$$a_{wi} = \left[\frac{1}{T} \int_0^T a_{wi}^2(t)dt \right]^{1/2} \quad ; i = x, y, z \tag{1}$$

$$a_{wv} = \sqrt{a_{wx}^2 + a_{wy}^2 + a_{wz}^2} \tag{2}$$

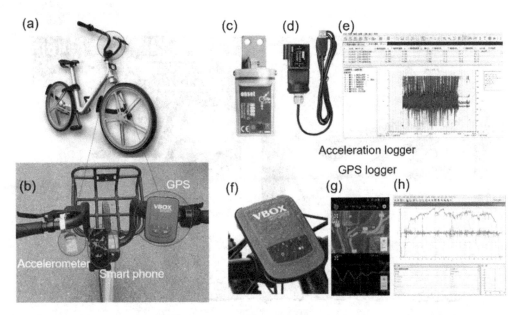

Figure 1. Set-up for dynamic cycling comfort measure system, DCC: (a) shared bicycle; (b) DCC installation; (c) acceleration logger; (d) controller and connecter of acceleration logger; (e) user interface of acceleration logger software; (f) GPS logger; (g) user interface of VBOX sport App installed on a smart phone; (h) user interface VBOX tools installed on computer for data displaying.

2.3 Cyclist' perception of vibration

Comfort is a subjective experience depending on individuals' perception. Therefore, a total of 17 volunteers were recruited to complete the questionnaire, and their assessments of vibration (and comfort) after cycling on each selected pavement section were recorded. Details of the questionnaire, volunteers and test procedures are described in our another publication (Gao et al., 2018). The key findings are summarised in section 3.2 of this paper.

2.4 3D laser scanning of asphalt pavements

A HandySCAN 300 3D scanner was used to capture the 3D digital data of the pavement surfaces. The used scanning parameters were 0.100 mm resolution, 0.040 mm precision, and 205,000 measurements/sec sampling rate, as shown in Figure 2(a). Cleaning off the dust from the selected area before the measurement starts, and scan the selected areas, as shown in Figure 2(b). Three area was scanned to represent a road section.

2.5 Characterization of pavement macro-texture

2.5.1 Height parameters
Four height parameters including the arithmetical mean height (*Sa*), height distribution symmetry (*Ssk*), height distribution kurtosis (*Sku*), and mean texture depth (MTD) are considered to describe the height characteristics of the pavement surface. The height parameters are defined and calculated using Eq.3 to Eq.8, in which *A* is the projection area of evaluation area; *z* (*x*, *y*) is the ordinate values and (*x*, *y*) is used to identify the position of calculation point.

$$Sa = \frac{1}{A}\int\int_{0}^{A}|z(x,y)|dxdy = \frac{\sum_{x=1}^{M}\sum_{y=1}^{N}|z(x,y)|}{M \times N} \tag{3}$$

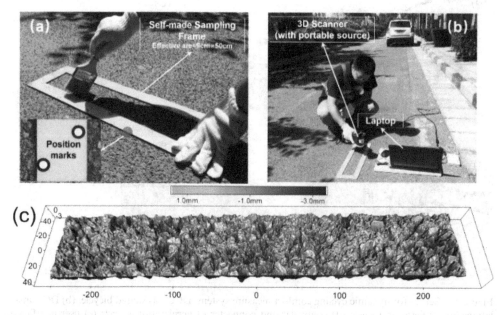

Figure 2. Field 3D scanning for the pavement surfaces: (a) sampling frame with position marks; (b) 3D scanning; (c) a typical 3D pavement surface model.

$$Ssk = \frac{1}{Sq^3}\frac{1}{A}\iint\limits_0^A z^3(x,y)dxdy = \frac{\sum_{x=1}^{M}\sum_{y=1}^{N} z^3(x,y)}{M \times N \times Sq^3} \tag{4}$$

$$Sq = \sqrt{\frac{1}{A}\iint\limits_0^A z(x,y)dxdy} = \sqrt{\frac{\sum_{x=1}^{M}\sum_{y=1}^{N} z(x,y)}{M \times N}} \tag{5}$$

$$Sku = \frac{1}{Sq^4}\frac{1}{A}\iint\limits_0^A z^4(x,y)dxdy = \frac{\sum_{x=1}^{M}\sum_{y=1}^{N} z^4(x,y)}{M \times N \times Sq^3} \tag{6}$$

$$MPD = \frac{Peak\ level(1^{st}) + Peak\ level(2^{nd})}{2} - Average \tag{7}$$

$$MTD = 0.8 \times MPD + 0.2 \tag{8}$$

2.5.2 *Hybrid parameters*

Height parameters involve only the statistical distribution of height values along the z axis, cannot describe the irregular texture of road surface. The hybrid parameters relate to the spatial shape of the surface data, can provide the necessary information for the total texture distribution. In this study, the root-mean-square slope (*Sdq*) and developed interfacial area ratio (*Sdr*) were introduced to describe the characteristic of total 3D macrotexture image. The parameters were calculated as follow, in which A is the projection area of evaluation area; $z(x, y)$ is the ordinate values and (x, y) is used to identify the position of calculation point.

$$Sdq = \sqrt{\frac{1}{A}\sum_A\left[\frac{\sigma^2 z(x,y)}{\sigma x} + \frac{\sigma^2 z(x,y)}{\sigma y}\right]} \tag{9}$$

$$Sdr = \frac{1}{A}\left\{\sum_A\left[\sqrt{\left[1+\left[\left(\frac{\sigma z(x,y)}{\sigma x}\right)\right]^2 + \left[\left(\frac{\sigma z(x,y)}{\sigma y}\right)\right]^2\right]} - 1\right]\right\} \tag{10}$$

2.5.3 *Feature parameters*

The peak characteristic parameters of 3D macrotexture images influence the contact types and areas between tire and pavement. Therefore, these parameters play important roles in cycling comfort of the pavement surface. In this study, the arithmetic mean of peak curvature (*Spc*) and average peak spacing (*Spa*) were used to describe the peak characteristic.

The calculation of *Spc* was conducted using Eq.11, in which A is the projection area of evaluation area; $z(x, y)$ is the ordinate values and (x, y) is used to identify the position of calculation point.

$$Spc = -\frac{1}{2}\frac{1}{n}\sum_1^n\left(\frac{\sigma^2 z(x,y)}{\sigma x^2} + \frac{\sigma^2 z(x,y)}{\sigma y^2}\right) \tag{11}$$

Average peak spacing (*Spa*) refers to the aggregate spacing that exposed on the pavement surface, to calculate the *Spa*, the profile line was extracted from the surface, and a filter with 2 mm cutoff wavelength was used, and was calculated via Eq.12. Where, p_i is distance between the $i+1^{th}$ and i^{th} peak (mm), and n is the total number of peaks.

$$\text{Spa} = \sum_{i=1}^{n} p_i \qquad (12)$$

3 RESULTS AND DISCUSSIONS

3.1 *Correlation between macro-texture and cycling comfort*

3.1.1 *Height parameters and cycling vibration*

The correlation between the height parameters and cycling vibration is shown in Figure 3. It can be found that there is no significant law between height distribution symmetry (*Ssk*), height distribution kurtosis (*Sku*) and cycling vibration level, as shown in Figure 3(a) and (b), indicating that the height distribution can not influence the cycling comfort. On the other hand, the arithmetical mean height (*Sa*), mean texture depth (MTD) and cycling vibration present a strong correlation, correlation coefficient for *Sa*-vibration and MTD-vibration is 0.60 and 0.41, respectively, higher than the 0.304, the critical value of correlation coefficient considering 42 samples.

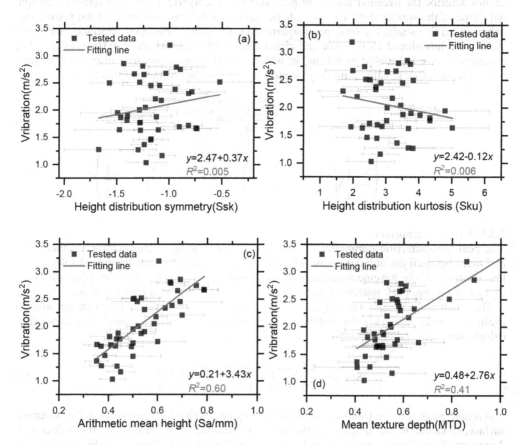

Figure 3. Correlation between the height parameters and cycling vibration: (a) *Sa*; (b) *Ssk*; (c) *Sku*; (d) *MTD*.

3.1.2 *Hybrid parameters and cycling vibration*

The correlation between the hybrid parameters and cycling vibration is shown in Figure 4. The root-mean-square slope (*Sdq*) and developed interfacial area ratio (*Sdr*) are not the key factors to cycling comfort since their Pearson's correlation coefficient are 0.04 and 0.06, respectively. Figure 4 confirm a fact that for the inclination of the pavement surface texture has no influence on cycling vibration.

3.1.3 *Feature parameters and cycling vibration*

The feature parameters essentially reflect the peak chrematistics of the aggregate peak distributed on pavement surface. By observing Figure 5, the average peak spacing (*Spa*) of tested pavement sections is in range of 4-8mm under a 2 mm cutoff length filter, and the *Spa* presents a strong positive correlation to the cycling vibration, indicating that the distance between aggregate, who is exposed on pavement surface, is a key factor to cycling vibration, the cycling comfort can be improved if a pavement surface with a lower *Spa*, this can be achieved by using smaller aggregate with a more dense gradation. On the other hand, the arithmetic mean of peak curvature (*Spc*) shows no clear relationship with cycling vibration, suggesting that the curvature of the aggregate peak can not determine the cycling vibration.

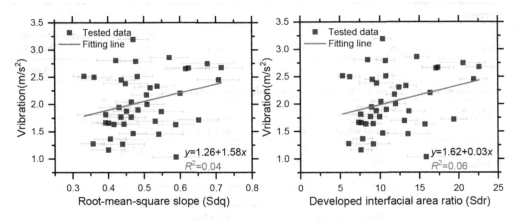

Figure 4.   Correlation between the hybrid parameters and cycling vibration: (a) *Sdq*; (b) *Sdr*.

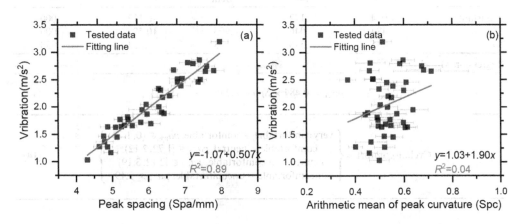

Figure 5.   Correlation between the feature parameters and cycling vibration: (a) *Spa*; (b) *Spc*.

3.2 Predicting cycling vibration based on the macro-texuture

The cycling comfort evaluation using the DCC system requires an on-site cycling to collect vibration data, apparently, this evaluation only suitable to field test. Therefore, it is necessary to develop a method to evaluate the cycling comfort of asphalt mixture in the laboratory stage. Based on the Pearson's correlation coefficient between the macro-texture and vibration, the arithmetical mean height (Sa) and the average peak spacing (Spa) are used as independent variable to build the regression model with cycling vibration. As such, the predicted cycling vibration pa_{wv} can be determined as well as the cycling comfort.

The results of ANOVA are shown in Table 1, in which the R^2 is 0.900, and Durbin-Watson is 1.359, indicating the regression model is well fitted and there is no problem of collinearity. Meanwhile, the significance of the variables below 0.05, indicating the fitting is valid. By involving the Sa and Spa to Eq.13, the predicted cycling vibration pa_{wv} can be calculated, comparing the pa_{wv} with the comfort threshold, as shown in Eq.14 which is developed in our previous study(Gao et al., 2018), the possible cycling comfort of a pavement surface can be predicted. Table 1 shows that the consistency of pa_{wv} and tested vibration, suggesting that the prediction is desirable.

Table 1.   The regression model between macro-texture and vibration.

Overview

Model	R	R^2	Adjusted R^2	Standard estimate error	Durbin-Watson
1	0.949	0.900	0.895	0.175	1.359

Independent variable: (Constant)、 Spa、 Sa;Dependent variable: a_{wv}

ANOVA

	Sum of squares	df	Mean square	F	Sig.
Regression	10.360	2	5.180	171.178	.000[b]
Residual	1.150	38	.030		
Total	11.510	40			

Coefficient

	Coefficient of non standardization		Standard coefficient		
	B	Std	Beta	t	Sig.
(Constant)	-1.054	.172		-6.117	.000
Spa	.484	.046	.905	10.527	.000
Sa	.235	.378	.053	.622	.038

Regression Model

$$pa_{wv} = 0.484 \times Spa + 0.235 \times Sa - 1.054 \qquad (13)$$

$$Cycling\ comfort = \begin{cases} very\ comfortable - comfortable, pa_{wv} \in (0, 1.72) \\ comfortable - neutral, pa_{wv} \in [1.72, 2.12) \\ neutral - uncomfortable, pa_{wv} \in [2.12, 3.19) \\ uncomfortable - uncomfortable, pa_{wv} \geq 3.19 \end{cases} \qquad (14)$$

4 CONCLUSIONS

This research has examined how pavement macrotexture influences the cycling comfort. It establishes a method for evaluating the cycling comfort, and provides recommendations for asphalt pavement design for bike lanes. Arithmetical mean height (Sa) and average peak spacing (Spa) are two main factors influencing the cycling comfort. Sa determines the amplitude of vibration, and Spa determines the vibration frequency. The flatter the height direction of the macro texture of the asphalt pavement and the denser the distribution of structural peaks, the lower the cycling vibration intensity and the better the comfort.

REFERENCES

CHOU, C.-P., LEE, W.-J., CHEN, A.-C., WANG, R.-Z., TSENG, I. C. & LEE, C.-C. 2015. Simulation of Bicycle-Riding Smoothness by Bicycle Motion Analysis Model. *Journal of Transportation Engineering*, 141, 04015031.

GAO, J., SHA, A., HUANG, Y., HU, L., TONG, Z. & JIANG, W. 2018. Evaluating the cycling comfort on urban roads based on cyclists' perception of vibration. *Journal of Cleaner Production*, 192, 531–541.

H LZEL, C., H CHTL, F. & SENNER, V. 2012. Cycling comfort on different road surfaces. *Procedia Engineering*, 34, 479–484.

ISO-2631 1997. Mechanical vibration and shock Evaluation of human exposure to whole-body vibration - Part 1 General requirements. International Organization for Standardization.

LI, H., HARVEY, J., THIGPEN, C. & WU, R. 2013a. Surface Treatment Macrotexture and Bicycle Ride Quality (Vol. Research Report: UCPRC-RR-2013-07): University of California Pavement Research Center: Davis, CA.

LI, H., HARVEY, J., WU, R., THIGPEN, C., LOUW, S., CHEN, Z. & REZAIE, A. 2013b. Preliminary Results: Measurement of Macrotexture on Surface Treatments and Survey of Bicyclist Ride Quality on Mon-198 and SLO-1 Test Sections (Vol. Technical Memorandum: UCPRC-TM-2013-07): University of California Pavement Research Center: Davis, CA.

LUO, W. 2018. Autonomous cars, shared bikes due for improvement. *Chinadaily*.

ZHANG, L., ZHANG, J., DUAN, Z.-Y. & BRYDE, D. 2015. Sustainable bike-sharing systems: characteristics and commonalities across cases in urban China. *Journal of Cleaner Production*, 97, 124–133.

Functional Pavements – Chen et al (eds)
© 2021 Taylor & Francis Group, London, ISBN 978-0-367-72610-2

Interlayer shear resistance and anti-skidding durability of asphalt wearing course overlay on tunnel cement pavement

Yi Zhang
Shanghai urban construction design and research institute (group) co., Ltd.
Shanghai, P.R. China

Mingjun Hu*, Daquan Sun, Tong Lu & Jianmin Ma
Key Laboratory of Road and Traffic Engineering of Ministry of Education Tongji University, Shanghai,
P.R. China

ABSTRACT: Asphalt wearing course overlay technology can prolong service life and improve service level of old cement pavement significantly, but few studies have focused on the interlayer shear resistance and long-term anti-skidding durability of asphalt wearing course overlay on cement pavement (AWCOCP). In order to study the interlayer shear resistance and long-term skidding resistance of AWCOCP, the cement concrete base, 2 types of tack coats, and 3 types of asphalt wearing courses were chosen to prepare AWCOCP. The direct shear test was conducted to study the effect of high temperature, water immersion, and freezing-thawing cycle on the interlayer shear property of AWCOCP. Then, the MMLS3 accelerated loading tester was used to carry out the long-term accelerated loading process of AWCOCP, and the anti-skidding durability of AWCOCP was evaluated by measuring BPN value at wheel track. The results show that AWCOCP has a decreasing interlayer shear strength with the effect of high-temperature, water immersion, and freezing-thawing cycle. SMA-10 and SMA-13 AWCOCP have higher immersion strength ratio and freezing-thawing strength ratio than SMA-5 AWCOCP, and the application of "waterproof binder+high viscosity modified asphalt" tack coat can improve the interlayer shear resistance of AWCOCP significantly. Moreover, the anti-skidding durability of SMA-10 and SMA-13 AWCOCP is better than that of SMA-5 AWCOCP, and the gap between the skidding resistance of them increases with the increase of loading times and temperature.

1 INTRODUCTION

Cement pavement is a common pavement form of tunnel pavement. However, due to the repeated action of high-speed, large-flow, heavy-axle traffic, the cement pavement is worn and polished constantly, and the skidding resistance of cement pavement is reduced gradually, which affects driving safety and driving comfort significantly (Li et al., 2016). Asphalt wearing course overlay technology can improve the service performance of old cement pavement effectively, which has shown broad application prospects in the maintenance of old cement pavement (Hu et al., 2019). However, the application of asphalt wearing course overlay on cement pavement (AWCOCP) also has some problems (Wang et al., 2012). It is difficult for cement pavement and asphalt wearing course overlay to work coordinately due to the different modulus between them (Lastra-Gonzalez et al., 2017). Therefore, AWCOCP should have excellent interlayer bonding properties to reduce interlayer displacement and ensure structure stability. Moreover, the long-term anti-skidding durability of AWCOCP should also be focused on to ensure driving safety (Wang et al., 2018, Wang et al., 2013).

---

* Corresponding author

The aim of this study was to investigate the interlayer shear resistance and long-term skidding resistance of AWCOCP based on laboratory tests. Cement concrete base, 2 types of tack coats (modified emulsified asphalt, waterproof binder+high viscosity modified asphalt), and 3 types of asphalt wearing courses (SMA-5, SMA-10, SMA-13) were chosen to prepare AWCOCP. Then, the direct shear test was conducted to study the effect of high-temperature, water immersion, freezing-thawing cycle on the interlayer shear resistance of AWCOCP. Finally, the long-term loading test of AWCOCP was conducted by MMLS3 accelerated loading tester, and the anti-skidding durability of AWCOCP was evaluated by measuring BPN value at wheel track.

2 MATERIALS AND METHODS

2.1 Materials and mix design

SBS modified asphalt (SBSMA), three grade basalt aggregates (3-5mm, 5-10mm, 10-15mm), one grade limestone aggregates (0-3mm) and limestone mineral power were used in the mix design of SMA asphalt mixtures. The basic properties of SBSMA were shown as follows: penetration (25°C): 61 (0.1mm), softening point: 72°C, ductility (5°C): 38 cm. The mix design results of SMA-5, SMA-10, and SMA-13 asphalt mixtures were shown in Table 1. The modified emulsified asphalt (M), waterproof binder+high viscosity modified asphalt (W+H) were used as the tack coats. C40 Portland cement concrete was used as the cement concrete base.

2.2 Specimen preparation

The specimen size of AWCOCP in direct shear test was 150mm (ϕ)×(50+80) mm (h). W+H and M were chosen as the tack coats, and SMA-5, SMA-10, SMA-13 asphalt mixtures were chosen as the wearing courses. The same specimens in direct shear test were prepared in three groups. The first group was kept at 25°C incubator for 2h before test. The second group was immersed at 25°C water for 48h and kept at 25°C incubator for 2h before test. The third group was treated for 4 freezing-thawing cycles and kept at 25°C incubator for 2h before test.

The preparation process of AWCOCP specimen in accelerated loading test was similar to direct shear specimen. W+H was chosen as the tack coat, and SMA-5, SMA-10, SMA-13 asphalt mixtures were chosen as the wearing courses.

2.3 Direct shear test

The direct shear test of AWCOCP was carried out by MTS 810 test machine, as shown in Figure 1(a). The loading rate was 2.5mm/min, the vertical load is 0.7MPa, and the test temperature was 10°C, 25°C, 60°C. According to the shear failure load obtained in test, the interlayer shear strength of AWCOCP can be calculated as shown in Eq. 1.

$$\tau = F/A \tag{1}$$

where F is shear failure load, kN; A is failure area, 0.0177 m^2; τ is shear strength, MPa.

Table 1. Mix design of SMA asphalt mixtures.

Mixture types	Mass ratio of following size of aggregates/%				Mineral powder	Asphalt content/%
	Basalt aggregates			Limestone aggregate		
	10-15mm	5-10mm	3-5mm	0-3mm		
SMA-5	-	-	55	35	10	6.7
SMA-10	-	70	5	13	12	6.3
SMA-13	60	15	-	14	11	6.2

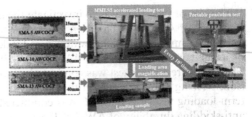

(a) Direct shear test	(b) Anti-skidding durability test

Figure 1.   Laboratory tests in this study.

2.4   *Anti-skidding durability test*

The MMLS3 accelerated loading test was conducted to simulate the wear of wheels on AWCOCP. The loading condition was 0.7MPa tire pressure, 5000r/h loading rate, and 10^6 loading times. The test temperature was 25°C and 70°C. The skidding resistance at wheel track was measured and recorded using portable pendulum tester for every 10^5 loading times. The loading and measurement process were repeated, until the BPN value during whole loading process were obtained correspondingly (Figure 1 (b)).

3   RESULTS AND DISCUSSION

3.1   *Interlayer shear strength results*

3.1.1   *Influence of water immersion and freezing-thawing cycle*
Figure 2 shows the effect of water immersion and freezing-thawing cycle on the interlayer shear strength of AWCOCP. It can be seen that the shear strength of AWCOCP shows a decreasing trend after water immersion, which indicating that the water immersion has an adverse effect on the interlayer shear property of AWCOCP. In addition, when using W+H tack coat, the immersion shear strength radio of AWCOCP is significantly higher than that with M tack coat, which demonstrates that the water sensitivity of W+H is less than that of M.

The interlayer shear strength of AWCOCP decreases significantly after freezing-thawing cycle, and the decreased degree is higher than that after water immersion. This result shows that the effect of freezing-thawing cycle is more significant than the effect of water immersion on the shear property of AWCOCP. Moreover, AWCOCP with W+H tack coat has a higher freezing-thawing strength ratio than that with M tack coat, which indicates that W+H tack

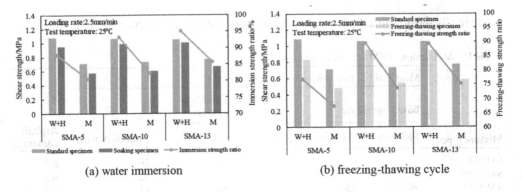

(a) water immersion	(b) freezing-thawing cycle

Figure 2.   Shear strength results of AWCOCP.

coat can prevent water intrusion and reduce the influence of freezing-thawing cycles effectively. SMA-10 and SMA-13 AWCOCP have higher immersion strength ratio and freezing-thawing strength ratio than SMA-5 AWCOCP. The skeleton structure of SMA-5 is relatively weak. Water immersion and freezing-thawing cycles can have a more serious impact on the skeleton friction at interface, which leads to a weak interlayer shear strength of SMA-5.

3.1.2 Influence of temperature

Figure 3 shows the effect of temperature on the interlayer shear strength of AWCOCP. According to Figure 3, it can be found that the increase of temperature leads to a significantly reduced shear strength of AWCOCP. The properties of tack coat play an important role in the interlayer shear strength of AWCOCP. With the increase of temperature, the viscosity of tack coat decreases gradually, which results in a serious interlayer sliding failure of AWCOCP.

3.2 Long term anti-skidding durability results

Figure 4 shows the changes in BPN value of AWCOCP during loading process at different temperatures. The initial BPN value of three AWCOCP are all higher than 60, and they present a decreasing trend with the increase of loading times. When loaded 10^6 times, the decreasing order of BPN value of AWCOCP is: SMA-13>SMA-10>SMA-5. This result indicates that SMA-5 AWCOCP is more sensitive to the long-term wear of wheels in comparison with SMA-10 and SMA-13 AWCOCP. SMA-5 asphalt mixture has less coarse aggregates and weak skeleton interactions. With the long-term loading of wheels, SMA-5 asphalt mixture is polished quickly, resulting in a decrease in anti-skidding performance. In contrast, SMA-10

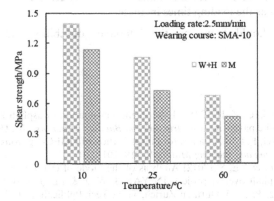

Figure 3.   Effect of temperature on the interlayer shear strength of AWCOCP.

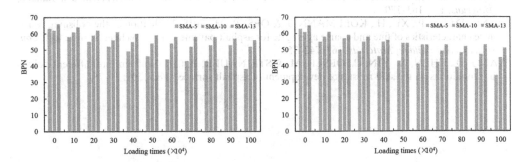

Figure 4.   Anti-skidding durability results of AWCOCP (left: 25°C, right: 70°C).

and SMA-13 have uniform coarse aggregates and fine aggregates, thus having excellent skeleton interaction and anti-wear ability. Therefore, with the long-term loading of wheels, SMA-10 and SMA-13 AWCOCP have better skidding resistance compared with SMA-5 AWCOCP.

Moreover, it can be found from Figure 4 that the increase of temperature leads to a deteriorated skidding resistance of AWCOCP. When loaded 10^6 times, the BPN values of SMA-5, SMA-10, and SMA-13 at 70°C decreases by 10.5%, 13.5%, and 8.9%, respectively in comparison to that at 25°C. With the increase of temperature, the viscosity of asphalt binder decreases, and the displacement of coarse aggregates under loading increases. This reduces the macroscopic texture of AWCOCP surface, thus leading to a deteriorated anti-skidding performance of AWCOCP.

4 CONCLUSIONS

The objective of this study was to study the interlayer shear resistance and anti-skidding durability of AWCOCP. The main conclusions were summarized as follows.

(1) With the effect of high temperature, water immersion, and freezing-thawing cycle, the interlayer shear strength of AWCOCP shows a decreasing trend. The effect of freezing-thawing cycle on the interlayer shear resistance of AWCOCP is greater than that of water immersion. The immersion strength ratio and freezing-thawing strength ratio of SMA-10 and SMA-13 AWCOCP are higher than that of SMA-5 AWCOCP. Furthermore, the use of W+H tack coat is more helpful to improve the interlayer shear resistance of AWCOCP compared with M tack coat.
(2) With the increase of loading times and temperature, the skidding resistance of AWCOCP decreases gradually. The anti-skidding durability of SMA-10 and SMA-13 AWCOCP is better than that of SMA-5 AWCOCP, and the gap between the anti-skidding performance of them increases as the loading times increase.

Future research should focus on the interlayer durability and fatigue property of AWCOCP.

REFERENCES

LI, S., HUANG, Y. & LIU, Z.-H. (2016) Experimental evaluation of asphalt material for interlayer in rigid-flexible composite pavement. *Construction and Building Materials*, 102, 699–705.
HU, M., Li, L. & PENG, F. (2019) Laboratory investigation of OGFC-5 porous asphalt ultra-thin wearing course. *Construction and Building Materials*, 219, 101–110.
WANG, H., WU, Y. F. & YE, S. (2012) Analysis of the mode of crushing and stability in old cement pavement during asphalt overlay project. IN YANG, W. J. & Li, Q. S. (Eds.) *Progress in Industrial and Civil Engineering, Pts. 1–5*. Durnten-Zurich, Trans Tech Publications Ltd.
LASTRA-GONZALEZ, P., INDACOECHEA-VEGA, I., CALZADA-PEREZ, M. A., CASTRO-FRESNO, D. & CARPIO-GARCIA, J. (2017) Analysis of the skid resistance and adherence between layers of asphalt concretes modified by dry way with polymeric waste. *Construction and Building Materials*, 133, 163–170.
WANG, D., Liu, P., XU, H., KOLLMANN, J. & OESER, M. (2018) Evaluation of the polishing resistance characteristics of fine and coarse aggregate for asphalt pavement using Wehner/Schulze test. *Construction and Building Materials*, 163, 742–750.
WANG, D., CHEN, X., YIN, C., OESER, M. & STEINAUER, B. (2013) Influence of different polishing conditions on the skid resistance development of asphalt surface. *Wear*, 308, 71–78.

Functional Pavements – Chen et al (eds)
© 2021 Taylor & Francis Group, London, ISBN 978-0-367-72610-2

3D reconstruction of pavement texture using single camera close-range photogrammetry

Ruoyu Zhou & Jun Chen
College of Civil and Transportation Engineering, Hohai University, Nanjing, China

Xijun Shi
Center for Infrastructure Renewal, Texas A&M University, Bryan, Texas, USA

ABSTRACT: This study developed a new approach to 3D reconstruction of pavement surface profile based on single camera close-range photogrammetry. Three asphalt concrete mixtures with different surface characteristics (i.e., AC-13, SMA-13 and OGFC-13) were examined using this new technique. Based on the reconstructed 3D surface profile database, surface property indexes including the mean texture depth and the standard deviation of texture depth were computed. The results showed that the OGFC-13 has the roughness surface while the AC-13 has the smoothest surface. This finding matches the existing knowledge of these asphalt concrete mixtures. To further validate the new method, a traditional pavement surface texture test (i.e., the sand patch test) were conducted using the same sample, and the test results yielded good agreement with those from the photogrammetry.

1 INTRODUCTION

There are two methods can be used to obtain pavement surface texture, contact measurement method and non-contact measurement method (Ueckermann et al., 2015). The contact measurement method refers to a field test using traditional equipment, which include the sand patch tester, British pendulum tester, dynamic friction tester, GripTester, etc. Despite the test methods are simple and straightforward, the results of the measurement are greatly influenced by human factors, and traffic could be interrupted during the measurement.

On the other hand, the non-contact measurement method is much more versatile. Non-contact measurement methods mainly include the use of laser technique, industrial computed tomography (CT), close-range photogrammetry (CRP), etc. to 3D reconstruct pavement surface. The laser method requires special equipment that is expensive and complex to operate (Zhang et al., 2018). One of the major drawbacks of the laser method is that it is only suitable for determining the texture depth of the dry surface of the asphalt pavement. The CT scanning method is only applicable to lab-sized specimens; this technique is not ready to be implemented in the field (Xu et al., 2018).

The close-range photogrammetry method, as a representative of the digital image reverse reconstruction technology, is to take pictures of the pavement surface from multiple angles and reconstruct the 3D texture pattern of pavement surface through digital image matching (Granshaw, 2010). Compared with the other non-contact measurement methods, it has the advantages of lower test time and higher measurement efficiency (Alamdarlo et al., 2018, Mçgowan et al., 2018, Sun et al., 2016). An automatic close-range photogrammetry system (ACRP System) was proposed and developed by Chen J. et al. (2019). The images taken from the three Basler industrial cameras were used to automatically reconstruct field pavement surface. The whole process was controlled and completed in real time, its efficiency and accuracy was greatly enhanced compared with the traditional close-range photogrammetry. However, the Basler cameras system is expensive and heavy, and the distance between the cameras and the measured object cannot be changed.

This study aims to improve the existing close-range photogrammetry system to develop an affordable and portable product for autonomous vehicle. The objectives of this study were to:

- Develop a 3D reconstruction method of asphalt pavement surface based on single-camera close-range photogrammetry and digital imaging technology.
- Obtain the surface properties (e.g., texture depth) of asphalt pavement from the reconstructed surface profile.
- Validate the technique by comparing the mean texture depth calculated from the reconstruction with that tested by the traditional sand patch method.

2 PRINCIPLES OF SURFACE TEXTURE 3D RECONSTRUCTION USING CLOSE-RANGE PHOTOGRAMMETRY

The workflow of asphalt pavement texture reconstruction using digital close-range photogrammetry technology is divided into three major steps: image acquisition, image processing, and 3D reconstruction based on point clouds.

During the image acquisition process, a single camera was used to capture images of pavement at different shooting angles. In a two-camera-position example (Figure 1), the camera captured two images (Image 1 and 2) of the pavement at two positions (Camera Position 1 and 2). For a point P on the surface of the pavement, its corresponding points in Image 1 and Image 2 were P' and P'', respectively.

The final and also the most critical step is the 3D reconstruction based on point clouds. The point clouds database of the surface can be calculated from the numerous amounts of images of the pavement which were photographed at different angles. The procedures are:

(1) Matching feature points: SIFT (Scale-Invariant Feature Transform) algorithm (Lowe, 2004) was used to extract all feature points in the two images and match same features of two photos. (e.g., P' and P'' in the two images are two feature points of P in Figure 1.)
(2) Coordinate system transformation: The coordinates of point P are inconsistent on camera coordinate system $o'-x'y'z'/o''-x''y''z''$ and pavement coordinate system $O\text{-}XYZ$. SFM (Structure from Motion) algorithm (Wu, 2013) was used to iteratively solving spatial transformation matrix between the coordinate systems based on large a large amount of image information and transform P' and P'''s coordinate from $o'-x'y'z'/o''-x''y''z''$ to $O\text{-}XYZ$.
(3) Solving the feature point coordinates: $o'P'P$ and $o''P''P$ have collinearity, so that there are two collinearity equations. The coordinate of point $P(X,Y,Z)$ on the pavement coordinate

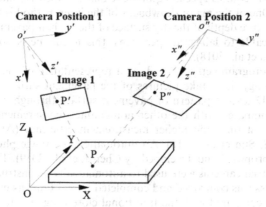

Figure 1.   Principles of close-range photogrammetry.

298

system *O-XYZ* can be readily solved by the two collinearity equations. Thereby, all the coordinates of feature points on the pavement surface can be obtained and formed as the 3D point clouds database.

3 ASPHALT CONCRETE SURFACE TEXTURE DETERMINATION

3.1 *Specimen preparation*

The conventionally used aggregate gradations of asphalt mixtures in China are AC-13, SMA-13 and OGFC-13, so these three aggregate gradations were adopted in this study to make lab specimens. Three 30 ×30 ×5 cm³ specimens were prepared according to the gradations and asphalt content in Table 1. The gradations and asphalt content are determined by the Marshall method.

3.2 *Image acquisition and processing*

A rectangular frame ruler with 24 ×15cm² inner measuring size was placed on the surface of the asphalt mixture specimen during the image acquisition. It would be used to adjust the coordinate system and measure the size of specimen during the reconstruction process. A digital camera with a resolution of 4608×3456 pixels and a 4mm focal length was used for image acquisition. The camera was fixed beside the specimen by a tripod (Figure 2). The tripod was placed in multiple positions around the specimen, and at least 10 photos were taken for each specimen.

3.3 *Surface texture reconstruction*

The VisualSFM open source platform was used to build up the spatial point clouds of asphalt concrete surface. The MeshLab software was used to delete irrelevant points in the point clouds and reconstruct the 3D surface profile. The Geomagic Design X software was used to

Table 1. Aggregate gradation and asphalt content of three asphalt mixtures.

Gradation type	Mass percentage (%) passing the sieve size (mm)										Asphalt content (%)
	16	13.2	9.5	4.75	2.36	1.18	0.6	0.3	0.15	0.075	
AC-13	100	96.5	76.0	52.7	33.6	26.5	20.2	13.5	10.3	7.2	5.1
SMA-13	100	94.7	63.0	27.7	20.3	18.7	17.0	15.0	13.3	10.9	6.0
OGFC-13	100	93.5	55.3	19.9	13.3	11.7	10.0	8.2	7.0	5.5	4.4

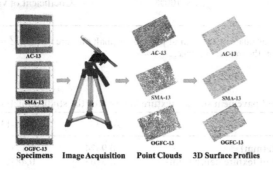

Figure 2. The workflow of surface reconstruction.

299

adjust the coordinate system and voxel size by using the rectangular frame ruler in model. After all the adjustments, the entire 3D surface profiles of the specimen within the rectangular frame of the three asphalt mixtures specimens were reconstructed (Figure 2).

4 ASPHALT CONCRETE SURFACE TEXTURE ANALYSIS

4.1 Calculation of texture characteristics

After the 3D surface profile was obtained, the mean texture depth (MTD), the standard deviation of texture depth (σ) and the relative standard deviation were chosen as the evaluation indexes of pavement surface texture. They are calculated in the equations listed in Table 2.

The MATLAB software was used for the data analysis. A 24 ×150 mm2 mesh with 0.1mm mesh size was created in the MATLAB code, and the profile data from the point cloud database was assigned to each mesh node by interpolation. Data outlier around the edge of the mesh due to improper camera sight was deleted. The MTD, σ and RSD of the surface were calculated based on 2.52 million data points on the surface. The results are shown in Table 3.

From Table 3, the mean texture depth and the standard deviation of texture depth are ranked as: OGFC-13> SAM-13> AC-13. This indicates that OGFC-13 has the highest surface roughness and depth fluctuation, while AC-13 has the lowest surface roughness and depth fluctuation. Relative standard deviation of SMA-13 is greater than AC-13 and OGFC-13, which means the ratio between the deviation and the mean of the surface texture is higher for the SMA-13 compared to AC-13 and OGFC-13.

4.2 Distribution of surface texture depth

The elevation distribution maps of the three different asphalt mixtures are output by the MATLAB (Figure 3), and each includes 2.52 million surface depth data. A 0 depth in the map means the highest elevation on the surface of asphalt mixture.

Figure 3 shows that the three asphalt mixtures all have uneven surface. The pattern matches extremely well with that in the raw images (Figure 2). This means the data processing was satisfactory. It can be seen from the depth scale color-bar that the variation range in the depth

Table 2. Evaluation indexes and equations of pavement surface texture.

Evaluation index	Equation	Illustration
Mean texture depth (MTD)	$\frac{1}{m \times n} \int_0^n \int_0^m (Z_p - z(x,y))dxdy$	Average depth of texture in the sampling area
Standard deviation of texture depth (σ)	$\sqrt{\frac{1}{m \times n} \int_0^n \int_0^m (z(x,y) - \overline{Z})^2 dxdy}$	Vertical deviation of texture depth
Relative standard deviation (RSD)	$\frac{\sigma}{MTD} \times 100\%$	Coefficient of variation

Note: Z_p is the vertical coordinate of the highest point on asphalt mixture surface; $\overline{Z}$ is the average vertical coordinate of all points on the mixture surface.

Table 3. Calculated pavement surface texture indexes for the studied mixtures.

Evaluation index	AC-13	SMA-13	OGFC-13
Mean texture depth (mm)	0.9424	1.1780	1.7944
Standard deviation of texture depth (mm)	0.3103	0.4879	0.5914
Relative standard deviation (%)	32.9	41.4	33.0

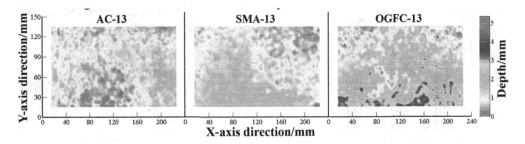

Figure 3.   Elevation colormap of three asphalt mixtures surface.

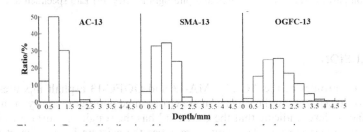

Figure 4.   Depth distribution histograms of three asphalt mixtures.

for AC-13, SMA-13, OGFC-13 is within 3mm, 4mm, 5mm, respectively. The surface roughness of the dense-graded asphalt mixture AC-13 is the smallest among the three mixtures, and
the surface roughness of open-graded asphalt mixture OGFC-13 is the largest. This finding
matches the existing knowledge of dense-graded and open-graded asphalt mixtures.

The 2.52 million depth data points on the surface of the mixture were statistically analyzed,
and their distributions are summarized in Figure 4.

Figure 4 shows that the majority of the AC-13 data are between 0.5 and 1.5 mm —50% of
the 2.52 million points are in the interval between 0.5 and 1mm, and 30% are between 1 and
1.5mm. SMA-13 has a wider texture depth distribution than AC-13, and 32%, 34%, and 25%
of the data are in the texture depth interval of 1- 1.5mm, 0.5-1mm, and 1.5-2mm, respectively.
Compared with AC-13 and SMA-13, OGFC-13 has the widest range of texture depth distribution, and the data generally obeys the normal distribution.

5   APPROACH VALIDATION

The texture depth of three asphalt mixtures were tested by the sand patch, and the result was
compared with the photogrammetry, which was shown in Figure 5(a). The texture depth
determined by the sand patch follows the trend of OGFC-13>SAM-13>AC-13, which
matches well with the photogrammetry results. The relative error in the depth value between
the photogrammetry and sand patch measurements for all three mixtures was less than 9.2%.
This verifies the accuracy of the 3D surface reconstruction based on close-range photogrammetry in this study.

The texture depths of six sites on pavement were measured using both the close-range
photogrammetry and sand patch testing to further verify the developed approach. Figure 5(b)
shows the linear correlation analysis between the two testing methods. The linear correlation
coefficient R^2 reaches up to 0.9984, which indicates that the measured texture by the close-
range photogrammetry has high linear relationship with the sand patch. The slope of the
regression line is 1.0077, which is almost equal to 1. This means that there is no obvious difference of the measured texture between the close-range photogrammetry and the sand patch.

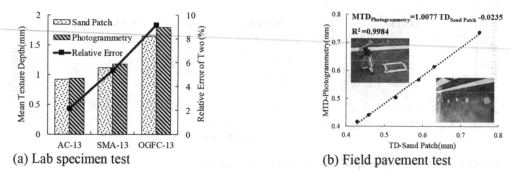

(a) Lab specimen test

(b) Field pavement test

Figure 5. Comparison between sand patch and photogrammetry. (a) Lab specimen test (b) Field pavement test.

6 CONCLUSION

(1) The surface profiles of the AC-13, SMA-13 and OGFC-13 asphalt mixtures were accurately reconstructed using the single camera close-range photogrammetry in this study.

(2) The surface indexes indicate that the OGFC-13 has the roughness surface while the AC-13 has the smoothest surface. This finding matches the existing knowledge of these asphalt concrete mixtures.

(3) The new approach was validated by comparing the mean texture depths calculated from the reconstruction with those tested by the traditional sand patch method for lab made specimens and field pavements. The new method may even be more efficient and accurate compared to the sand patch method.

REFERENCES

Alamdarlo M.N. & Hasami S. 2018. Optimization of the photometric stereo method for measuring pavement texture properties. *Measurement* 127: 406–413.

Chen J., Huang X. & Zheng B., et al. 2019. Real-time identification system of asphalt pavement texture based on the close-range photogrammetry. *Construction & Building Materials* 226: 910–919.

Granshaw S.I. 2010. Close range photogrammetry: principles, methods and applications. *Photogrammetric Record* 25(130): 203–204.

Lowe D.G. 2004. Distinctive image features from scale-invariant keypoints. *International Journal of Computer Vision* 60(2): 91–110.

Mçgowan R., Feighan K., & Mulry B., et al. 2018. Use of close-range photogrammetry to characterize texture in a pavement surfacing material. *The Transportation Research Board(TRB) 97th Annual Meeting*, pp.18–27.

Sun L., Abolhasannejad V. & Gao L., et al. 2016. Non-contact optical sensing of asphalt mixture deformation using 3D stereo vision. *Measurement* 85: 100–117.

Ueckermann A., Wang D. & Oeser M., et al. 2015. Calculation of skid resistance from texture measurements. *Journal of Traffic and Transactions Engineering(English Edition)* 2(1): 3–16.

Wu C. 2013. Towards linear-time incremental structure from motion. *International Conference on 3d Vision, IEEE Computer Society*, pp.127–134.

Xu H., Chen F. & Yao X., et al. 2018. Micro-scale moisture distribution and hydrologically active pores in partially saturated asphalt mixtures by X-ray computed tomography. *Construction & Building Materials* 160:653–667.

Zhang D., Zou Q. & Lin H., et al. 2018. Automatic pavement defect detection using 3D laser profiling technology. *Automation in Construction* 96: 350–365.

Functional Pavements – Chen et al (eds)
© 2021 Taylor & Francis Group, London, ISBN 978-0-367-72610-2

Research on the texture orientation characteristics of pavement surface

W. Ren
School of Civil and Transportation Engineering, Beijing University of Civil Engineering and Architecture, Beijing, P.R. China

S. Han
School of Highway, Chang'an University, Xi'an, Shaanxi, P.R. China

J. Ji
School of Civil and Transportation Engineering, Beijing University of Civil Engineering and Architecture, Beijing, P.R. China

X. Han
School of Highway, Chang'an University, Xi'an, Shaanxi, P.R. China

T. Jiang & Y. Yang
School of Civil and Transportation Engineering, Beijing University of Civil Engineering and Architecture, Beijing, P.R. China

ABSTRACT: Pavement surface texture affects tire/pavement interaction related performances. More attention has been paid to texture depth and spectra. Texture orientation is one of important evaluation indexes affecting related performances. Whereas, it is hard to find related research. In this research, three characterization indexes, skewnees R_{sk}, the ratio of MPD to Root Mean Square (R_{ms}) and shape factor g were investigated. The Exposed Aggregate Cement Concrete (EACC) and Porous Asphalt (PA) pavement were acknowledged as positive and negative texture orientation, respectively. They were characterized by the three orientation evaluation indexes. The results were compared with the acknowledged cognition. The three indexes showed that EACC was positive. The R_{sk} and shape factor g showed that the PA texture was negative. This was consistent with the acknowledged cognition. Whereas, the ratio of MPD to Rms of PA texture showed opposite results. This means R_{sk} and g could be more widely used to characterize texture orientation.

1 INTRODUCTION

Road surface texture can influence many aspects of tire/pavement interaction, such as anti-skidding especially on wet roads, tire/pavement noise, rolling resistance, splash and spray,etc (Snyder, 2007, Henry, 2000, ISO13473-1, 1997). So it is essential to focus on surface texture.

Many researchers have conducted research on the relation between the pavement surface texture and tire-pavement interaction related performances. Tire/pavement noise of asphalt pavements could be even predicted from the surface texture evaluated by Mean Profile Depth (MPD) and International Roughness Index (IRI), and texture spectra.(Mak et al., 2012, Anfosso-Lédée and Do, 2002, Sandberg and Descornet, 1980, Li et al., 2015, Liao et al., 2014). A number of studies have also built relation between friction and road surface texture characterized by the MPD, skewness, kurtosis, fractal dimension, etc. (Britton et al., 1974, Miao et al., 2011).

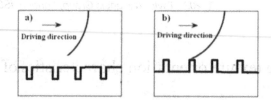

Figure 1.   Positive and negative texture shape for the surface texture with the same spectra and MPD. a): negative texture shape; b): positive texture shape.

Many kinds of indexes were used to characterize pavement texture from different aspects. Whereas, when some texture indexes were the same, the actual pavement texture maybe different, thus resulting in different performances correspondingly. The orientation of surface texture is such kind of characteristics, as shown in Figure 1. In this research, we used three different method to characterize the surface texture orientation and aimed to find a suitable one. This may contribute to understanding related performances.

2   REARCH SCOPE AND OBJECTIVES

The overall objective of the research was to characterize surface texture orientation based on the surface texture data. Different characterization methods were attempted, and the results were compared in order to find a suitable one. This study consists of two main sections, texture orientation characterization indexes calculation and comparison analysis.

3   MATERIALS AND EXPERIMENTAL PROGRAM

3.1   Materials and specimen fabrication

In this research, the texture orientation of two types of road surface, the Exposed Aggregate Cement Concrete (EACC) pavement (EACC) and Porous Asphalt (PA) pavement, were investigated. For the former, the surface texture of EACC pavement in Highway and Airport Pavement Research Center (HAPRC) in Chang'an University was measured. For the latter, the PA laboratory specimens were used, with a nominal maximum aggregate size (NMAS) of 19.0 mm, 16.0 mm,13.2 mm, 9.5 mm respectively. They were fabricated via a single-drum walking-behind roller. The specimens used had the dimensions of 50 cm × 50 cm ×5 cm.

3.2   Road surface measurement

A Laser Texture Scanner (LTS) was used to obtain the elevation data of the road surface texture. Its vertical resolution was 0.005 mm. The maximum scanning area is 104 mm× 72.01 mm. In each scan, 163 profiles with a length of 104 mm were obtained, which were evenly dis-tributed. The sample interval in each line was 0.0095 mm.

4   ACQUISITION OF DIFFERENT CHARACTERIZATION INDEXES OF SURFACE TEXTURE ORIENTATION

4.1   Preprocessing of road surface texture data

The surface texture data obtained directly using the LTS was surface height data. Before using the data to calculate related texture indexes, several pretreatments were needed. For the dropouts, the invalid points were replaced with the lowest valid value before or after them in the same scanned profile line. Slope and offset suppression were also dealt with.

4.2 *Calculation of different characterization indexes of road surface texture orientation*

Skewness (named R_{sk}), the ratio of MPD to Root Mean Square (R_{ms}) value (named MPD/R_{ms}) and shape factor (named g), were used to characterize the pavement surface texture orientation. They were all calculated based on the road surface texture data obtained by the LTS. After preprocessing, texture orientation characterization indexes could be calculated directly.

For the R_{sk}, it is a statistical index. It could be used to describe the height symmetry characteristics. It could be calculated according to Equation (1) and Equation (2).

$$R_{sk} = \frac{1}{NR_{ms}^3} \sum_{i=0}^{N-1} z_i^3 \tag{1}$$

$$R_{ms} = \sqrt{\frac{1}{N} \sum_{i=0}^{N-1} z_i^2} \tag{2}$$

where Z_i=profile height after preprocessing; N=data number on the scanned profile line.

For the MPD/R_{ms}, it was used based on height data obtained using the Circular Texture Meter (CTM) at first. In our research, it was calculated using the data acquired using the LTS. MPD was a common surface texture characterization index. The R_{ms} could be calculated according to Equation (2), so the MPD/R_{ms} could be calculated.

For the g, it was also statistical index to characterize the surface height distribution. It could be calculated by the height data. Its definition was shown in Figure 2. It could be calculated according to the cumulative distribution function.

5 TEST RESULTS AND DISCUSSION

5.1 *Determination of critical values for different characterization methods*

In order to investigate the texture orientation characteristics, it was necessary to determine the critical values for the three different characterization indexes. Then they could be used to describe whether the surface texture was positive or negative. From previous research, the critical values for the three texture orientation characterization indexes were summarized, as shown in Table 1.

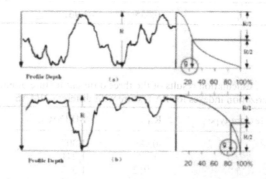

Figure 2.   The definition of shape factor g and its calculation (Beckenbauer and Kuijpers, 2001).

Table 1. Critical values for the three different texture orientation characterization indexes.

Texture orientation	Critical values for the three characterization indexes for texture orientation		
	R_{sk}	MPD/R_{ms}	g (%)
Positive texture	>0	>1.050	<50
Neutral texture	Near 0; Do not have exact range	Near 1.05; Do not have exact range	Near 50; Do not have exact range
Negative textue	<0	<1/1.05	>50

5.2 Comparative analysis of different texture indexes

What was well acknowledged was that the EACC was a typical type of positive texture and the PA surface texture was typical negative texture. So we use the three different texture orientation characterization indexes to characterize the EACC and PA respectively to check whether the evaluation results was consistent with the well acknowledged fact.

For the EACC texture, the three different texture orientation characterization indexes were tested, and results were shown in Table 2. It could be found that the R_{sk} was much more than 0, the MPD/R_{ms} was much more than 1.05 and shape factor g was less than 50%. According to Table 1, the three characterization results of three indexes were the same and EACC was positive texture, which was consistent with acknowledged cognition.

For the PA texture with different NMAS, the three different texture orientation characterization indexes were tested, and results were shown in Table 3. As shown in Table 3, R_{sk} of PA texture with different NMAS were all much smaller than 0, and the shape factor g were all much more than 50%. According to Table 1, this indicated that the PA texture with different NMAS were negative. Whereas, the MPD/R_{ms} of PA texture with different NMAS were all much more than 1.05 and it indicated that the PA texture were positive. The characterization results of R_{sk} and shape factor g were opposite to MPD/Rms. As we know, the PA texture is negative and texture orientation characterization result was opposite to the fact. So MPD/Rms could not be used widely. Further research was needed.

Table 2. Calculation results of the three different texture orientation evaluation indexes for EACC texture.

Surface texture type	R_{sk}	MPD/R_{ms}	g (%)
EACC	0.435	1.957	40.3

Table 3. Calculation results of the three different texture orientation characterization indexes for PA texture with different NMAS.

Surface texture type	R_{sk}	MPD/R_{ms}	g (%)
PA-20	-0.986	1.307	81.2
PA-16	-1.184	1.241	83.4
PA-13	-1.012	1.323	80.6
PA-10	-0.624	1.444	77.5

6 CONCLUSIONS

In this research, three different evaluation methods of surface texture orientation were investigated. R_{sk}, MPD/R_{ms} and shape factor g were calculated using the surface texture height data obtained by the LTS. They were used to characterize the EACC surface texture and PA surface texture with different NMAS. Different characterization results were compared based on the acknowledged fact that the EACC texture is positive and the PA texture is negative. Although further research was need, it seemed that that R_{sk} and shape factor g could be more widely used.

REFERENCES

ANFOSSO-LéDéE, F. & DO, M. T. (2002) Geometric Descriptors of Road Surface Texture in Relation to Tire-Road Noise. *Transportation Research Record Journal of the Transportation Research Board*, 1806, 160–167.

BECKENBAUER, T. & KUIJPERS, A. (2001) Prediction of pass-by levels depending on road surface parameters by means of a hybrid model.

BRITTON, S., LEDBETTER, W. & GALLAWAY, B. (1974) Estimation of skid numbers from surface texture parameters in the rational design of standard reference pavements for test equipment calibration. *Journal of Testing and Evaluation*, 2, 73–83.

HENRY, J. J. (2000) *Evaluation of pavement friction characteristics*, Transportation Research Board.

ISO13473-1 (1997) Characterization of pavement texture by use of surface profiles–Part 1: Determination of mean profile depth. *Eur Stand ICS 1714030, Eur Comm Stand Brussels 2004*.

LI, M., KEULEN, W. V., CEYLAN, H., TANG, G., VEN, M. V. D. & MOLENAAR, A. (2015) Influence of Road Surface Characteristics on Tire–Road Noise for Thin-Layer Surfacings. *Journal of Transportation Engineering*, 141, 04015024.

LIAO, G., SAKHAEIFAR, M. S., HEITZMAN, M., WEST, R., WALLER, B., WANG, S. & DING, Y. (2014) The effects of pavement surface characteristics on tire/pavement noise. *Applied Acoustics*, 76, 14–23.

MAK, K. L., HUNG, W. T. & LEE, S. H. (2012) Exploring the impacts of road surface texture on tyre/road noise - A case study in Hong Kong. *Transportation Research Part D-Transport and Environment*, 17, 104–107.

MIAO, Y., CAO, D. & LIU, Q. (2011) Relationship Between Surface Macrotexture and Skid Resistance of Asphalt Pavement. *Journal of Beijing University of Technology*, 37, 547–553.

SANDBERG, U. & DESCORNET, G. (1980) Road surface influence on tire/road noise-Part1 and Part 2. *INTER-NOISE and NOISE-CON Congress and Conference Proceedings*.

SNYDER, M. (2007) Current perspectives on pavement surface characteristics. *R&T Update: Concrete Pavement Research and Technology*.

Functional Pavements – Chen et al (eds)
© 2021 Taylor & Francis Group, London, ISBN 978-0-367-72610-2

Enhanced prediction of the tire-road-interaction by considering the surface texture

J. Friederichs & L. Eckstein
Institute for Automotive Engineering RWTH Aachen University, Aachen, Germany

ABSTRACT: This paper describes a method to describe the tire-road-interaction more accurately. It is shown that the parameterization of physical tire models can be improved using measured shear forces in the tire contact area. This additional step in the model parameterization process enables an accurate investigation of the tire-road-interaction on different surface textures. It is further shown that the resolution of the road model used in the simulation needs to be matched not only to the maneuver being simulated, but also to the parameterization process applied. Only when considering the road model and the tire model together during the parametrization process a realistic modelling of the tire-road-interaction can be achieved.

1 INTRODUCTION

The automotive and road construction industry uses simulation tools to achieve shorter development cycles for complex products, such as tires or the asphalt surface layer. New functions and a higher accuracy of the tools help to specify the vehicle and road design more easily in early stages of the product development. As the contact patch represents the interaction layer between the tire and the road surface, both models need to be taken into consideration to predict a more realistic tire-road-interaction.

Numerical tire model approaches (Scarpas, 2005; Liu et al., 2017) include a discrete contact patch, which is able to interact with a high-resolution road model. The high parametrization effort and computation time is disadvantageous, especially in full vehicle simulation. Standardized, simplified, physical tire models such as FTire (Gipser, 2016), RMOD-K (Oertel, 2011) or CDTire (Gallrein, 2004) show high potential to analyze the vehicle-tire-road-interaction on asphalt texture due to lower computation time and good connectivity to road and vehicle simulation software.

It is common practice to use pressure sensitive mats for a non-rolling tire under varying loads to validate the pressure distribution within the parametrization process of physical tire models. Relevant shear stresses on the other hand can only be measured for the rolling tire. In previous work from (Anghelache & Moisescu, 2012; Fernando et al., 2006), triaxial force transducer pins arranged in an array are used for the measurement of dynamic footprints of free rolling truck tires. However, these systems are not suitable for high shear stresses generated by rolling tires with high slip angles. To validate the shear stresses of a physical tire model under several slip angles, tire measurements in stationary conditions are performed on a force matrix sensor (A&D, 2018) that can measure high dynamic pressure and shear stresses in the tire contact patch locally via a cluster of strain gauges. Using a mobile tire test rig, the sensor is crossed in a straight line with varying rolling and slip conditions. The resulting data is used as input for parameterization to increase the level of detail of the tire model. Subsequent simulations will show the effect of an extended model in combination with a detailed road surface texture.

2 TIRE MODEL OPTIMIZATION FOR THE TIRE-ROAD-INTERACTION

2.1 *Measurement of dynamic footprints*

The experimental setup to measure pressure and shear stress distribution in the contact patch consists of two independent measurement systems, which are shown in Figure 1. The mobile tire test rig (Bachmann, 2018) is used to perform straight-line approaches crossing the sensor module "FMS" (Force-Matrix-Sensor) at constant velocity, slip angle and camber angle.

The wheel guidance unit in the middle of the semitrailer is able to decouple the measuring wheel from the trailer movement with an active control of the wheel load, the track guidance and the camber angle. Numerous sensors, such as an inertial measurement unit, a measuring hub or laser sensors, are used to determine the forces, moments, movements and several component temperatures.

The FMS (A&D, USA) consists of 40 sensor-pins with an area of 7.5 mm x 7.5 mm each and a pin spacing of 0.5 mm allowing a maximum contact patch width of 319.5 mm. The sensor pins have spring elements and strain gauges with a realizable nominal pressure range of 1.77 MPa in all three directions making it possible to measure a dynamic footprint when braking torques or slip angles occur. Two laser barriers trigger the measurements and determine the rolling velocity.

In the evaluation the time-dependent sensor pin data is transferred into a path dependent map with longitudinal and lateral extend. Trivially, the map's lateral resolution is the sensor pin width, while the longitudinal resolution is dependent on the sampling frequency, which can be raised up to 100 kHz. The tests were carried out with several passenger car tire dimensions at 2.5 bar with varying wheel loads, slip angles and camber angles. A discussion on the results of this measurement study can be found in (Friederichs, 2019).

2.2 *Measurement of friction coefficients*

The shear forces between the sensor surface and the tread surface are dependent on the friction behavior. In this study, a linear friction tester (Figure 2) is used to determine the friction

Figure 1.   Mobile tire test rig (Bachmann, 2018) and the Force-Matrix-Sensor (Friederichs, 2019).

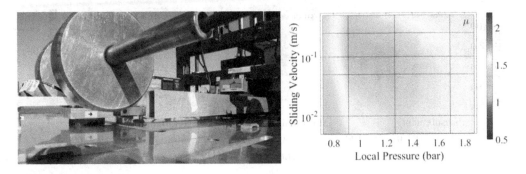

Figure 2.   Linear friction tester on the FMS-surface and the corresponding sliding friction map.

coefficients for varying sliding velocities and loads. It consists of a linear motion actuator, a sledge construction and a supporting frame. A three-dimensional force transducer connects the sledge construction with a rubber specimen, which is cut out of the tire tread with an area of 60 mm x 60 mm.

The specimen is pulled over the test surface with a steady translational motion. The division of the longitudinal force and the vertical load is defined as the friction coefficient. In the following investigations, only the steady-state sliding friction is taken into consideration. Figure 2 exemplarily shows the linear friction tester and the friction map of the tested tread specimen on the FMS-surface. The local contact pressure is assumed consistent on a flat surface due to a flattening subsequent treatment of the rubber specimen. The sliding friction coefficients for this material pairing varies between 1.1 and 1.7.

2.3 *Model optimization through simulation of the dynamic tire footprints*

The tire model FTire from (Gipser, 2016) is used to simulate tire-road-interaction. The mechanical model describes the tire belt as an extensible and flexible ring, which carries bending stiffnesses and is elastically bounded on the rim by distributed spring and damper elements in radial, tangential and lateral direction. A finite number of belt elements numerically approximates the ring. Mass-less tread blocks are associated to each belt element, which interact with the belt and the road model. These contact elements are aligned according to a defined outer contour.

During the standard parameterization process, the structural stiffnesses and the footprint contour are measured with a non-rolling tire. The transient behavior of the rolling tire is determined on a drum test rig with cleat excitation. The dynamic force and moment behavior is measured on a drum or flat track test rig on corundum paper. In the tire model parametrization process, these measurements are approximated by iteratively optimized parameter identification of the spring and damper parameters and the friction coefficients for the contact elements for varying relative sliding speed, load and optionally temperature.

The new approach in this study additionally uses the FMS-data for the iterative parameterization process to achieve a higher level of detail for the tread model. In this way the identification of the tire model's integrated friction characteristics can be validated not only with the resulting forces and moments on the rim, but also locally with the dynamic footprints. Furthermore, a portable hand scanner (HandySCAN 700™, AMETEK GmbH, Germany) is used to measure the outer contour of the tire. Geometric parameters, such as the tread curvature radius, can be derived from the cross-section. With the help of the tread pattern, the position of the contact elements of the tread model can be determined in FTire. In result, the geometry of the tire can be better approximated in the model.

Figure 3 exemplarily shows the ground pressure distribution of a passenger car tire at a rolling speed of 30 kph, 6 kN wheel load and 8° slip angle in measurement and simulation. The outer side of the tire shows a significant increase in pressure up to five times the tire

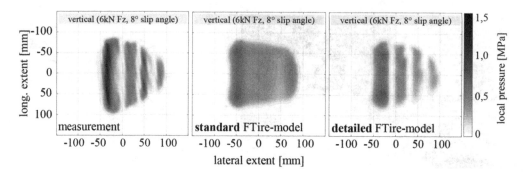

Figure 3.   Ground pressure distribution in the contact patch in measurement and simulation at 30 kph.

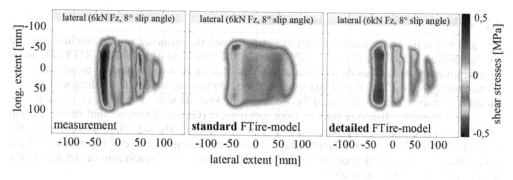

Figure 4. Lateral shear stress distribution in the contact patch in measurement and simulation at 30 kph.

pressure of 2.5 bar. The simulation with the standard FTire has a very homogeneous pressure distribution with lower pressure values. The simulation with the detailed FTire model shows a similar trapezoidal contour to the one in the measurement with higher ground pressure on the outer side.

Due to the integration of the longitudinal tread pattern and optimized geometry parameters, the parameter set of the standard tire model is no longer valid for the detailed model. A new parameterization with additional integration of the measured data of the ground pressure and shear stress distribution into the iterative parameterization process results in a changed friction map for the tire model. For example, the friction coefficients for high ground pressure have been increased.

Consequently, the changes between the "standard" and the "detailed" model are bigger for the lateral shear stresses than for the ground pressure distribution in the same maneuver. Figure 4 indicates that the shear stresses of the detailed model show higher correlation to the measurement, especially on the outer side, where most of the side force is transmitted.

It is to mention, that averaged friction levels of the rubber-road-interaction are calculated using the friction measurements on the sensor surface and the corundum paper, which is used in the drum tests. The friction deviation is considered in the footprint simulations by scaling of the friction coefficients. For the detailed model, the total forces acting on the rim are similar to the standard model and correspond to the measurements on a sufficient level. As the model is validated for the wheel forces and the contact patch behavior, the detailed model shall be used for further investigation on realistic asphalt textures.

3 SIMULATION OF THE TIRE-ROAD-INTERACTION ON ASPHALT

3.1 Preparation of digital roads with asphalt texture

The benefits from digitalizing a road section in the micrometer range do not justify the effort because even in the quasi-static component simulation the road length needs to be up to 100 m depending on rolling velocity and the settings of the wheel load controller. Therefore, the digital road is composed of different rectangular asphalt samples with the aim of the lowest possible periodicity. One or several cut samples (300 mm x 100 mm) of one asphalt type are measured with the handheld scanner from chapter 2.3, which operates with a minimum grid width of c. 200 µm. A compensation plane is projected through the 3D point cloud and the small inclination angle error to the horizontal is compensated so that the texture can be used as a two-dimensional matrix for further processing in MATLAB. Randomly rectangular sections measuring 400 mm² to 1,600 mm² are lined up in longitudinal and lateral direction. They are connected by linear interpolation in a transition distance of twice the grid width of the scan data. In a next step, the matrix is integrated into the OpenCRG format, which is able to connect road textures and courses by simple means, cf. (Rauh, 2010).

3.2 *Influence of tire and road model resolutions*

The aim of the following simulation study is to evaluate the influence of grid resolution of the two models and to adjust them to each other. In the following example, a CRG road model representing fine grained asphalt with a grid width of 0.5 mm is resampled up to grid width of 10 mm using the MATLAB griddata interpolation. On the other side the resolution of the tire model's contact elements is varied between 2 mm and 10 mm. The contact elements are arranged rather in a fishbone pattern than rectangular (Gipser, 2016), so that the resolution specification is rather an approximation. Figure 5 illustrates the lateral shear stresses in the contact patch of the detailed FTire-model with contact element distance of c. 2 mm at 50 kph, 6 kN wheel load and 6 degree slip angle on asphalt with a grid resolution of 0.5 mm (left), 4 mm (middle) and 10 mm (right).

The grid interpolation works like a low-pass filter and smoothens the texture. As a result, the overall contact area increases under this condition by 4.03 % for the road model with 4 mm grid resolution and increases by 8.51 % for the road model with 10 mm grid resolution. Due to the flattening of the texture peaks, the local ground pressure also decreases. In result, the maximum shear stress decreases by 9.95 % for the road model with 4 mm grid resolution and 12.47 % according to the reduced friction values at higher ground pressures in comparison to the values at low ground pressures within the friction map of the tire model. The lateral deflection of the longitudinal tread profile does not change significantly.

These effects also have an impact on the total transmitted side forces of the tire model. Figure 6 exemplarily shows the tire side forces for 3 kN wheel load (left) and 6 kN wheel load (right) at 50 kph. While the side forces of the standard and detailed model have a clear offset, the grid resolution of the tire model only shows little impact on the resulting side forces under these conditions, which allows a reduction of the computation time regarding the distances between the contact elements. As the grid width of the road model decreases, also the side forces decrease for all models within a range of c. 700 N for 3 kN and c. 1000 N for 6 kN.

Using an FTire-model and an OpenCRG model, the grid resolution of the asphalt texture within the road model has the main influence on the resulting tire forces. Therefore, tire measurements need to be carried out on the asphalt, which has been digitalized via the hand scanner, to find the best fitting CRG grid resolution. The mobile tire test rig from chapter 2.1 is used to perform slip angle sweeps at varying wheel loads, which is shown in Figure 7 as an example for a wheel load of 6 kN. The measurements show a low frequency fluctuation in the tire side force of up to 1000 N. This effect can rather be attributed to the road unevenness and the wheel load control than to the texture and is therefore not existent in the simulations. The data from the real road tests correlate best with the tire simulations at 6° slip angle with a road grid width of 0.5 mm (cf. Figure 6 right and real road test in Figure 7). Therefore, the simulations on asphalt texture are carried out on a CRG grid width of 0.5 mm in Figure 7.

Regarding the full real road measurement and simulation on fine grained asphalt, the side force maximum as well as the cornering stiffness show only little deviation. The simulation is

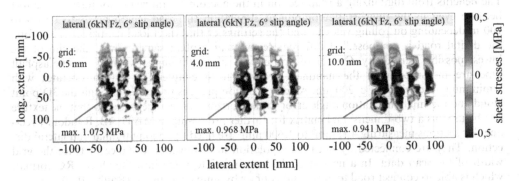

Figure 5. Lateral shear stress distribution depending on the grid resolution of the asphalt texture.

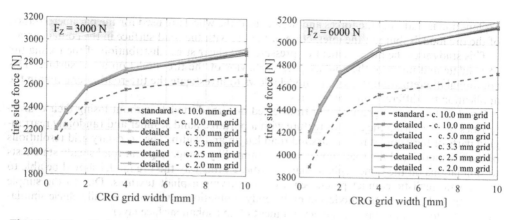

Figure 6. Tire side forces at 6° slip angle on fine grained asphalt for different model resolutions.

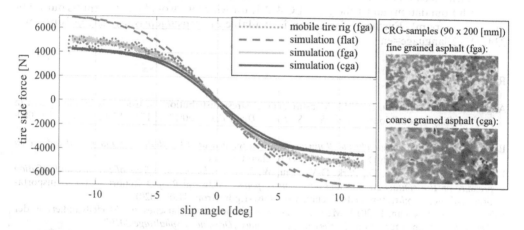

Figure 7. Slip angle sweeps in simulation and measurement on different surface textures.

able to extrapolate the side force behavior for the full slip angle range at a constant road grid width of 0.5 mm. Consequently, one needs to accept higher computation time when simulating the tire-road-interaction on realistic asphalt texture due to the necessary high grid resolution.

After the validation of the tire model on basis of the wheel and the contact patch, it is now possible to simulate on different asphalt textures to predict the wheel forces and moments, but also to analyze the shear stress and ground pressure distribution for the design of the asphalt surface layer. As an example, Figure 7 also shows the side forces during a slip angle sweep on coarse grained asphalt. As the texture peaks are higher, the ground pressure peaks increase and the contact area within the simulated tire footprint decreases, which results in a smaller cornering stiffness and side force potential. The little high frequency excitations, which can be seen on the fine grained asphalt, do not occur on the coarse grained asphalt in the same extent. This phenomenon still needs to be investigated.

4 SUMMARY AND OUTLOOK

Tire simulation models are parameterized based on test bench measurements. These are performed under laboratory conditions on homogeneous corundum paper with fine grain sizes. When comparing tire measurements on a real asphalt surface with the simulation results of test rig models on asphalt texture, conspicuous differences are found. One possible explanation is that only the forces and moments at the wheel are used for the parameterization of the tire models, but not the interaction of the tire with the road surface in the contact area.

This study uses the measurement of pressure and shear stress distribution of the rolling tire to validate and improve the contact patch behavior of the tire model on a flat surface. This additional discretization step of the model allows to investigate the tire-road-interaction in the application for different surface textures.

Road surfaces are digitalized and transferred into road models without losing their characteristics by measuring the asphalt texture with a portable hand scanner and randomly rearranging parts of the samples for the desired road length and width. The necessary grid resolutions of the tire model and the road model are determined by comparing real road steady state tests to simulations on the same asphalt characteristic. In this way the tire model should be able to extrapolate realistic contact patch behavior on varying asphalt textures. Due to the simple integration of physical tire models in multi-body simulation environments, full vehicle simulations can additionally support the development of an asphalt surface layer.

The research work documented here is part of the German research group FOR 2089 - "Permanent Road Pavements for Future Traffic Loads" financed by the DFG (German Research Foundation) under the grant EC 412/1, for which we would like to express our gratitude at this point. Further information is available at https://gepris.dfg.de/gepris/projekt/239224712?language=en.

REFERENCES

Anghelache, G., Moisescu, R. 2012. Measurement of stress distributions in truck tyre contact patch in real rolling conditions, *Vehicle System Dynamics*. 50:12, 1747–1760. doi: 10.1080/00423114.2012.674143.

Bachmann, C. 2018. *Vergleichende Rollwiderstandsmessungen an Lkw-Reifen im Labor und auf realen Fahrbahnen*. Aachen: Ph.D. Thesis - RWTH Aachen University

Fernando, E. G., Musani, D., Park, D-W., Liu, W. 2006. *Evaluation of effects of tire size and inflation pressure on tire contact stresses and pavement response*. Technical Project Report, Texas Transportation Institute, http://tti.tamu.edu/documents/0-4361-1.pdf. (from 31.07.2020).

Friederichs, J.; Eckstein, L 2019. Messung des dynamischen Reifenlatsches zur Modellvalidierung der Reifen-Fahrbahn-Interaktion, *Conference proceedings „Dresdner Asphalttage 2019"*

Gallrein, A. 2004. CDTire: *State-of-the-art Tire Models for Full Vehicle Simulation*. Kaiserslautern: Fraunhofer ITWM.

Gipser, M. 2016. *The FTire Tire Model Family*. Esslingen: FH Esslingen. https://www.cosin.eu/wp-content/uploads/ftire_eng_3.pdf. (from 31.07.2020).

Liu, P., Xing, Q., Wang, D., Oeser, M. 2017. Application of dynamic analysis in semi-analytical finite element method. *Materials*, 2017, 10, 1010, doi:10.3390/ma10091010.

Oertel, C. 2011. *RMOD-K Formula Documentation*. Brandenburg an der Havel: FH Brandenburg.

Rauh, J. 2010. *OpenCRG -The open standard to represent high precision 3D road data in vehicle simulation tasks on rough roads for handling, ride comfort, and durability load analyses*. http://www.opencrg.org.

Scarpas, A. 2005. CAPA-3D: *A Mechanics Based Computational Platform for Pavement Engineering*. Delft: Ph.D. Thesis - TU Delft.

Functional Pavements – Chen et al (eds)
© 2021 Taylor & Francis Group, London, ISBN 978-0-367-72610-2

Multiscale and multiphysical tire-pavement analysis: A mesostructure inspired material model for the short- and long-term material behavior of asphalt

R. Behnke, F. Hartung, I. Wollny & M. Kaliske
Institute for Structural Analysis, Technische Universität Dresden, Dresden, Germany

ABSTRACT: For the detailed analysis of tire-pavement interaction, its structural components (tire, pavement, tire-pavement interface) have been represented in finite-element-discretized form within a thermo-mechanical simulation approach. The tire-pavement model captures short-term phenomena (rolling tire, day-night temperature change) as well as long-term phenomena (annual temperature change, change of traffic intensity). Several submodels (tire, multiscale friction law, pavement structure) have been developed considering different length and time scales in combination with an arbitrary Lagrangian Eulerian (ALE) formulation for tire and pavement (steady state rolling condition). In this contribution, a mesostructure inspired material model for asphalt is proposed to realistically capture the material behavior within the described simulation approach. The material model consists of a short- and a long-term part connected in series and represents the material's short- and long-term response to repetitive loading.

1 INTRODUCTION

With the help of a thermo-mechanical simulation approach considering several length and time scales, complex interaction phenomena between rolling tire and asphalt pavement are investigated in a closed system with displacement and temperature as solution fields. For tire and pavement, finite element (FE) models have been developed during the last years in the framework of the Research Unit FOR 2089 funded by the Deutsche Forschungsgemeinschaft (DFG), see Figure 1. The FE-discretized approach of the system allows from a macroscopical point of view the detailed but still efficient analysis of tire (Behnke & Kaliske 2015) and pavement (Wollny et al. 2016) as well as the tire-pavement interface (Hartung et al. 2018) during the service life of a pavement, e.g. to assess the alteration of its functional properties, see e.g. Alber et al. (2020). With the help of the coupled structural analysis (Behnke et al. 2019), new insights into the interaction phenomena and the identification of influence quantities are expected. Furthermore, mechanistic-empirical pavement predictions can be replaced by more design specific numerical analyses. In this case, the number of unknown model and system parameters increases drastically, which requires a detailed experimental-numerical study of each subcomponent and its representation over several length and time scales regarding the service life of pavements. Recent advances in materials and pavement predictions are reported e.g. in Masad et al. (2018). Especially in the future, more resource- and energy-efficient tire-pavement systems facing the challenges of climate change, finite availability of resources and environmental constraints are needed by providing comfort and safety at the same time during the entire service life.

In this contribution, the focus is on a constitutive material model for asphalt material, which can be used in the aforementioned simulation approach to describe in a numerically efficient way from a macroscopical point of view the constitutive behavior of asphalt

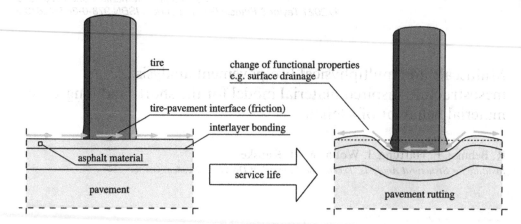

Figure 1. Overview of the tire-pavement system and associated submodels developed within the FOR 2089, where the continuum mechanical description of asphalt material is discussed in this contribution.

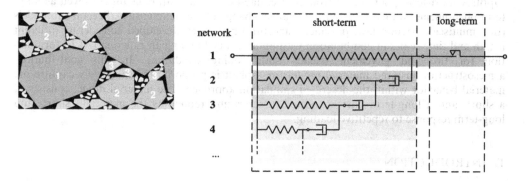

Figure 2. From the mesostructure of asphalt material inspired generalized continuum mechanical model consisting of a short-term and a long-term part: aggregates and binder connected in series form idealized networks of different size fractions in the short-term part of the model.

materials. Differences in the material's response to short- and long-term loading are captured in a short- and a long-term part of the model, which are connected in series, see Figure 2.

2 MESOSTRUCTURE INSPIRED MATERIAL MODEL

For the representation of large strains in the context of asphalt materials (see e.g. Chen et al. (2017), Shahsavari et al. (2016), Zopf et al. (2015)), a multiplicative split of the deformation gradient

$$\mathbf{F} = \nabla_{\mathbf{X}} \, \varphi(\mathbf{X}, t) \ \text{with} \ \mathbf{u}(\mathbf{X}, t) = \varphi(\mathbf{X}, t) - \mathbf{X} = \mathbf{x} - \mathbf{X} \tag{1}$$

into volumetric and isochoric parts, i.e.

$$\mathbf{F} = \bar{\mathbf{F}} \, \mathbf{F}_{\text{vol}} \ \text{with} \ \bar{\mathbf{F}} = J^{-1/3} \, \mathbf{F} \ \text{and} \ \mathbf{F}_{\text{vol}} = J^{1/3} \, \mathbf{1}, \tag{2}$$

is employed. In this case, $J = \det \mathbf{F} > 0$ denotes the determinant of the deformation gradient. To distinguish between short-term and long-term deformations, the isochoric part

$$\bar{\mathbf{F}} = \bar{\mathbf{F}}_{\text{short}} \, \bar{\mathbf{F}}_{\text{long}} \quad \text{with} \quad \bar{\mathbf{F}}_{\text{short}} = \bar{\mathbf{F}}_e \, \bar{\mathbf{F}}_i \tag{3}$$

is further split into a short-term and a long-term part, where inelastic features (viscoelasticity) are included in the short-term part. The multiplicative decomposition is depicted in Figure 3 for the representative example of two rheological branches. In this case, the isothermal reference potential per unit reference volume is

$$\psi = U(J) + \bar{\psi}^s(\bar{\mathbf{F}}_{\text{short}}) + \bar{\psi}^v(\bar{\mathbf{F}}_e) \tag{4}$$

and includes the volumetric energy function $U(J)$ (modeled as purely elastic) and isochoric energy functions depending on the number of branches of the short-term part,

$$U(J) = \kappa_0 \, (J - \ln J - 1) \text{ as well as} \tag{5}$$

$$\bar{\psi}^s(\bar{\mathbf{F}}_{\text{short}}) = \frac{G_c^s}{2} \, (\bar{I}_1^{\text{short}} - 3) \text{ and } \bar{\psi}^v(\bar{\mathbf{F}}_e) = \frac{G_c^v}{2} \, (\bar{I}_1^e - 3), \tag{6}$$

where $\bar{I}_1^{\text{short}}$ and $\bar{I}_1^e$ are the first invariants of the associated Cauchy-Green deformation tensors computed from $\bar{\mathbf{F}}_{\text{short}}$ and $\bar{\mathbf{F}}_e$, respectively. κ_0 (bulk modulus) and G_c^s, G_c^v (shear moduli) are material parameters. The inelastic rheological elements are governed by evolution equations in terms of the inelastic creep rates $\dot{\gamma}_i$ and $\dot{\gamma}_{\text{long}}$. For the identification of the model parameters, information on the asphalt mixture (binder content, void content, particle size distribution) as well as uniaxial tests (compacted mixture) are used. From the particle size distribution, load-bearing networks are identified for which the effective shear moduli are computed. Further details with respect to the numerical implementation and the identification of model parameters from information of the asphalt material's mesostructure and mixture are provided in Behnke et al. (2020).

The material response is evaluated for different discrete testing temperatures for which a set of model parameters is derived. For an arbitrary temperature state of the material, the material response is computed by interpolating the sets of material parameters.

For cylindrical specimens of a stone mastic asphalt (SMA11S B5070) with binder type B5070, the long- and short-term response at 20°C in terms of the engineering strain ε has been computed with the help of the material model (7 branches) and is compared to experimental data in Figure 4. The uniaxial loading (force controlled test) with stress amplitude $\hat{\sigma}_1 = -0.25$ MPa consists of a repeated cyclic loading, where each load cycle T consists of a series of 6 sinusoidal pulses at a frequency of $f = 8$ Hz and a subsequent relaxation time of 2 s. It can be observed

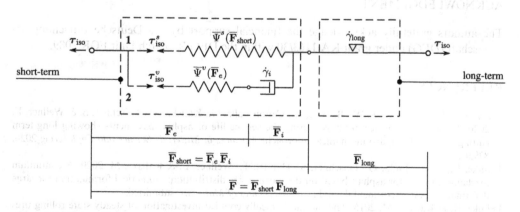

Figure 3. Isochoric part of the material model with isochoric stresses τ_{iso}, exemplarily depicted with two branches 1 and 2 in the short-term part of the model, compare Figure 2.

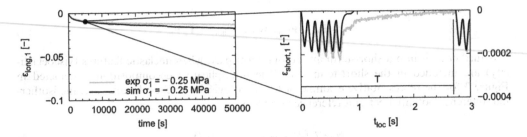

Figure 4.   Simulated material response of SMA11S B5070 at 20°C compared to experimental data, see Behnke et al. (2020), for a force controlled uniaxial test performed on cylindrical specimens (stress amplitude $\hat{\sigma}_1 = -0.25$ MPa, total cycle duration $T = 2.75$ s): evolution of engineering strains $\varepsilon_{long,1}$ and $\varepsilon_{short,1}$ in loading direction during repetitive loading (local time t_{loc} at total time $t = 5000$ s).

that the experimentally recorded material response is qualitatively well captured by the proposed material model. Only minor deviations between measurement and simulation results occur at the end of the test, where further material phenomena (fatigue, healing, damage and failure) become predominant. These phenomena are not captured by the actual material model. The short-term response of the material (strain amplitude, viscoelastic features) is well captured by the model. Only during the removal of the cyclic loading (relaxation phase), the simulated and experimentally recorded responses slightly diverge. This might be improved by a more detailed study of the relaxation behavior (frequency spectrum) of the material and is judged as negligible for the quality of the long-term predictions regarding rutting on the structural scale.

Further details of the experimental-numerical study (material response at different temperatures and for different stress amplitudes) are reported in Behnke et al. (2020).

3   CONCLUSIONS

The mesostructure inspired material model enabled to capture different time scales and features of the asphalt material within a continuum mechanical model for large deformations. With the help of this model, the material behavior of different asphalt materials can be captured on the structural scale, e.g. for the identification of influence factors during sensitivity analyses of a representative part of the tire-pavement system for the improved design of future smart roads.

ACKNOWLEDGEMENT

The authors gratefully acknowledge the financial support by the Deutsche Forschungsgemeinschaft (DFG) under grant KA 1163/30 within the DFG Research Unit FOR 2089.

REFERENCES

Alber, S., Schuck, B., Ressel, W., Behnke, R., Canon Falla, G., Kaliske, M., Leischner, S. & Wellner, F. 2020. Modeling of surface drainage during the service life of asphalt pavements showing long-term rutting: A modular hydro-mechanical approach. *Advances in Materials Science and Engineering* 2020: 8793652.

Behnke, R., Canon Falla, G., Leischner, S., Händel, T., Wellner, F. & Kaliske, M. 2020. A continuum mechanical model for asphalt based on the particle size distribution: Numerical formulation for large deformations and experimental validation. *Mechanics of Materials*, submitted.

Behnke, R. & Kaliske, M. 2015. Thermo-mechanically coupled investigation of steady state rolling tires by numerical simulation and experiment. *International Journal of Non-Linear Mechanics* 68: 101–131.

Behnke, R., Wollny, I., Hartung, F. & Kaliske, M. 2019. Thermo-mechanical finite element prediction of the structural long-term response of asphalt pavements subjected to periodic traffic load: Tire-pavement interaction and rutting. *Computers and Structures* 218: 9–31.

Chen, F., Balieu, R. & Kringos, N. 2017. Thermodynamics-based finite strain viscoelastic-viscoplastic model coupled with damage for asphalt material. *International Journal of Solids and Structures* 129: 61–73.

Hartung, F., Kienle, R., Götz, T., Winkler, T., Ressel, W., Eckstein, L. & Kaliske, M. 2018. Numerical determination of hysteresis friction on different length scales and comparison to experiments. *Tribology International* 127: 165–176.

Masad, E., Bhasin, A., Scarpas, T., Menapace, I. & Kumar, A. (eds.) 2018. *Advances in Materials and Pavement Prediction*. London: CRC Press/ Balkema,Taylor & Francis Group.

Shahsavari, H., Naghdabadi, R., Baghani, M. & Sohrabpour, S. 2016. A finite deformation viscoelastic-viscoplastic constitutive model for self-healing materials. *Smart Materials and Structures* 25: 125027.

Wollny, I., Hartung, F. & Kaliske, M. 2016. Numerical modeling of inelastic structures at loading of steady state rolling – Thermo-mechanical asphalt pavement computation. *Computational Mechanics* 57: 867–886.

Zopf, C., Garcia, M. & Kaliske, M. 2015. A continuum mechanical approach to model asphalt. *International Journal of Pavement Engineering* 16: 105–124.

Functional Pavements – Chen et al (eds)
© *2021 Taylor & Francis Group, London, ISBN 978-0-367-72610-2*

Study on porous asphalt concrete clogging and cleaning mechanism based on numerical simulation

W.J. Wen & B. Li

National and Provincial Joint Engineering Laboratory of Road & Bridge Disaster Prevention and Control, Lanzhou Jiaotong University, Lanzhou, China

ABSTRACT: In this paper, specimens of porous asphalt mixture with design porosities of 18%, 20%, and 25% are produced. Use wind-blown sand to clog these PAC specimens and clean them with vacuum pump and high-pressure water. Use X-ray CT to get CT images of PAC in the initial and after cleaning, and process these CT images to build 3D digital model of pore. Use Fluent software to numerically simulate of water seepage characteristics. Comparing the changes in the seepage rate of the pore model under vacuum pump and high-pressure water cleaning, the results show: the vacuum pump cleaning has a better effect on the water seepage rate recovery of the small porosity specimens. The high-pressure water cleaning effect increases with the increase of porosity, the average water seepage rate recovery rate of the 18% porosity model after the vacuum pump cleaning is 49.90% and the 25% porosity model after high-pressure water cleaning is 53.25%.

1 INTRODUCTION

As an environmentally friendly paving material, porous asphalt mixture is increasingly applied in pavement engineering. Porous asphalt pavement (PAC) are designed to reduce road surface runoff and driving noise due to pores. And the rough surface texture of PAC can provide enough friction to ensure driving safety, especially in rainy days. PAC has the above advantages and has been increasingly used in pavement engineering recently. However, the pore of PAC will be clogged by soil, sand and tire debris, etc. during service, and the water seepage performance of the pavement is seriously attenuated. In order to alleviate the attenuation of water seepage performance of PAC caused by the clog of pore, cleaning equipment can restore its function. The common cleaning and maintenance methods include high-pressure water cleaning, vacuum pumping and manual sweeping.

PAC due to its pore being clogged during service and the water seepage performance of the pavement is seriously attenuated, which has become a critical issue restricting the development of PAC (Moriyoshi A. 2013. Deo O 2010.). Kandhal et al (Kandhal, P. S. 2001.) found that the permeability of asphalt mixture is related to factors such as porosity, aggregate gradation, and aggregate shape. Wang et al (Wang, C. 2020.) uses the second-generation aggregate image measurement system to study the effect of crushing operation on the three-dimensional control characteristics of coarse aggregates and its influence on the shape distribution characteristics of aggregates (i.e. angularity, sphericity, and flatness & elongation ratio). Vancura M tried to use an electron microscope to observe the distribution of the clogging material on the cross-section of the specimens by injecting epoxy resin into the permeable concrete and then use an electron microscope to study the depth of movement of the clogging material inside the permeable concrete (Vancura M. 2012.). Mahsa Amirjani found that wind-blown sand is the main clogging material for PAC in semi-arid areas (Mahsa Amirjani. 2010.). With the advent of high-resolution X-ray computed tomography (CT), it is possible to obtain the characteristics of the microscopic pore inside the PAC (Tashman, L. 2010. Kassem, E. 2008). Kayhanian used CT scanning technology to study permeable concrete specimens and found the relationship between clog and permeability (Kayhanian M. 2012.). Shirke used the back-wash method to study the effects of water pressure, clogging material, porosity, and backwashing times on the recovery effect of PAC water permeability after clogging (Shirke N A. 2009.). Gruber tried to numerical modeling of flow in the pore network is presented based on actual pore geometry reconstructed employing X-ray computer tomography (Gruber, I. 2012.). In order to study the cleaning effect and cleaning mechanism of different cleaning methods on the clogged PAC. The 3D model of pore is constructed after CT scanning the PAC initial and after cleaning respectively. Use computational fluid dynamics (CFD) software to numerically simulate the pore model to obtain water-flow information and compare the seepage permeability under vacuum pump cleaning and high-pressure water cleaning. So as to analyze the cleaning mechanism and cleaning effect.

The research route can be seen in Figure 1

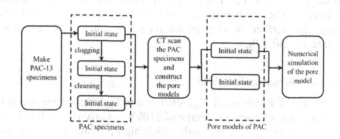

Figure 1. Research route.

2 MATERIALS AND METHODS

2.1 Porous asphalt pavement

The aggregate used in the test is limestone. Cylindrical specimens were prepared in standard Marshall moulds (diameter is 101.6mm). by applying 50 blows on both faces using the Marshall hammer. The mixture gradation and porosity used in the test are listed in Table 1. asphalt-aggregate ratio is determined by drain-down test and raveling test. Porosity is determined by volume method.

2.2 Gradation of clogging material

The clogging material is selected from wind-blown sand in Northwest China. The gradation clogging materials shown in Table 2.

Table 1. Gradation and porosity of porous asphalt concrete.

| Gradation number | Passing ratio (by mass) of different sieve size (mm)/% | | | | | | | | | | asphalt-aggregate ratio/% | porosity/% |
	16.0	13.2	9.5	4.75	2.36	1.18	0.6	0.3	0.15	0.075		
A	100.0	95.0	55.0	19.5	10.8	9.0	7.4	6.2	5.6	4.6	4.4	25.0
B	100.0	95.2	56.8	23.8	16.4	13.2	10.0	7.4	6.3	4.7	4.9	20.0
C	100.0	95.3	57.6	25.7	18.1	14.4	10.7	7.8	6.5	4.7	5.1	18.0

Table 2. Gradation of clogging material.

	Passing ratio (by mass) of different sieve size (mm)/%						
	4.75	2.36	1.18	0.6	0.3	0.15	0.075
Clogging material							
Wind-blown sand	100	95.5	76.7	56.2	39.7	21.1	14.5

2.3 Clogging and cleaning process

During the clogging process of the specimens, in order to avoid the accumulation and clog of the specimens, each time 5g of clogging material is evenly spread on the surface of the specimens, and then the water seepage time is measured until the water seepage time is greater than 5 minutes. Vacuum pump cleaning and high-pressure water cleaning are used respectively. The vacuum pump cleaning use the HC-T2103Y vacuum cleaner with a vacuum degree of ≥14kPa (Figure 2a) to uniformly suck the surface of the specimens for 3 minutes. The high-pressure water cleaning uses a high-pressure cleaner (Figure 2b) Rinse the surface of the specimens for 3 minutes with water pressure of 5 MPa and flow rate of 330 L/h.

2.4 3D model reconstruction and meshing

This paper uses the Y. CT Precision II high-energy and high-resolution industrial CT scanner of Tongji University, with a scanning voltage of 210kV, a current of 0.36mA, and three different porosity under the two cleaning methods which high-pressure water cleaning and vacuum pumping at 0.1mm intervals. Scan the PAC13 specimens in two states (initial state, after cleaning) to obtain CT images (see Figure 3a). Use image processing software Avizo to construct 3D pore model of PAC specimens (Figure 3b).

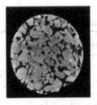 

(a) Vacuum cleaner (b) High pressure cleaner

Figure 2. Cleaning equipment used in the laboratory.

(a) CT image (b) pore model of 18% porosity specimen

Figure 3. CT image and pore model.

2.5 Numerical simulation

The Navier-Stokes governing equations were used to simulate flow of water in-side the pore model of the specimens. By constructing the connected pore 3D model (eliminate non-connected pores) of the PAC with design porosity of 18%, 20%, and 25%, and imported into the ICEM CFD software for model meshing, and then imported into Fluent for numerical simulation. Due to the large geometric size of the PAC (101.6mm in diameter and 70mm in height), the use of this full-size model will make the numerical simulation calculations very large, so it is necessary to extract the pore model. In order to ensure the PAC model in initial state and after cleaning state, the pore model of the cylindrical area with a diameter of 45mm and a height of 30mm was extracted from the above two states for numerical simulation research. Figure 4 demonstrates that the pore structure is relatively complete in the initial state and some pore structures are missing after cleaning in the same porosity model. This is because cleaning can restore the clogged pore structure, but not completely. The model is meshed in ICEM CFD and imported into Fluent for numerical simulation. The boundary conditions and model gird are shown in Figure 5.

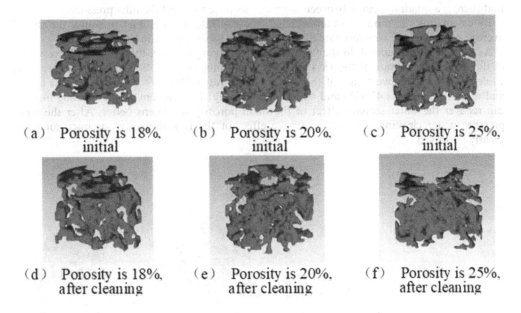

(a) Porosity is 18%, initial

(b) Porosity is 20%, initial

(c) Porosity is 25%, initial

(d) Porosity is 18%, after cleaning

(e) Porosity is 20%, after cleaning

(f) Porosity is 25%, after cleaning

Figure 4.   Extracted model of each porosity of vacuum pump cleaning group.

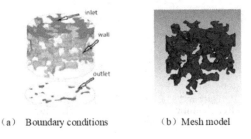

(a) Boundary conditions

(b) Mesh model

Figure 5.   Boundary conditions and Mesh model.

323

3 RESULTS AND DISCUSSION

Processing Fluent calculation results can get the contour of seepage velocity vector and water seepage rate value of each model. Figure 6 shows the vector contour of seepage velocity of vacuum pump cleaning model.

Figure 6 shows that as the porosity increases, the water seepage traces are more distributed, which means that there are more water flow channels. There are 2-3 thicker seepage traces in the 18% porosity model at the initial state. As the porosity increases, the number of seepage traces increases rapidly and the diameter gradually becomes thinner. When the porosity reaches 25%, the seepage traces cover the entire pore model. The water seepage traces of the pore model of the same porosity specimens at the after cleaning are less than the initial. These results may be attributed to the cleaning can restore the pore structure of the PAC, but cannot fully restore it. By observing the void model and the water seepage trace, it can be found that there are some pores (red area) with a significant increase in velocity in the pore structure model. This is because when the water flows through these pore throats, the pore size decreases and the seepage rate increases sharply, and the water seepage rate begins to decrease after passing through the pore throats.

Figure 7 demonstrate the 20% porosity model has a higher water seepage rate in the initial state than after cleaning. It shows that the water seepage rate of the clogged specimens after cleaning does not reach the initial water seepage performance. Meanwhile, it is also found that there is a quadratic curve between the water seepage rate and the inlet pressure.

By calculating the model's water seepage recovery rate at each inlet pressure, the model average water seepage recovery rate can be calculated as shown in the Figure 8.

As illustrated in Figure 8, in the vacuum pump cleaning method, the average recovery rate of the water seepage rate of the 18% model after cleaning reached 49.90% of the initial state, while the average recovery rate of the water seepage rate of the 20% and 25% models before and after cleaning was 47.96% and 44.12% respectively. The vacuum pump cleaning method can recover the water seepage effect of the small porosity specimens better. After the high-pressure water cleaning, the water seepage effect of the large-porosity specimen model is greatly improved.

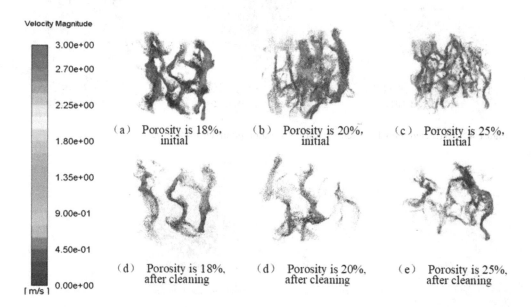

Figure 6.   The contour of seepage velocity vector of the model under vacuum pump cleaning.

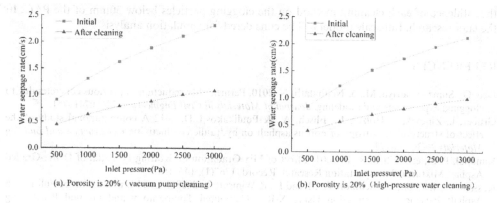

(a). Porosity is 20% (vacuum pump cleaning)   (b). Porosity is 20% (high-pressure water cleaning)

Figure 7.   The relationship of Inlet pressure - Water seepage rate under different cleaning methods.

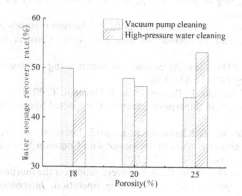

Figure 8.   Pore model average seepage recovery rate.

4   CONCLUSION

The results of this study show that vacuum pump cleaning has better cleaning effect on small porosity specimens, while high pressure water cleaning has better cleaning effect on large porosity specimens. Under the vacuum pumping, the cleaning mechanism is mainly to suck out the clogged particles in the pore through the pumping, and the cleaning effect decreases as the porosity of the PAC increases. The possible reason is that the depth of movement of the clogging material in the small porosity specimen is smaller than that of the large porosity specimen, and the vacuum pump cleaning can only remove the clogging particles stay in the shallow area of PAC specimen. The effect of high-pressure water cleaning is mainly divided into three parts. First, the larger clogging particles that have not entered the inside of the road surface can be removed from the surface of the specimen under the cleaning action of high-pressure water Secondly, due to the connecting gap, part of the water can be reversed. The particles inside the specimen are flushed out; the last part of the clogging particles staying inside the specimen is cleaned to the bottom under the action of high-pressure water.

However, there are deficiencies in this research: since the full-scale model is not used in the numerical simulation, the extracted model cannot well represent the overall performance of the PAC. The part of the pore model in the area within 30mm of the extracted model ignores the influence of each cleaning method on the clogging particles below 30mm of the PAC. In the later research, full-scale model will be considered for simulation analysis.

REFERENCES

Deo, O., Sumanasooriya, M., & Neithalath, N. 2010. Permeability reduction in pervious concretes due to clogging: experiments and modeling. *Journal of Materials in Civil Engineering*: 22(7):741–751.

Gruber, l., Zinovik, l., Holzer, L., Flisch, A., & Poulikakos,L.D. 2012 A computational study of the effect of structural anisotropy of porous asphalt on hydraulic conductivity. *Construction and Building Materials* 36(36): 66–77.

Kandhal, P. S., & Mallick, R. B. 2001. Effect of Mix Gradation on Rutting Potential of Dense-Graded Asphalt Mixtures. Transportation Research Record: 1767(1), 146–151.

Kassem, E., Walubita L., Scullion T., Masad E., & Wimsatt A. 2008. Masad E Evaluation of Full-Depth Asphalt Pavement Construction Using X-Ray Computed Tomography and Ground Penetrating Radar. *Journal of Performance of Constructed Facilities* 22(6): 408–416.

Kayhanian, M., Anderson, D., Harvey, J.T., Jones, D., & Muhunthan, B. 2012. Permeability measurement and scan imaging to assess clogging of pervious concrete pavements in parking lots. *Journal of Environmental Management* 95(1): 114–123.

Mahsa Amirjani. 2010. Clogging of permeable pavements in semi-arid areas. Master of Science thesis. *Delft University of Technology*.

Moriyoshi, A., Jin, T., Nakai, T., & Ishikawa, H. 2013. Evaluation methods 2013. Evaluation methods for porous asphalt pavement in service for fourteen years. *Construction & Building Materials*: 42:190–195.

Shirke N. A., & Shuler, S. 2009. Cleaning porous pavements using a reverse flush process. *Journal of Transportation Engineering* 135(11): 832–838.

Tashman, L., Masad, E., D' Angelo, J., Bukowski, J., & Harman, T. 2010. X-ray Tomography to Characterize Air Void Distribution in Superpave Gyratory Compacted Specimens. *International Journal of Pavement Engineering*: 19–28

Vancura, M. E., MacDonald, K,. & Khazanovich, L. 2012. Location and Depth of Pervious Concrete Clogging Material before and after Void Maintenance with Common Municipal Utility Vehicles. *Journal of Transportation Engineering* 138(3):332–338.

Wang, C., Wang, H., & Mohd Hasan, M. R. 2020. Investigation on the morphological and mineralogical properties of coarse aggregates under VSI crushing operation. *International Journal of Pavement Engineering* 2020 Jan 24: 1–14.

Study on aggregate strengthening performance and principle of recycled concrete based on macro and micro tests

Xuqiu Teng*, Xiaolong Yang & Pengcheng Yang
College of civil engineering, Lanzhou Jiaotong University, Lanzhou, China

ABSTRACT: Recycled concrete aggregates have a large number of voids, micro-cracks and interface transition areas, resulting in low apparent density and high water absorption and crushing values. If the recycled concrete aggregate is directly applied to the semi-rigid base, it will easily cause a series of problems such as cracks in the semi-rigid base and insufficient bearing capacity during service. In view of this, different concentrations of water glass, nano-SiO_2 and infiltration crystalline materials were used to soak the regenerated aggregates, and the macro-strengthening effect was evaluated using the apparent density, water absorption rate and crushing value of the regenerated aggregates after chemical strengthening. The strengthening effect of the microstructure of the recycled aggregate is evaluated by SEM, EDS and pore size analysis test. The experimental results show that the water absorption and crushing value of the modified recycled aggregate soaked with three kinds of strengthening agents are significantly decreased in macro aspect. The micro cracks and interface transition zone of recycled aggregate are repaired, and the total pore diameter and maximum pore diameter are reduced. Compared with the strengthening effect of three kinds of chemical agents on recycled concrete aggregate, it is found that nano-SiO_2 has a significant effect on macro index and microstructure of recycled aggregate.

1 INTRODUCTION

In recent years, with the rapid development of China's economy, urbanization construction is also proceeding rapidly. The environmental pollution problem caused by waste concrete is becoming more and more serious. A large amount of waste concrete is directly transported to the suburbs and directly piled up without treatment, causing environmental pollution, which increases the cost of cleaning and takes up farmland (Xu et al. 2009; Wang et al. 2003). Therefore, recycling waste concrete prepared by crushing and sieving into a certain size of recycled aggregate is one of the effective methods to solve such problems. Recycled aggregates generally have defects such as low apparent density, high water absorption rate and crushing value. If it is directly applied to the projects, it will cause a series of problems such as low strength and cracking of the structure. Therefore, it needs to be strengthened before application. Physical modification and chemical modification are generally two ways to increase the strength of recycled aggregates. The physical modification is mainly achieved by improving the shape of the recycled aggregate particles. In order to reduce the micro-cracks on the surface of the regenerated aggregates, the cement mortar and aggregate particles, partially attached to the surface of the recycled aggregate, are removed by the abrasion impact between the recycled aggregate and the abrasive. Through physical modification, the recycled aggregate particles are cleaner and denser. The common physical modification methods are grinding method, including vertical eccentric high-speed grinding method, horizontal rotating grinding method and heating grinding method (Li et al. 2006). The physical modification effect is poor, and it will increase the production cost, so it is generally used as an auxiliary method of

---

* Corresponding author

modification. Chemical modification is to soak the recycled aggregate with chemical slurry, and produce a kind of cementitious material through the hydration reaction between the chemical slurry and the mortar on the surface of the recycled aggregate. This cementitious material fills the pores and interface transition zone of the recycled aggregate, which improves the strength of the recycled aggregate and reduces the water absorption rate of the recycled aggregate, thus playing a modified role. Xiong Feng used dilute HCl, cement mortar mixed with fly ash or silica fume to chemically strengthen the recycled aggregate, and used water absorption as an index to evaluate the strengthening effect (Xiong 2017). The regenerated coarse aggregate treated with dilute HCl has the best modification effect when the immersion time is 60 min and the concentration is 2%. At this time, the water absorption decreases the most. Vivian used HCl, H_2SO_4 and H_3PO_4 to soak the recycled aggregates, which can remove the old mortar on the surface of the recycled aggregates and reduce the water absorption rate (Vivian et al. 2006). Purushothaman soaked recycled coarse aggregate with HCl, H_2SO_4, washing, and heat washing (Purushothaman et al. 2015). The results showed that the H_2SO_4 treatment and hot washing of the recycled aggregate can greatly reduce its water absorption, and the performance of the prepared recycled concrete is not much different from that of natural concrete. Suo Lun designed five methods to improve the water absorption of recycled aggregates, and the results showed that 5% dilute HCl can better improve the water absorption of recycled aggregates (Suo et al. 2015). After chemically soaking recycled coarse aggregate with $H_2C_2O_4$ solution and HCl solution, Wang Jianghao can effectively remove the hardened cement mortar attached to the surface of recycled coarse aggregate and reduce water absorption (Wang et al. 2016). Both Hu Zhonghui and Liu Lingqing used silicone resin solution, silane coupling agent and phthalic acid coupling agent to treat the recycled aggregate. The test results show that the silicone resin solution has the best effect on improving the crushing value and water absorption performance of recycled aggregates (Hu 2016; Liu 2014). Chun-Ran Wu used a new biological precipitation method to improve the performance of recycled concrete aggregate (RCA). The new biological precipitation method is based on the respiration of bacterial precipitation of calcium carbonate. It is different from the traditional biological precipitation method. The results show that the RCA treated by the biological precipitation method has the advantages of higher density, crushing value and lower water absorption rate than untreated RCA (Wu et al. 2018). Li Wengui conducted a comparative test on the impact performance of recycled concrete and recycled concrete with 1% to 2% nano-SiO_2. The test results show that recycled concrete with nano-SiO_2 has higher impact strength (Li 2017). Tao Ji studied water permeability resistant behavior and microstructure of concrete with nano-SiO_2 by experiments. A water permeability test shows that, for concretes of similar 28-day strength, the incorporation of nano-SiO_2 can improve the resistance of water penetration of concrete. An ESEM test reveals that the microstructure of concrete with nano-SiO_2 is more uniform and compact than that of normal concrete (Tao 2005).

Although some modification effects are good, the economic cost is high. It is difficult to apply them to actual production. At present, the evaluation of the modification effect is only to carry out some macroscopic performance tests such as apparent density, water absorption and crushing value, and there is no microscopic analysis on the changes of microstructure of recycled aggregate before and after modification. Therefore, in this paper, three kinds of modifiers are used to modify the recycled aggregate, and the modification effect is evaluated from macro and micro aspects respectively, and the modification mechanism is revealed at the same time.

2 TEST

2.1 Test materials

Recycled aggregate: The original recycled aggregate comes from the abandoned concrete block of the demolished house in Lanzhou City. The concrete block is crushed into particles with a particle size of less than 30mm, and particles with a particle size of more than 30mm

Table 1. Technical parameters of water glass.

Density(20°C) (g/mL)	Sodium oxide mass fraction ≥	Silica mass fraction ≥	Modulus (M)	Concentration (%)
1.395	8.2	26	32	38

Table 2. Technical parameters of nano-SiO$_2$.

SiO$_2$ content (≥)	The average particle size(nm)	Specific surface area(≥)(m^2/g)	Density (g/cm^3)
98%	20	120	0.03 ~ 0.05

Table 3. Technical parameters of infiltration crystalline waterproofing agent.

Water content (%)	Fineness 0.63mm sieve residue(%)	Chloride ion content (%)	Wet base bonding strength (MPa)
1.2	≤2.5	≤0.06	1.3

are screened out and continue to be crushed until they pass the 30mm sieve completely. Since this article only studies recycled coarse aggregates, fine recycled aggregates with a particle size of less than 5mm are directly removed during the screening process.

Water: ordinary tap water.

Modifier: In this paper, three chemical reagents such as water glass, nano-SiO$_2$ and infiltration crystalline material are used. The parameters of each reagent are shown in Table 1~Table 3.

2.2 *Modification of recycled aggregate*

Before the test, the recycled aggregate of each grade should be cleaned, put into the oven with the temperature of 105 °C ± 5 °C to constant weight, and then taken out and cooled to room temperature for use. Three kinds of modifiers were prepared into solutions with mass concentration of 2%, 4% and 6%. After soaking the recycled aggregate with different concentrations of modifier for 24 hours, the modifier solution was filtered out and dried naturally for 3 days at room temperature. Then the apparent density, water absorption and crushing value of the modified recycled aggregate were tested. After measuring the physical indexes of recycled aggregate, the surface of recycled aggregate was observed by scanning electron microscope (SEM), and the composition of surface products was analyzed by energy spectrum analysis. Finally, the inner pore structure of the modified recycled aggregate was analyzed by mercury intrusion method (MIP), so as to reveal the modification mechanism at a deeper level.

3 RESULTS AND DISCUSSION

3.1 *Macro test results and analysis*

Figure 1 to Figure 3 show the test results of the apparent density, water absorption and crushing value of the recycled aggregate after soaked by three reagents with three different mass concentrations.

It can be seen from Figure 1 that after soaked the recycled aggregate by the three reagents with three different mass concentrations for 24h, the effects on the apparent density are as follows. Firstly, water glass and permeable crystalline materials show a trend of first

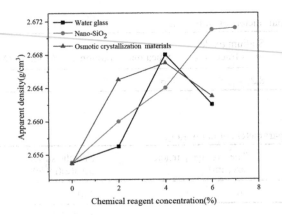

Figure 1.   Apparent density.

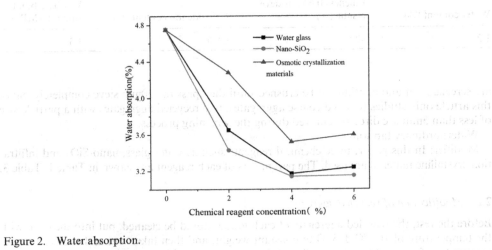

Figure 2.   Water absorption.

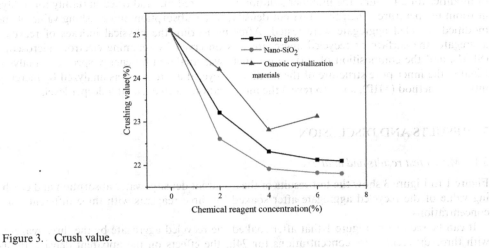

Figure 3.   Crush value.

increasing and then decreasing with the increase of concentration, both of which have a peak value when the concentration is 4%. Secondly, the apparent density of recycled aggregates after water glass treatment is $2.668g/cm^3$, and the improvement effect is better. Thirdly, the apparent density of recycled aggregate soaked with nano-SiO_2 increases with the increase of nano-SiO_2 concentration. However, when the concentration of nano-SiO_2 is higher than 6%, the apparent density of recycled aggregates did not increase significantly. The apparent density of recycled aggregates after nano-SiO_2 treatment is $2.672g/cm^3$, and the improvement effect is best.

It can be seen from Figure 2 that after soaked the recycled aggregate by the three reagents with three different mass concentrations for 24h, the effects on the water absorption are as follows. Firstly, the three reagents have a trend of first decreasing and then rising with the increase of concentration, and the peak value appears when the chemical reagent concentration is 4%. secondly, the water absorption of recycled aggregate after three kinds of reagent treatment are 3.17%, 3.14% and 3.52%, respectively. Finally, nano-SiO_2 treatment is the best, followed by water glass, but the difference between them is not obvious.

It can be seen from Figure 3 that after soaked the recycled aggregate with three different mass concentrations for 24h, the effect of the three reagents on the crushing value is as follows. Firstly, the crushing value of recycled aggregate soaked with water glass and nano-SiO_2 showed a downward trend with the increase of chemical reagent concentration. Secondly, the recycled aggregate soaked by permeable crystalline material showed a trend of first decreasing and then increasing. The crushing value of modified recycled aggregate was the lowest when the concentration of water glass and nano-SiO_2 was 7%, but the crushing value of recycled aggregate soaked with 4% reagent was already enough to meet the specification requirements.

3.2 Microscopic test results and analysis

Figure 4~Figure 7 respectively show the 2000 times SEM scanning images of recycled aggregate soaked by different concentrations of three reagents.

It can be seen from Figure 4~Figure 7, that the surface roughness of the recycled aggregate is obviously improved. The C-S-H gel material, which is produced by soaking recycled aggregates in chemical reagents, filled the pores, repaired the micro cracks, and optimized the interfacial transition zone. Water glass and nano-SiO_2 have good filling effect on pores and microcracks. After strengthening, the surface of recycled aggregate can hardly see the existence of pores and microcracks. Therefore, the modification is more thorough. In contrast, the strengthening effect of permeable crystallization material on recycled aggregate is slightly poor. After treatment, there are still cracks on the surface of aggregate, and the strengthening effect is general.

Figure 4. SEM image of unmodified recycled aggregate.

Figure 5.   SEM image of recycled aggregate soaked by water glass.

Figure 6.   SEM image of recycled aggregate soaked by nano-SiO₂.

Figure 7.   SEM image of recycled aggregate soaked by permeable crystalline materials.

Figures 8 to 10 respectively show the energy spectrum analysis chart of the surface products of recycled aggregate after soaking with different concentrations of modifiers.

Figure 8 shows the energy spectrum analysis of the 21# point in the diagram. The results show that the content of Oxygen and Silicon is very high in the recycled aggregate after soaking through the water glass solution. It is obvious that after the regenerated aggregate and sodium silicate acted, a large amount of cementitious substances containing Oxygen and Silicon were generated at that position, and the gel material filled the pores of recycled aggregate.

Figure 9 shows the energy spectrum analysis of the 7# point in the diagram. From the energy spectrum analysis, it can be seen that the content of Oxygen and Silicon is very high in the recycled aggregate after soaking by nano-SiO_2 solution, which indicates that a large amount of C-S-H gel material is generated, and C-S-H gel material fills the pores of recycled aggregate very well.

Figure 10 shows the energy spectrum analysis diagram of point 20 in the figure. It can be seen that the content of Aluminum and Silicon in the recycled aggregate soaked in the solution of permeable crystalline material is very high, which indicates that a new cementitious material with Aluminum and Silicon as the main components is formed at this point, which has a certain filling effect on the pores of recycled aggregate.

Figure 11 to Figure 12 show the pore size analysis chart of recycled aggregate after soaking with different concentrations of reagent.

It can be seen from Figure 11 that compared with the recycled aggregate before modification, the total pore area, average pore diameter and porosity of the recycled aggregate after 4% sodium silicate treatment decreased by 19.7%, 27.5% and 15.2% respectively. After 6% nano-SiO_2 treatment, the total pore area, average pore size and porosity of recycled aggregate

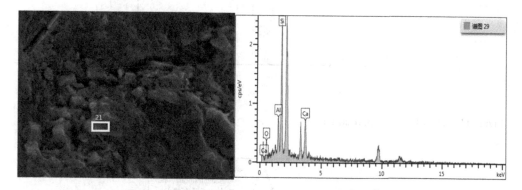

Figure 8.　Energy spectrum analysis diagram of recycled aggregate soaked with water glass solution.

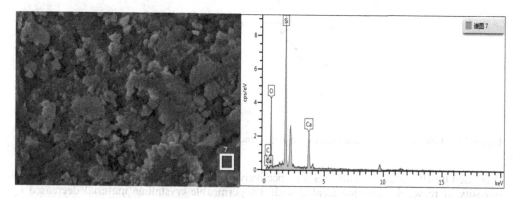

Figure 9.　Energy spectrum analysis diagram of recycled aggregate soaked with nano-SiO_2 solution.

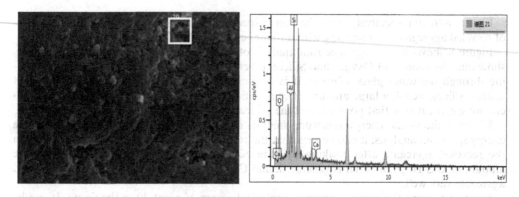

Figure 10. Energy spectrum analysis diagram of recycled aggregate soaked with permeable crystalline material.

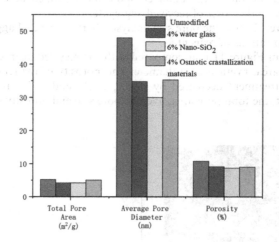

Figure 11. Distribution of pore structure.

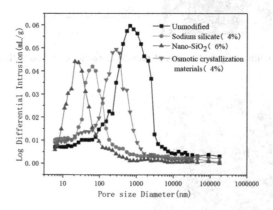

Figure 12. Relationship between mercury injection and pore size.

decreased by 19.3%, 37.5% and 18.9% respectively. The total pore area, average pore size and porosity of recycled aggregate treated with 4% permeable crystalline material decreased by 2.7%, 26.5% and 16.1% respectively. The results of pore structure analysis of recycled aggregate show that the pores of recycled aggregate treated with different chemical reagents are improved, and the pore size is reduced in different degrees. The modified effect of recycled aggregate treated with nano-SiO$_2$ is the best.

Figure 12 shows the relationship between mercury injection rate and pore size distribution of the sample. There are peaks in all four curves, and the pore size corresponding to the peak value is called the most probable pore size. When the most probable pore size is smaller, that is, the curve shifts to the direction of small pore size (left), which means that the probability of small pore size is greater. It can be seen from the figure that the peak value of the curve after treatment with three reagents moves to the left, which means that the most probable pore size of the samples corresponding to the three curves is reduced, that is, the pore size of recycled aggregate is reduced. On the other hand, the pore size distribution of the unmodified recycled aggregate is about 100nm ~ 5000nm, while the pore size distribution of the recycled aggregate treated with water glass, nano-SiO$_2$ and permeable crystalline material is between 20nm ~ 300nm, 10nm ~ 200nm and 80nm ~ 2000nm respectively. It is obvious that the pore size of recycled aggregate treated by three kinds of chemical reagents is greatly reduced. The modification effect of Nano-SiO$_2$ is the best.

4 CONCLUSION

In this paper, different concentrations of water glass, nano-SiO$_2$ and infiltration crystalline materials were used to soak the recycled aggregates respectively, and the macro-strengthening effect was evaluated using the apparent density, water absorption rate and crushing value of the regenerated aggregates after chemical strengthening. The strengthening effect of the micro-structure of the recycled aggregate is evaluated by SEM, EDS and pore size analysis test. The conclusions are as follows:

- The test results show that the effect of 4% water glass and 6% nano-SiO$_2$ is better, the apparent density of recycled aggregate is increased by 0.49% and 0.6% respectively, the water absorption rate is decreased by 33.3% and 34.3% respectively, and the crushing value is decreased by 11.1% and 13.1% respectively. The reason for this test result is that nano-SiO$_2$ and water glass filled the voids on the surface of the recycled concrete aggregate, and reacted with Ca(OH)$_2$ in the cement paste on the surface of the aggregate to form C-S-H gel. Therefore, the recycled concrete aggregates soaked by nano-SiO$_2$ and water glass become denser, the ability to resist water damage is improved, and the strength increases. It can be seen from the above test results that the strengthening effect of recycled aggregate soaked in 6% nano-SiO$_2$ is the best. Compared with 6% nano-SiO$_2$, the strengthening effect of recycled aggregate soaked in 4% sodium silicate is slightly worse, but its indicators also meet the requirements of the specification.
- The surface microstructure of the recycled aggregate was tested by scanning electron microscopy and energy dispersive spectrometry. It was found that the micro cracks and weak interface transition areas of the recycled aggregate were all repaired. The pore size of the recycled aggregate was also studied by MIP method. The experimental results showed that the large void existed in the recycled aggregate was filled with the gel material. The total pore size decreases.
- The strengthening mechanisms of water glass, nano-SiO$_2$ and infiltration crystalline materials can be summarized as: filling pores, optimizing interface transition zone and wrapping recycled aggregate.

REFERENCES

Chun-Ran Wu,Ya-Guang Zhu,Xiao-Tong Zhang,Shi-Cong Kou. 2018. Improving the properties of recycled concrete aggregate with bio-deposition approach. *Cement and Concrete Composites*, 94.
Hu Zhonghui. 2016.Experimental research on the performance of large particle size recycled cement stabilized macadam base. Shandong University of Technology.

Li Qiuyi, Li Yunxia, Zhu Chongji, Tian Li. 2006.Research on Reinforced Concrete Aggregate Strengthening Technology. *Concrete*, Vol:01. 74–77.

LI Wengui, LUO Zhiyu, LONG Chu, HUANG Liang. 2017.Experimental Study on the Dynamical Mechanical Performance of Nanomodified Recycled Aggregate Concrete. Journal of Hunan University. Vol.44: 92–99.

Liu Lingqing. 2014.Performance and design study of cement stabilized recycled mixture. Chang'An University.

Revathi Purushothaman, Ramesh Ruthirapathy Amirthavalli, Lavanya Karan. 2015. Influence of Treatment Methods on the Strength and Performance Characteristics of Recycled Aggregate Concrete. *Journal of Materials in Civil Engineering*. Vol.27: 1–7

Suo Lun, Peng Peng, Zhao Yanru. 2015. Reinforced experimental study of recycled coarse aggregates. *Materials Herald*, 29(S1): 362–365.

Tao Ji.2005. Preliminary study on the water permeability and microstructure of concrete incorporating nano-SiO$_2$. *Cement and Concrete Research*. Vol:35.1943–1947.

Vivian W.Y. Tam, C.M. Tam, K.N. Le. 2006. Removal of cement mortar remains from recycled aggregate using pre-soaking approaches. Resources, *Conservation &amp Recycling*, 50(1).

Wang Jian, Li Yi. 2003. Research on Treatment and Recycling of Construction Waste. *Environmental Engineering*, (06): 49-52+4.

Wang Jianghao, Geng Ou, Li Fumin. 2016. Test of the effect of various modified methods of recycled coarse aggregate on the compressive strength of concrete. *Construction Science and Engineering*, 33(02): 91–97.

Xiong Feng. 2017. Study on the modification of construction waste recycled coarse aggregate and the performance of recycled concrete. Chongqing University.

Xu Ping, Zhang Minxia. 2009. Analysis of recycling of construction waste in my country. *Energy Environmental Protection*, 23(01): 24–26.

Pavement geotechnics & cementitious materials

Functional Pavements – Chen et al (eds)
© *2021 Taylor & Francis Group, London, ISBN 978-0-367-72610-2*

Erosion-resistance performance of SG-1 soil stabilizer-stabilized gravel soil compared with cement-stabilized gravel soil

Yulong Zhao & Zhizhong Zhao
Shandong Jiaotong University, China

Shuya Xie
Hohai University, China

Yiluo Zhang
Chang'an University, China

Shaoquan Wang
Southeast University, China

ABSTRACT: In order to reduce the use of crushed stone and improve erosion resistance performance of road base, SG-1 Soil Stabilizer is used to stabilize natural gravel soil instead of cement. In this study, the erosion resistance performance of SG-1 Soil Stabilizer-Stabilized Gravel Soil-Natural Gradation (SG-1 SSSGS-NG), SG-1 Soil Stabilizer-Stabilized Gravel Soil-Optimized Gradation (SG-1 SSSGS-OG) and Cement-Stabilized Gravel Soil-Natural Gradation (CSGS-NG) were tested and studied. The results reveal that SG-1 Soil Stabilizer has better bonding effect than cement, and SG-1 SSSGS-NG has the best erosion resistance performance. That is to say, the use of soil stabilizer can not only improve the erosion resistance performance, but also reduce the use of crushed stone. Overall, it is feasible to use a soil stabilizer instead of cement to stabilize gravel soil as road material.

1 INTRODUCTION

A large number of road condition surveys show that the erosion of the road base is one of the main reasons for the decline of the function of the pavement and the structural fracture of the surface course (Zhu et al., 2012). The semi-rigid base with cement as the cementing material for road construction in China is widely used in all levels of roads due to its good integrity, high rigidity and good water stability, but its erosion resistance ability is insufficient, which is likely to cause pavement cracks and slurry pumping(Guo et al., 2019; Cong et al., 2011).

On the other hand, stone is an important raw material of the base course in highway pavement structure, and its quality will affect the service life of the road directly. A large number of stone mining will also pose a threat to the ecological environment; long-distance transportation will increase the cost of the project. Therefore, local materials should be fully utilized to save the use of crushed stone and reduce transportation costs.

In some mountainous areas of China, there are abundant gravel soil resources. A large number of gravel soil can be used as a new source of road construction material. Moreover, soil is everywhere and cheap in our lives. It is also possible to mix soil and crushed stone at a certain ratio to compound into artificial crushed stone soil to save the use of stone.

Inorganic binder stabilized materials such as cement and lime, which are commonly used in road bases in our country, have poor anti-scouring ability and are easy to cause road water damage after being washed by water (Nie, 2009). Studies have shown that adding fly ash to cement stabilized gravel soil can reduce the amount of cement when the same soil stabilizing effect is obtained (Zhang et al., 2011; Jongpratist et al., 2009). Moreover, the mechanical properties and durability of composite soil stabilizer stabilized gravel soil are better than those of cement stabilized macadam (Zhao et al., 2016). Using soil stabilizer to stabilize gravel soil instead of a certain amount of cement and lime to stabilize natural gravel soil as road material can reduce the cost of construction and reduce ecological damage. It is necessary to study the ability of erosion-resistance of the soil stabilizer stabilized gravel soil. Therefore, the purpose of this study is to study the erosion-resistance performance of stabilized gravel soil with different content of binder and different gradation through erosion-resistance test, and the comparative results are given.

2 EXPERIMENTAL PROGRAMME

2.1 Raw materials

SG-1 soil stabilizer is used as binder. The main components are cement clinker, quicklime, fly ash and polymer materials. Basic technical properties of SG-1 soil stabilizer are shown in Table 1. In this research, 32.5# Portland cement, which was produced in the city of Tongchuan, Shaanxi Province, played a comparative role with SG-1 soil stabilizer.

Natural gravel soil utilized in tests was taken from the county of Shenmu, Shaanxi Province. According to the requirements of T 0302-2005 (JTG E42-2005), the Grain-size distribution curve is shown in Figure 1.

The optimized gradation are shown in Figure 2.

In the process of mix proportion design, based on the secondary and lower highway base, three kinds of mix proportion were set: (1) SG-1 Soil Stabilizer-Stabilized Gravel Soil-Natural Gradation (SG-1 SSSGS-NG); (2) SG-1 Soil Stabilizer-Stabilized Gravel Soil-Optimized Gradation (SG-1 SSSGS-OG); (3) Cement-Stabilized Gravel Soil-Natural Gradation (CSGS-NG). The crushed stone used in the experiment in this paper was granite, which was produced in Inner Mongolia, and had four grades. With the unconfined compressive strength of 2.8 Mpa as the target, the dosages of the three groups of materials are shown in Table 2.

Table 1. Basic technical properties of SG-1 soil stabilizer.

Item	Test value	Specified value	Standard
Appearance	Gray powder	—	—
Fineness (total percentage of material retained on 0.074 mm sieve) (%)	0.5%	<15%	CJ/T3073-1998
Initial setting time (min)	290	>210	CJ/T3073-1998
Stability	Qualified	Qualified	
Conclusion	Qualified		

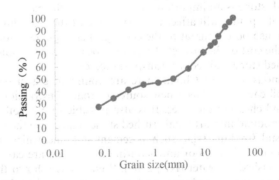

Figure 1. Grain-size distribution curve of the natural gravel soil.

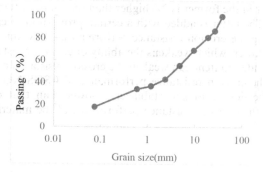

Figure 2. Grain-size distribution curve of the optimized gradation.

Table 2. The dosages of the three groups of materials.

	SG-1 SSSGS-NG	SG-1 SSSGS-OG	CSGS-NG
Binders Dosage (%)	7	5	7

2.2 Specimens preparation and testing procedures

The specimen was cylindrical with a diameter to height ratio of 1:1 (150 mm in diameter and 150 mm in height). For the three research objects, six specimens in each group were cured at standard temperature and humidity for 27 days, saturated with water for 24 hours.The erosion mass loss of the specimen can be calculated as follows:

$$P = \frac{mf}{m0} \times 100 \qquad (1)$$

where P is the erosion mass loss (%); mf is the mass of erosion; $m0$ is the mass of the specimen.

3 RESULTS AND DISCUSSION

The results of erosion resistance test are shown in Figure 3. The erosion-resistance perform-ance of SG-1 SSSGS-NG is better than that of CSGS-NG. From the test results, the bonding ability of SG-1 soil stabilizer to soil particles is better than that of cement. The erosion-resistance performance of SG-1 SSSGS-NG is better than that of SG-1 SSSGS-OG. The

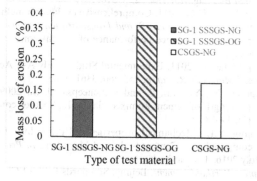

Figure 3. Histogram of erosion mass loss.

341

dosage of soil stabilizer in the former is 2% higher than that in the latter, and the latter adopts optimized gradation. The latter is added with a certain proportion of crushed stone, which is conducive to enhancing the erosion-resistance performance. However, the content of binder dosage in the latter is lower, which weakens the ability of erosion-resistance. According to the test, the decrease of binder content can weaken the erosion-resistance more than the improvement of gradation. The erosion resistance performance of CSGS-NG is better than that of SG-1 SSSGS-OG. The dosage of soil stabilizer is lower than that of cement, which has a stronger impact on the erosion resistance performance of the material than the gradation improvement.

4 CONCLUSIONS

Through erosion-resistance test, the reinforcement effect of SG-1 soil stabilizer on natural gravel soil was studied. The relationship between the erosion-resistance results of SG-1 soil stabilizer stabilized gravel soil and cement stabilized gravel soil with different gradations was analyzed to evaluate the suitability of SG-1 soil stabilizer stabilized gravel soil as base or subbase. Based on the test results, the following conclusions are drawn:

(1) Under the same composition of natural gravel soil materials, the binding effect of the same dosage of binder on soil particles is different, and the binding effect of SG-1 soil stabilizer is better than that of cement.
(2) Under different composition of gravel soil gradation and binder dosage, the change of binder content has a stronger effect on the erosion resistance than the improvement of the gradation.
(3) While using soil stabilizer to improve the erosion-resistance performance of road, it can also reduce the use of crushed stone. Therefore, it is feasible to use soil stabilizer instead of a certain amount of cement and lime to stabilize gravel soil as a road material in terms of erosion resistance.

ACKNOWLEDGEMENTS

This study was supported by Science and Technology Plan of Shandong Transportation Department (2019B63 and 2020B93).

REFERENCES

Zhu, T.L., Tan, Z. M. and Zhou, Y. M., 2012. Experimental research on erosion-resistance performances of cement stabilized base materials. *Journal of Building Materials*, 15(4):565–569.
Guo, R., et al.,2019. Anti-erosion performance of asphalt pavement with a sub-base of cement-treated mixtures. *Construction and Building Materials*, 223: 278–287.
Cong, Z.H., Zheng, N.X.and Yan, H.G.,2011. Comprehensive evaluation method of semi-rigid base mixture's pavement performance. *Journal of Traffic and Transportation Engineering*, 2011,11(4):23–28.
Nie, W. L.,2009. Study on erosion resistance performance of cement concrete pavement base material. Thesis. Chang'an university.
Zhang, X.D., Liang, Y. and Yue, Y., 2011. Experimental Study of Fly Ash Activator in the Application of Aeolian Soil. *Advanced Materials Research*, 255-260: 3361–3365.
Jongpradist, P., Jumlongrach, N., Youwai, S. and Chucheepsakul, S., 2009. Influence of fly ash on unconfined compressive strength of cement-admixed clay at high water content. *Journal of Materials in Civil Engineering*, 22(1): 49–58.
Zhao, Y.L., Gao, Y., Zhang, Y.L., Mechanical properties and durability of Composite Soil Stabilizer Stabilized Gravel Soil compared with Cement Stabilized Macadam, *In Proceedings for 4th CEW Conference*, 27 June to 3 July 2016, The Netherlands.
GB 175-2007, 2007. *Common Portland Cement*. Beijing: Standards Press of China.

JTG E42-2005, 2005. *Test Methods of Aggregate for Highway Engineering*. Beijing: China Communications Press.

JTG E40-2007, 2007. *Test Methods of Soils for Highway Engineering*. Beijing: China Communications Press.

JTG/T F20-2015, 2015. *Technical Guidelines for Construction of Highway Roadbases*. Beijing: China Communications Press.

JTG E51-2009, 2009. *Test Methods of Material Stabilized with Inorganic Binders for Highway Engineering*. Beijing: China Communications Press.

Functional Pavements – Chen et al (eds)
© 2021 Taylor & Francis Group, London, ISBN 978-0-367-72610-2

Study on the law of compressive rebound modulus of liquid soil materials

Yi Zhang
Shanghai Urban Construction Design and Research Institute (Group) Co., Ltd, Shanghai

Dandan Guo, Chongwei Huang* & Meixuan Zhu
Department of Transportation Engineering, University of Shanghai for Science and Technology, Shanghai

ABSTRACT: Aiming at the unconfined compressive strength of the new self-flow platform backfill material which is called liquid soil. The influence of different water content, cement content, admixture type and additive content on the compressive resilient modulus of liquid soil are comprehensively analyzed. The results show the moisture content has a significant effect on it, and the relationship between them is a quadratic function. With the increase of cement content, the compressive resilience modulus of liquid soil increases significantly. The effect of different types of admixtures is quite different: the strength of liquid soil can be improved by reactive mineral powder, and the compressive modulus of resilience is almost unchanged; the compressive resilient modulus of liquid soil can be increased by about one time by water reducer; and the compressive resilient modulus of liquid soil can be increased by two to three times by medium sand.

1 INTRODUCTION

The back filling of abutment refers to the backfill at the structure. After the completion of the structure, the left part between the structure and the subgrade is filled with the materials that meet certain requirements in layers[1]. There are many factors such as the performance and compactness of abutment backfill material, abutment type and construction method. Among these factors, the material, construction method and quality of abutment backfill are the key factors leading to vehicle bump at bridge head[2–4]. The key to solve the differential settlement between subgrade and structure is to adopt appropriate backfill materials and improve the compactness of abutment backfill[3]. In the practical application of abutment backfill, there is a complex demand for the compressive modulus of resilience of liquid soil. The liquid soil material studied in this paper is a new type of backfill material[4]. Its main components are cement, fly ash, water and additives. Through theoretical analysis and calculation, the limit value of compressive resilient modulus of new self flowing platform backfill material liquid soil is defined[5,6]. At the same time, the compressive resilient modulus test of liquid soil with different mixture ratio is carried out to obtain the contribution law of various raw materials and their mixture ratio of liquid soil, which provides a theoretical basis for the backfill engineering of liquid soil[7].

2 ANALYSIS OF MAIN COMPONENTS OF LIQUID SOIL

Liquid soil is a mixture of fly ash, cement, admixture and water. The content of its compressive resilience modulus is the ability to resist the vertical deformation under the action of external vertical load in the elastic deformation stage.

---

* Corresponding author

Table 1. Chemical composition of fly ash.

Chemical composition	Fe_2O_3	Al_2O_3	SiO_2	CaO	MgO
The content (%)	4.74	24.84	51.15	7.71	2.39

2.1 Fly ash

Fly ash, also known as soot, is very similar to cement in appearance. Its main components are SiO_2 and Al_2O_3, as shown in Table 1. Its color can reflect the difference of carbon content and fineness of fly ash to a certain extent.

2.2 Cement

Cement is a kind of powdery, gray black inorganic cementitious material, which hardens when meeting water or air. The chemical composition of cement is shown in Table 2. After adding water to the cement, the slurry is formed. Sand and other materials are added into the paste and mixed again. If it is left standing, the cement will harden in air or water. Among them, cement can improve the strength and strengthen the setting of liquid soil materials.

3 TEST ON COMPRESSIVE RESILIENT MODULUS OF LIQUID SOIL

3.1 Test materials

In the liquid soil materials, fly ash is an important part of the mixture, and cement is a hydraulic inorganic cementitious material. On the premise of determining the amount of fly ash and cement unchanged, the use of additives has a strong catalytic effect on the cement molecules and can improve the early strength of liquid soil.

(1) Cement

As one of the most important raw materials of liquid soil, the main contribution of cement to liquid soil is to enhance its strength. When the content of cement is high and the strength of cement itself is high, the strength of liquid soil will be greatly improved. The cement used in this paper is P.O. 32.5 conch composite Portland cement produced by Shanghai Conch Cement Co., Ltd.

(2) Fly ash

Coal is a kind of solid waste produced after high temperature combustion. Its form is similar to pozzolanic ash. We call it fly ash. In the power generation process of thermal power plant, a large amount of fly ash will be produced, which can be collected by dust collection device. The parameters of fly ash are shown in Table 3 below.

Table 2. Chemical composition of fly ash.

Chemical composition	CaO	SiO_2	Al_2O_3	MgO	Fe_2O_3
The content (%)	60.6	22.3	6.1	1.46	3.25

Table 3. Parameters of fly ash.

Project	Fineness	Loss on ignition	Specific surface area	Apparent density	The compaction density	Water content
Parameters	≤10%	7.9%	≥450m²/kg	700kg/m³	1150Kg/m³	3% ~ 7%

Table 4.	Standard curing conditions.	
Temperature	20±2°C
Relative humidity	≥90%

(3) Admixtures

In the test liquid soil, the addition mainly includes water reducing agent, medium sand and active mineral powder.

① Water reducing agent

The cement additive used in this test is a kind of nano additive.

② Medium sand

The sand used in the test is the sand whose full particle size is less than 5mm and the content of particles is more than 90%.

③ Active mineral powder

Mineral powder and water will react with water to form hydrated calcium silicate, which will increase its activity and disperse in the pores of liquid soil materials. It will greatly increase the density of liquid soil materials. At the same time, $Ca(OH)_2$ crystals with low compressive strength will be transformed into calcium silicate gel with high compressive strength.

(4) Water

The liquid soil can be used as drinking water for human and livestock.

3.2 *Test content*

In this paper, according to the method of "highway engineering inorganic binder stabilized material test specification" (JTJ057-94), the compressive resilient modulus test of liquid soil specimens at 28 days age was carried out. In the test, the liquid soil specimen was molded by the triple test mold with 100 × 100 × 100cm test mold, and the mold was removed after curing for 28 days under the standard curing conditions (as shown in Table 4). The 810 MTS material testing system was used as the testing instrument.

At the beginning of the test, first select the measuring point, place the mixture sample on the horizontal surface and level the surface, then place the instrument; apply the pressure to 0.05Mpa in advance, then give the stable pressure for one minute, unload, and then apply the stable pressure for one minute. Finally, adjust the pointer to zero scale or directly record the initial reading. Step by step loading and unloading method is adopted in this test. The preset unit pressure is divided into 5 levels as the pressure value applied each time. At the beginning of the test, the preset level 1 load is applied. After the load lasts for 1 minute, record the reading at this time and unload at the same time to restore the elastic deformation of liquid soil specimen. Record the reading again at 0.5 min, and then apply the second level load. After the same load lasts for 1 min, record the reading and unload the load. Continue to apply the third stage load according to the above method. Repeat the test above until the last level of load is applied, record the rebound deformation of liquid soil specimen, and finally calculate the rebound deformation and total deformation of the specimen. Repeat three parallel tests according to the above steps, and the difference between the results of each test and the mean value of compressive modulus of resilience should not exceed 5%.

4 TEST RESULTS AND DISCUSSION

Three groups of parallel experiments were done in each group of tests. The Figure 1 and 2 showed that the compressive modulus of resilience of liquid soil decreased with the increase of

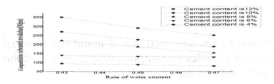

Figure 1.　Test results of compressive modulus of resilience under different moisture content.

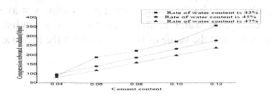

Figure 2.　Test results of compressive modulus of resilience under different cement contents.

water content under different cement content, and increased significantly with the increase of cement content under different water content.

The Table 5 and 6 show that: (1) the compressive resilient modulus and water content of liquid soil meet the equation y = aexp (BX), and the values of a and B are different greatly, which change with the change of cement content, but there is no obvious change rule. In practical application, it is necessary to carry out appropriate tests to obtain these two parameters. (2) The relationship between compressive resilience modulus of liquid soil and cement content is linear, i.e., y = ax + B. The values of a and B change with the change of water content, and have certain regularity. The value of a decreases with the increase of water content, while the value of B is opposite.

Different admixtures have different effects on the improvement of compressive resilience modulus of liquid soil. The effect of medium sand is the most obvious, it can fill the tiny pores in liquid soil to make it more dense; and adding medium sand can improve its compressive resilience modulus without a lot of water. Naphthalene series water reducing agent (β - naphthalene sulfonate sodium formaldehyde condensate) can significantly enhance the activity of cement molecules, and the effect of improving the compressive resilience modulus of liquid soil is acceptable. Although the active mineral powder has potential hydration activity, but from the test results, the effect of active mineral powder on improving the compressive resilient modulus of liquid soil is not as expected.

Table 5.　Fitting equation of compressive resilient modulus under different moisture content.

Cement content%	4	6	8	10	12
Strength equation	$y=592.1exp$ $(-4.198x)$	$y=330.3exp$ $(-1.982x)$	$y=15810exp$ $(-9.861x)$	$y=14000exp$ $(-9.166x)$	$y=18370exp$ $(-9.218x)$
R^2	0.9834	0.8355	0.8412	0.9985	0.9104

Table 6.　Fitting equation of compressive resilient modulus under different cement content.

Cement content%	43	45	47
Strength equation	$y=3232x-41.46$	$y=2320x+0.8207$	$y=1967x+1.435$
R^2	0.9861	0.9863	0.9571

5 EPILOGUE

In this paper, the compressive resilient modulus of liquid soil material in abutment backfill engineering is analyzed, and an index of compressive resilience modulus of liquid soil is studied by means of laboratory test and data analysis.

(1) The compressive resilient modulus and moisture content of liquid soil meet the equation y = aexp (bx), and the equation y = ax + b is satisfied with the cement content.
(2) In the equation y = aexp (bx), parameters a and b will change with the change of cement content, but the regularity of change is not strong; in equation y = ax + b, parameters a and b change with the change of water content, the value of a gradually decreases with the increase of water content, and the value of b increases with the increase of water content.

REFERENCES

[1] Liu Guoyuan. 2007. Application of fluid cement fly ash slurry in abutment backfill [D].
[2] Tang Lizhi. 2010. Causes and prevention measures of vehicle bump at bridgehead [J]. Shanxi architecture, 36 (19): 274–275.
[3] Zhang Keqian. 2003. Key measures to control vehicle bump at bridge head [J]. Shanxi architecture, 29 (17): 135–136.
[4] Xiao Lijing. 2003. Application of foam mixed lightweight fill technology to solve the problem of bumping at the bridge head of Soft Foundation Embankment on high grade highway [J]. Chinese foreign highway, 23 (5): 121–123.
[5] Li Yunfei. 2002. Study on mechanical properties of EPS and reduction of embankment settlement [D].
[6] Ge Zhesheng, Huang Xiaoming & Zhang Xiaoning. 2006. Study on the application of ceramsite fly ash concrete in the back filling of bridge and culvert [J]. Geotechnical mechanics, 27 (11).
[7] Bao Kongbo, Gu Fengyun. 2019. Relationship between compactness and resilient modulus of gravel soil subgrade [J]. Urban road and bridge and flood control, 2019 (08): 240–241.

Functional Pavements – Chen et al (eds)
© 2021 Taylor & Francis Group, London, ISBN 978-0-367-72610-2

Study on mechanical properties of cement stabilized macadam with different admixtures

M. Zhang, K. Zhong & M. Sun
Research Institute of Highway Ministry of Transport, Beijing, China
Key Laboratory of Transport Industry of Road Structure and Material, Beijing, China

H. Wang
Delft University of Technology, Beijing, The Netherlands

X. Tian
Chongqing Jiaotong University, Chongqing, China

Q. Huang
Southeast University, Beijing, China

ABSTRACT: In order to evaluate the influence of different admixtures on the performance of cement stabilized macadam, the influence of fly ash, expansive agent, brucite fiber and alkali salt early strength agent on dry shrinkage, temperature shrinkage, compression and tensile properties of cement stabilized macadam mixture was studied by unconfined compressive strength test, dry shrinkage test, temperature shrinkage test and splitting test, respectively. The results show that the temperature shrinkage, dry shrinkage, unconfined compressive strength and Splitting strength of cement stabilized macadam with fly ash, expansive agent, brucite fiber and basic salt are greatly improved.

Cement stabilized macadam materials have the typical damages of shrinkage cracks caused by the cooling and water loss of base materials, and the secondary reflection cracks extending from the bottom of the base layer to the surface layer under the action of traffic load (Wang et al. 2018, Gao et al. 2020, Guo et al. 2019). Therefore, it's significant to improve the physical and mechanical property of cement stabilized macadam materials, so that to prevent base layer from typical damages (Zhang et al. 2017, Ma et al. 2015, He et al. 2010, Xiong et al. 2015).

In this paper, four kinds of admixture materials, brucite fiber, alkali salt, fly ash and expansive agent, were selected to investigate the influence of admixtures on dry shrinkage, temperature shrinkage, unconfined compressive strength and splitting strength of cement stabilized macadam through unconfined compressive strength test, dry shrinkage test, temperature shrinkage test and splitting test, respectively.

1 RAW MATERIALS

The cement is grade 32.5 ordinary Portland cement produced in Xuancheng, Anhui Province. Its basic technical properties are shown in Table 1. The aggregate is made of Nanyang Limestone, Henan Province, and its basic performance indexes meet the requirements of JTJ 034-2018"Technical Specifications for Construction of Highway Road-bases". Combined with the existing research results of cement stabilized macadam gradation (Liu et al. 2019), the composite gradation of cement stabilized macadam is finally determined by controlling the content

Table 1. Basic technical properties of portland cement.

Initial setting time/Min	Final setting time/Min	Soundness	Bending strength/MPa		Compressive strength/MPa	
			3d	28d	3d	28d
240	400	Qualified	16.3	33.4	4.6	6.2

Table 2. Technical properties of uea-iv concrete expansive agent.

Projects	Sieveallowance/%		Setting time/min		Total alkali content/%
	0.08mm	1.25mm	Initial setting	Final setting	
Results	10	0	115	195	0.31
Standards	≤12	≤0.5	≥45	≤600	≤0.75

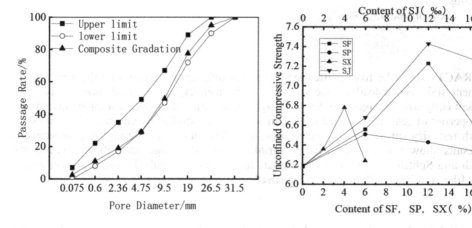

Figure 1. Composite gradation.

Figure 2. 28d unconfined compressive strength.

of coarse and fine aggregate and according to the grading of each grade, with the aim of improving the crack resistance and erosion resistance of the mixture, as shown in Figure 1.

Fly ash produced by Luoyang thermal power plant is selected as fly ash, with loss on ignition of 4.2%; UEA-IV low alkali concrete expansion agent produced by Nanjing Teheng building materials Co., Ltd. is selected as expansion agent, and its properties are shown in Table 2; Mineral brucite fiber is used as fiber; Inorganic basic salt is selected as early strength agent, and the effective component anhydrous sodium sulfate content is more than 99%.

2 TEST RESULTS AND ANALYSIS

2.1 Additional material

The cement content was fixed as 4%. As for additional materials, the content of expansive agent accounts were 0%, 6%, 12%, 18% of cement by weight respectively. The brucite fiber content were 0%, 2%, 4%, 6% of cement by weight respectively. The content of basic salt accounts were 0‰, 6‰, 12‰, 18‰ of cement by weight respectively. Meanwhile, 0%, 6%, 12%, 18% of limestone aggregate were replaced by fly ash respectively. According to the types of admixtures, the cement stabilized macadam added with additional materials is recorded as expansion agent SP, brucite fiber SX, basic salt SJ and fly ash SF.

As some of the additional materials may have negative influence on the early strength of the mixture, the 28d unconfined compressive strength is taken as the best selection standard of the added materials (Lv et al. 2015). The unconfined compressive strength test results of cement stabilized macadam with various dosage of additional materials are shown in Figure 2.

It can be found from Figure 2 that the 28d compressive strength of cement stabilized macadam specimen increases at first and then decreases with the increase of the amount of admixture. When the amount of fly ash reached 12%, the compressive strength of SF reached a maximum value of 7.23 MPa. When the content of expansion agent reached 6%, the compressive strength of SP reached a maximum value of 6.51 MPA. When the content of the brucite fiber reached 4%, the compressive strength of the specimens reached a maximum value of 6.78 MPa. When the content of the basic salt reached 12‰, the compressive strength of the specimens reached a maximum value of 7.43 MPa. Finally, determine the admixture content at the peak of the compressive strength of cement stabilized macadam to form new specimens. The optimum content of fly ash, expansion agent, brucite fiber and basic salt early strength agent is 12% (SF-12), 6% (SP-6), 4% (SX-4) and 12‰ (SJ-12), respectively.

2.2 Dry shrinkage

The instrument method in T0855-2009 was used in the dry shrinkage test. The saturated specimen was dried for 7 days, the Vernier Caliper was used to measure the length of the initial specimen, and then the specimen was put into the dry shrinkage chamber for curing. The dial indicator is placed on the top of the test piece to measure the vertical deformation. The test results are shown in Figure 3 ~ Figure 4.

As can be seen from Figure 3 ~ 4, the dry shrinkage strain rate of the first 12h cement stabilized macadam specimen is higher than that of the first 12h cement stabilized macadam specimen, the cumulative drying shrinkage strain increase rate of cement stabilized macadam with addition material is lower than that of reference group. The average dry shrinkage coefficients of fly ash, expansive agent and brucite fiber were decreased by 21.2%, 19.5%, 30.2%, 5.6%, 22.2%, 24.0%, 33.6% and 11.6%, respectively.

2.3 Thermosetting properties

In the same way, the instrument method is used to measure the vertical temperature contraction deformation of the specimen in the range of 20 °C ~-20 °C for 7 days.

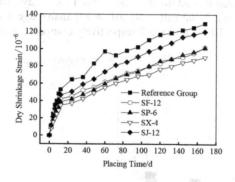
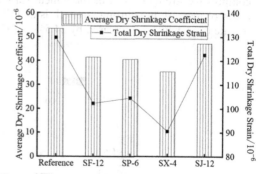

Figure 3. Relationship between cumulative dry shrinkage strain and placing time.

Figure 4. Average dry shrinkage coefficient and total dry shrinkage strain.

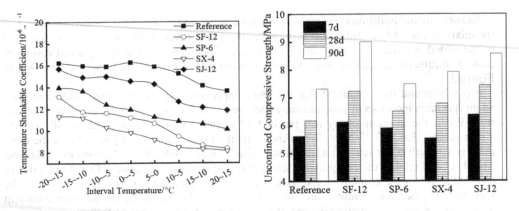

Figure 5. Interval temperature shrinkable coefficient.

Figure 6. Unconfined compressive strength.

As shown in Figure 5, the addition of additional materials can improve the temperature shrinkage performance of crushed stone, reduce its temperature sensitivity, and to a certain extent inhibit the cracking problem of base course caused by temperature change.

2.4 Unconfined compressive strength

Unconfined compressive strength test was conducted according to T0805-1994. The specimens were cured to the specified age of 7d, 28d and 90d, and the specimens soaked for 1d were tested. The results are shown in Figure 6.

As can be seen from Figure 6, the order of increase of unconfined compressive strength from curing to 7d is: basic salt > fly ash > expansive agent > brucite fiber; For the unconfined compressive strength of the specimens from curing to 28d and 90d, fly ash > basic salt > brucite fiber > expansive agent.

2.5 Splitting strength

The splitting test was carried out according to T0852-2009 test. The specimens were cured to the specified age of 7d, 28d and 90d, and the specimens soaked for 1d were subjected to splitting test. The results are shown in Figure 7.

It can be seen from Figure 7 that the variation law of the splitting strength with age of cement stabilized macadam with different kinds of additional materials is similar to that of compressive strength test. For the cement stabilized macadam specimens cured for 7 days, the order of increase in splitting strength is as follows: basic salt > fly ash > expansive agent > brucite fiber, which is increased by 34.5%, 13.8%, 10.3% and 6.9% respectively compared with

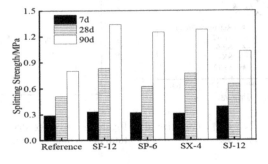

Figure 7. Splitting strength.

the reference group; The order of improvement degree of 90d splitting strength of specimens is: fly ash > brucite fiber > expansive agent > basic salt, and the increase range is 67.5%, 60%, 56.3% and 28.7% respectively compared with the reference group.

3 CONCLUSION

(1) Adding fly ash and brucite fiber can reduce the early forming strength and rigidity of cement stabilized macadam, and the strength of the specimen with the standard curing period of 28 days is taken as the evaluation index, according to the principle of peak strength, the optimum fly ash, expansion agent, brucite fiber and basic salt early strength agent content of cement stabilized macadam is 12%, 6%, 4% and 12‰, respectively.

(2) Fly ash, expansive agent, brucite fiber and basic salt early strength agent can effectively improve the shrinkage performance, unconfined compressive strength and Splitting strength of cement stabilized macadam.

REFERENCES

Gao, J.Q. & Jin, P.P. & Sheng, Y.X. & An, P. 2020. A case study on crack propagation law of cement stabilised macadam base. *International Journal of Pavement Engineering* 21(4):516–523.

Guo, R. & Yang, X.J. & Jiang, H. & Liu, C. 2019. Experimental Analysis of Influencing Factors on Anti-erosion Performance of Cement Stabilized Macadam Base for Asphalt Pavement. *Advanced Engineering Sciences* 51(2):78–84

He, X.B. & Yang, Q.G. & He, G.J. 2010. Anti-erosion Property of Polypropylene Fiber Reinforced Cement-Stabilized Macadam Base Material. *Journal of Building Materials* 13 (02):263–267+276.

Liu, Z.J. & Wei, X.B. & Wang, D.Q. & Wang, L.L. 2019. Performance of cement-stabilized macadam roads based on aggregate gradation interpolation tests. *Mathematical Biosciences and Engineering* 16(4):2371–2390.

Lv, S.T. & Zheng, J.L. & Zhong, W.L. 2015. Characteristics of Strength, Modulus and Fatigue Damage for Cement Stabilized Macadam in Curing Period. *China Journal of Highway and Transport* 28(09):9–15+45.

Ma, Y.H. & Gu, J.Y. & Li, Y. & Li, Y.C. 2015. The bending fatigue performance of cement-stabilized aggregate reinforced with polypropylene filament fiber. *Construction and Building Materials* 83:230–236.

Wang, X.Y. & Li, K. & Zhong, Y. & Xu, Q. & Li, C.Z. 2018. XFEM simulation of reflective crack in asphalt pavement structure under cyclic temperature. *Construction and Building Materials* 189:1035–1044.

Xiong, R. & Fang, J.H. & Xu, A.H. & Guan, B.W. & Liu, Z.Z. 2015. Laboratory investigation on the brucite fiber reinforced asphalt binder and asphalt concrete. *Construction and Building Materials* 100:318–318.

Zhang, X.Z. & Liu, M.L. & Du, X.H. & Yang, X.Z. & Zhou, Z.H. 2017. Synergistic Effect of Nano Silica and Fly Ash on the Cement-based Materials. *Materials Review* 31(24):50–55+62.

Functional Pavements – Chen et al (eds)
© 2021 Taylor & Francis Group, London, ISBN 978-0-367-72610-2

Analysis of factors affecting durability of concrete sleepers

Min Li
Dezhou University

Yulong Zhao
ShanDong Jiaotong University

ABSTRACT: The reasons for the lack of durability of concrete sleepers can be divided into two categories: one is caused by the characteristics of concrete itself, and the other is related to the environmental status of sleepers when they are used. The durability of concrete sleeper can be improved when mineral admixture and air entrainment agent are added at the same time; The durability of concrete sleepers is the best when fly ash and slag powder are mixed in proportion. If the curing temperature is too high, the carbonization resistance of concrete sleeper will be reduced.

1 INTRODUCTION

The service life of most concrete sleepers is much lower than the design service life. The lack of durability of concrete sleepers caused a lot of waste of resources [1]. And there is no reliable evaluation system to predict the durability and the remaining service life of concrete members. Therefore, it is very important to improve the durability of concrete sleeper from the source. This article mainly analyzes the various factors that cause the failure of concrete sleepers, and prepares concrete specimens by compounding mineral admixtures, and adopts two curing methods of steam curing and ordinary curing. The frost resistance, chloride ion permeability resistance and carbonization resistance of the specimens are tested.

2 INTERNAL FACTORS THAT AFFECT THE DURABILITY OF CONCRETE SLEEPERS

2.1 Influence of mineral admixtures

An appropriate amount of fly ash or slag powder could improve the chloride ion permeability resistance and corrosion resistance in the early stage, but it were not significantly improved the compressive strength, porosity and dynamic elastic modulus [2–4]. The effect of mineral admixture on improving the frost resistance of the concrete is limited. Without air entraining agent, the addition of mineral admixtures will reduce the frost resistance of concrete sleepers. When air entraining agent and mineral admixture are added at the same time, the resistance of concrete sleepers to freeze-thaw cycle is significantly enhanced. When fly ash and slag powder are added together, there is an optimal dosage for the improvement of chloride ion permeability resistance, and it is about 30%. When adding fly ash alone, the durability index is the best when the dosage reaches about 15% [5]. The addition of an appropriate amount of mineral admixtures can reduce the strength loss caused by steam curing and ensure the durability of concrete sleepers in the later stage [6].

In this paper, concrete specimens are prepared. The proportion of C60 high strength concrete is shown in the Table 1. The cement is ordinary Portland cement P.O.52.5, the cement property is shown in the Table 2. The mineral admixture dosage scale is shown in the Table 3.

Table 1. The proportion of C60 high strength concrete of C60 high strength concrete.

strength	Water-cement ratio (%)	Quantity of single square concrete (kg/m³)						
		water	cement	medium sand	macadam	Fly ash	slag powder	superplasticizer
C60	30	124	380	653	1389	50	50	6.24

Table 2. The cement property.

P.O52.5 Ordinary Portland cement

28-day compressive strength value	28 day value of flexural strength	loss on ignition	chlorine ion content	alkali content	specific surface area	initial setting time	final setting time
58.4MPa	9.2MPa	3.37%	0.016%	0.58%	348 m²/kg	148min	231min

Table 3. Mineral admixtures table.

grouping	E1	E2	E3	F
fly ash	10%	15%	20%	15%
slag powder	20%	15%	10%	15%
adding amount	30%	30%	30%	30%
Strength(MPa)	67.6	76.1	69.8	48.2

Standard curing for 28 days to test the freeze resistance, chloride ion Permeability resistance and carbonization resistance of the specimens.

The durability of specimens was evaluated by testing their frost resistance, chloride ion permeability resistance and carbonization resistance. The frost resistance of the specimen was tested by rapid freeze-thaw method. The lower the mass loss rate, the better the frost resistance. The chloride ion permeability resistance was tested by the rapid chloride penetrability test. The smaller the electric flux value, the better the chloride ion permeability resistance. The carbonization resistance of the specimen was tested by alcohol phenolphthalein solution titration. The smaller the carbonization depth, the better the carbonization resistance.

The quality loss rate of the specimen after 200 freeze-thaw cycles is shown in the Figure 1. The comparison diagram of the electric flux value of the specimen after 6 hours of electrification is shown in Figure 2. The comparison of residual strength of the specimen after 28 days of carbonization is shown in Figure 3.

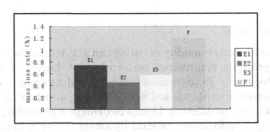

Figure 1. Comparison of mass loss rate after 200 times freeze-thaw.

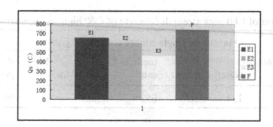

Figure 2.   Electric flux value of specimens after six hours of power.

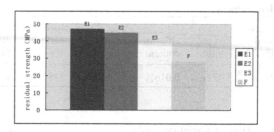

Figure 3.   Comparison of residual strength of specimens after carbonization for 28 days.

As can be seen from the figure that the mass loss rate of E2 mix ratio is the lowest after 200 freeze-thaw cycles for specimens with mineral admixtures. This shows that adding fly ash and slag powder in equal proportions has the best frost resistance. After the rapid chloride pene-trability test, the electric flux value of the E3 mix ratio is the lowest. This indicates that increasing the content of fly ash can improve the chloride ion permeability of specimens. The carbonization depth of F mix ratio is the lowest after 28 days of carbonization. It shows that mineral admixtures have little effect on ordinary strength concrete specimens, and the addition of mineral admixtures is not conductive to the carbonization resistance of high-strength con-crete specimens.

It can be seen from Figure 1, Mass loss rate of E2 is the lowest after 200 times of freeze-thaw. that is, the frost resistance performance is the best when fly ash and slag powder are mixed in proportion; It can be seen from Figure 2 that the E3 blending ratio has the smallest electrical flux value after 6 hours of energization, that is, the more fly ash content, the better the resistance of the specimen to chloride ion penetration; It can be seen from Figure 3 that the E3 mix ratio has the lowest residual strength value after 28 days of carbonization, that is, the more the amount of slag powder, the better the carbonization resistance of the test specimen.

When the content of mineral admixture is 30%, the addition of fly ash is beneficial to improve the chloride permeability of concrete sleepers, but it is unfavorable to carbonization resistance; The more slag powder is added, the better carbonization resistance of concrete will be; The concrete sleeper has the best frost resistance when fly ash and slag powder are mixed in equal proportions. It can be seen that the mixture of fly ash and slag powder is the most beneficial to the durability of concrete sleepers.

2.2   Influence of curing mode

Improving the temperature and humidity of curing can also reduce the shrinkage and cracking of specimens, and improve the deformation performance of sleepers. A survey of some bridges that have been used for nearly 20 years shows that the bridges under the condition of natural curing have strong corrosion resistance and hardly need repairing, while the bridges under the condition of steam curing have different degrees of damage and need repairing. Therefore, the improvement of concrete performance indexes by steam curing is limited to the early stage, which may be detrimental to the long-term durability of concrete.

In this paper, concrete specimens with strength of C40 and C60 were prepared. Steam curing (60°C) and standard curing (20°C) were carried out. Standard curing is shown in Figure 4, and steam curing is shown in Figure 5. The frost resistance, chloride ion permeability resistance and carbonization resistance to be tested. The comparison of the mass loss rate of the specimen under 200 times of freeze-thaw is shown in Figure 6. Comparison of the electric flux value of the specimen is shown in Figure 7. The comparison of residual strength of specimens after 28 days of carbonization is shown in Figure 8. After 200 times of freeze-thaw cycles, it can be seen from the figures, the mass loss rate of steam curing is less than that of standard curing, the electric flux of steam curing is less than that of standard curing, and the carbonization residual strength value of steam curing is greater than that of standard curing. The specimens cured by steam at 60°C have better frost resistance and chloride ion permeability resistance than those cured by standard condition, but their carbonization resistance is lower than that cured by standard condition.

Under the condition of steam curing, the concrete sleeper can reach 75% of the designed strength within 9-12h, forming a relatively dense internal structure. Therefore, in the early stage, the freeze-thaw resistance performance and chloride ion permeability resistance of the specimen are better. Steam curing is also easy to cause internal micro-cracks in concrete. Because the thermal expansion coefficient of solid materials in concrete differs a lot from that of substances such as water and air in the internal capillary channels, uneven expansion and deformation will lead to internal micro-cracks, which provides the conditions for the invasion of CO_2 and reduced the carbonization resistance of the specimen. Therefore, under the condition

Figure 5.   Steam curing.

Figure 4.   Standard curing.

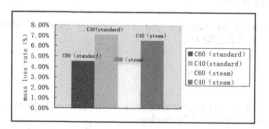

Figure 6.   Comparison of mass loss rate of each group of specimens after 200 times freeze-thaw.

357

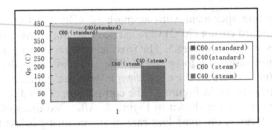

Figure 7. Electric flux value of specimens under different curing conditions.

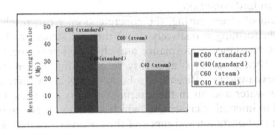

Figure 8. Comparison of residual strength of specimens after carbonization for 28 days under two curing conditions.

of adding mineral admixtures, appropriately increasing the temperature and humidity during curing of concrete sleepers is conducive to the formation of early strength and durability of concrete sleepers, but the curing environment with high temperature and humidity may damage its long-term carbonization resistance.

3 CONCLUSION

(1) When the content of mineral admixture is 30%, the addition of fly ash is beneficial to improve the chloride permeability of concrete sleepers, but it is unfavorable to carbonization resistance; The more slag powder is added, the better carbonization resistance of concrete will be; The concrete sleeper has the best frost resistance when fly ash and slag powder are mixed in equal proportions.
(2) Properly increasing the temperature and humidity of curing can reduce the shrinkage and cracking of concrete sleepers. However, if the curing temperature is too high, the internal micro-cracks are likely to occur in concrete, which reduces the long-term durability of concrete sleepers.

ACKNOWLEDGEMENTS

This study was supported by Science and Technology Plan of Shandong Transportation Department (2019B63 and 2020B93).

REFERENCES

[1] Penghuan Liu. Study on the Effect of several salts on the early self-shrinkage and Anti-crack properties of Concrete. Harbin Institute of Technology, 2015.
[2] Hover K. C., Evaporation of water from concrete surface. ACI Materials Journal 2006, 5(103), 384–389.

[3] Dhanya Sathyan; Kalpathy Balakrishnan Anand, Inflfluence of superplasticizer family on the durability characteristics of fly ash incorporated cement concrete. 2019,204,864–874.

[4] Memon S. A.; Shah, S. F. A., Khushnood R. A.; Baloch W. L., Durability of sustainable concrete subjected to elevated temperature–A review. Construction and Building Materials 2019, (199), 435–455.

[5] Mohammed A.; Rita N., Long-term durability of self-compacting high-performance concrete produced with waste materials. Construction and Building Materials 2019, (212), 350–361.

[6] Qiang jin, Hu DE, et al.Study on the Influence of mineral admixtures on the mechanics and erosion properties of steel slag sleeper concrete. Concrete and Cement Products 2017, (9), 34–38.

Functional Pavements – Chen et al (eds)
© 2021 Taylor & Francis Group, London, ISBN 978-0-367-72610-2

Study on road performance of liquid soil materials

Chongwei Huang & Juanjuan Chen
Department of Transportation Engineering, University of Shanghai for Science and Technology, Shanghai

Yi Zhang*
Shanghai Urban Construction Design and Research Institute (Group) Co., Ltd, Shanghai

Dandan Guo
Department of Transportation Engineering, University of Shanghai for Science and Technology, Shanghai

ABSTRACT: In this paper, a new type of backfill material for gravity platform, that is, liquid soil, is proposed, considering the influence of different water content, cement content, age and additives, the road performance of liquid soil material was tested in laboratory, including fluidity, density and shrinkage tests. According to the test results, the density range of the liquid soil material is 1.35g/cm³ ~ 1.41g/cm³, which has good light weight; the water content has a relatively large influence on the shrinkage of the liquid soil. In contrast, the cement content has a little influence on the shrinkage; the flow performance of liquid soil with water content of more than 45% meets the construction requirements, and the cement content and mineral powder content have little impact on the fluidity. Medium sand can greatly improve the fluidity of the liquid soil and reduce the water content at the same time. The water content is fixed between 40% and 41% after using the water reducing agent.

1 INTRODUCTION

With the continuous advancement of urban road engineering in my country, new fillers that have been applied and studied more include foamed lightweight soil, EPS, ceramsite concrete, coal gangue, etc. [1–3]. In addition, some treatment techniques for backfilling of abutment back have also been researched and applied, such as approach slab at bridge head, compaction grouting, geocell, etc[4]. Although the above-mentioned backfilling materials and treatment techniques have some effect on solving the problem of backfilling on the back of the abutment, there are also many problems that follow, such as higher prices, and inconvenient construction.

The American Concrete Institute (ACI) has proposed a new type of lightweight, low-strength, and alternative to traditional backfill materials, called CLSM [5]. Therefore, in this paper, on the basis of CLSM raw material ratio, water reducing agent, activated slag, medium sand and other additives that can change its properties are added to form a new road backfill material, called liquid soil. Its density, shrinkage and fluidity were studied in detail.

2 TEST SCHEME

2.1 Test materials

Liuid soil slurry mixture is mainly composed of fly ash, cement, admixture and water. The propeties and chemical characteristics of fly ash play an important role in the performance of

---

* Corresponding author

mixture; cement provides early strength and polymerization force for mixture; under the condition of keeping fluidity and cement dosage unchanged, additives can reduce the amount of water in mixture and improve the early strength.

2.2 Test content

2.2.1 Density
Use a 100 mm × 100 mm × 100 mm cube test mold for molding. After molding, measure its side length with a ruler and calculate the volume, then weigh the test piece to find its density.

2.2.2 Contractility
In the shrinkage test, 100mm×100mm×400mm prism test mold was used for the mixture sample. After pouring, put the mixed material with mold into the curing box, cure for 7 days under standard curing conditions, then remove the mold, number it, indicate the direction of the test, and let it stand at room temperature for 3-6 hours then measure. After measuring the initial length, place it in a room with a temperature of (20 ± 2)°C and a relative humidity of (60 ± 5)%, and measure the length of the test piece at 7d, 14d, 21d, and 28d.

2.2.3 Liquidity
Fluidity is a measure of the fluidity of a fluid mixture. Liquid soil contains a lot of water. On the one hand, it is used as the raw material for the physical and chemical reaction of fly ash and cement, on the other hand, it makes the early mixing, transportation and pouring convenient and feasible, and plays a great role in the later molding and maintenance.In the test and actual construction, how to ensure the fluidity of liquid soil mixture is very important.

In this paper, sz145 mortar consistency meter is used to measure the fluidity of liquid soil. The operation is simple, the results are intuitive, and it is easy to carry.

3   TEST RESULTS AND DISCUSSION

In the following data analysis, the average values of three groups of parallel tests were discussedfor each group of tests.

3.1   Density

As shown in Figure 1 and figure 2:
It can be seen that under the same cement content, with the increase of water content, the density of liquid soil gradually decreases. At the same moisture content, the density of liquid soil increases with the increase of cement content.

Since in the composition of liquid soil, the density of water is the smallest; and the density of cement is the largest. The formula for calculating the density of the mixture is:

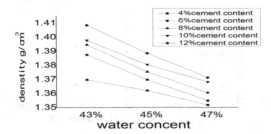 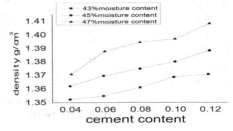

Figure 1.   The density of soil specimens with different moisture content.

Figure 2.   Liquid soil specimen density with different cement content.

$$\rho = \frac{\rho_1 \rho_2 \rho_3}{a\rho_2\rho_3 + b\rho_1\rho_3 + c\rho_1\rho_2}$$

ρ_1, ρ_2, ρ_3 are the density of water, fly ash and cement, respectively; a, b, c are the corresponding percentage.

Therefore, when the content of water or cement increases, the density of liquid soil will change as expected. On the other hand, because the change of cement content refers to the change of the mass ratio of cement and fly ash, and the solid composition only accounts for about 60% of the mass ratio of liquid soil, so the change range of density with cement content is small. At the same time, according to the test results, the density of liquid soil material is less than that of traditional backfill material. It has a certain lightness and can reduce the settlement problem caused by the self weight of backfill material. At the same time, according to the test results, the density range of liquid soil material is 1.35g/cm3 ~ 1.41g/cm3. It has a certain lightness and can reduce the settlement problem caused by the self weight of backfill material.

3.2 Contractile

The shrinkage rate of liquid soil is taken as the evaluation index, the calculation is as follows:

$$\gamma_t = \frac{L_0 - L_t}{L}$$

L_0: The initial length of the specimen;
L_t: test length of the specimen at t days;
L: effective length of the specimen; $L = 400$ mm.
the test results are shown in Figure 3 and Figure 4.

It can be seen that the dry shrinkage rate of liquid soil increases with the increase of water content and cement content. The higher the water content is the higher the cement content is, and the easier the liquid soil is to produce drying shrinkage cracking.

When the moisture content changes from 38% to 47%, the shrinkage increases by about 40%; when the cement content increases from 4% to 10%, the dry shrinkage of liquid soil increases by about 6%. Therefore, it is an important method to control the shrinkage of liquid soil with lower moisture content while keeping other properties of liquid soil (such as fluidity) meet the requirements. When the strength meets the requirements, the cement content of liquid soil should be controlled as much as possible to prevent the shrinkage and cracking of liquid soil. On the other hand, it can also be seen from the two figures that the shrinkage rate of liquid soil changes greatly in the early stage (between 7d and 14d), and small change in the later stage (between 21d and 28d). It shows that the shrinkage rate of liquid soil tends to be stable after 14 days.

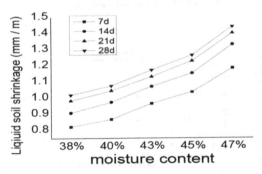

Figure 3. Liquid soil fluidity with different moisture content.

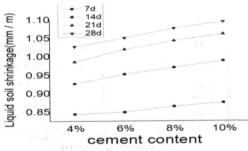

Figure 4. Liquid soil fluidity with cement content.

3.3 Liquidity

In the road abutment backfill project, the fluidity of liquid soil in the range of 11.0 cm ~ 13.0 cm can achieve good self filling performance and improve its compactness

The fluidity of liquid soil under different moisture content and different cement content is shown in Figure 5 and Figure 6.

According to the test results, the fluidity of liquid soil has a very obvious positive correlation with water content. When the water content is 43%, the fluidity of liquid soil obviously fails to meet the requirements of construction. In fact, when the moisture content is less than 40%, the mixture is too thick to be mixed, and accompanied by a large number of bubbles, voids can not be eliminated. When the water content is 45%, the mixture becomes thinner obviously, with the increase of water content, the fluidity changes evenly, and the mixture tends to the ideal state; when the water content is higher than 47%, the consistency of the mixture is too large, and the mixture can not be formed.

With the change of cement content, the fluidity of liquid soil only fluctuates slightly, and the relationship between them is not obvious. The reason for this phenomenon is that only the cement content is changed, which is based on the mass ratio of fly ash to cement. When the cement content changes from 4% to 12%, the fly ash content also changes from 96% to 88%. In this process, the water cement ratio of liquid soil does not change. The fluidity of the liquid soil under different admixture types and applied admixtures is shown in Figure 7.

Dosage 1:No additives added. Dosage 2: The content of mineral powder water-reducing agent is 1.5% of the soild, and the mass ratio of medium sand and fly ash is 75:25; Dosage 3: The content of mineral powder and water reducer is 2.5% of solid material, and the mass ratio of medium sand and fly ash is 70:30.

The fluidity of the three admixtures increased positively, and the fluidity of liquid soil increased with the increase of the content. In fact, in the medium sand test group, the moisture content is only 25%, and the fluidity reaches the requirement of construction; in contrast, the medium sand with the amount of 1 is basically

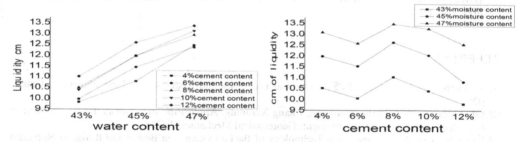

Figure 5. Liquid soil fluidity with different moisture contentdifferent moisture content.

Figure 6. Liquid soil fluidity with cement content.

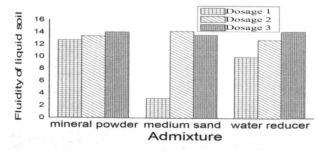

Figure 7. Effect of admixture on liquid soil liquidity.

unable to mix at 25% moisture content because it is not mixed with the admixture. In the water-reducing agent group, we find that when the moisture content is less than 40%, the mixture will thicken sharply, and the mixture with moisture content more than 41% will quickly become thin and can not be formed. Relatively speaking, mineral powder is not as effective as the former two in improving the fluidity of liquid soil.

3.4 Principle of optimum mix proportion of liquid soil

Based on the analysis of the above test results, it can be concluded that:

(1) Based on the consideration of light weight, the moisture content of liquid soil should be increased and the cement content should be reduced.
(2) Based on the consideration of shrinkage, the moisture content and cement content of liquid soil should be minimized.
(3) The water content of liquid soil should be increased, and the admixture should be mixed with medium sand or water reducing agent.

4 KNOTS

(1) The density range of liquid soil material is 1.35 g/cm³~1.41 g/cm³, which has good light-weight property.
(2) The effect of water content on the shrinkage of liquid soil is greater than that of cement, In practical application, the effect of moisture content on shrinkage can be considered more.
(3) The flow performance of liquid soil with water content of more than 45% meets the construction demand, and the cement content and mineral powder content have little influence on the fluidity. Medium sand can greatly improve the fluidity of liquid soil and reduce water content. After using water reducing agent, the water content is fixed between 40% and 41%.

REFERENCES

[1] Li Yunfei. 2002. Study on EPS Mechanical Properties and Reduction of Embankment Settlement [D].
[2] GE Fusheng. 2006.Huang Xiaoming, Zhang Xiaoning. Application of ceramsite fly ash concrete in backfill of bridge and culvert abutment .Geotechnical Mechanics. 27(11).
[3] Dong Jianxun. 2013. Application Technology of Backfill Gangue in Bridge and Bridge D.Northeast Forestry University.
[4] Ding Fang. 2015. A Study on the Integration of Bridgehead Board and Subgrade and Pavement Structure.
[5] ACI Committee 229R, 1999. Controlled low strength materials(CLSM), ACI 229R—99.

Functional Pavements – Chen et al (eds)
© 2021 Taylor & Francis Group, London, ISBN 978-0-367-72610-2

FEM analysis of heat transfer in porous asphalt pavement

Muhammad Waheed Abid, Ali Raza Khan & Bin Yu*
School of Transportation Engineering, Southeast University, Nanjing, China

ABSTRACT: This study examines the cooling performance of porous asphalt pavement and the effectiveness of an addressed type of pavement for the mitigation of urban heat island effect by using FEM modeling. The city of Lahore is considered as a case study keeping in view the trends of changing temperature due to the invigorating urbanization process in Lahore. A FEM model was developed using material thermal properties and hot summer environmental conditions. The temperature data was collected from the Pakistan Meteorological Department (PMD). The developed model was used to identify the cooling efficiency of porous asphalt pavements. The model indicates a temperature drop of 6°C at the surface when porous asphalt pavement was used instead of conventional dense-graded pavement. The Sensitivity analysis performed for all the factors (the wind velocity, solar radiations, daily peak air temperature, and daily lowest air temperature) indicates that increasing the intensity of solar radiation highly affect the pavement surface temperature.

Keywords: Porous asphalt, Heat transfer, Urban heat island effect, Finite element modeling

1 INTRODUCTION

From the last few decades, the world population is increasing dramatically. To meet the demand of the population there has been a huge addition in the construction of road and residential projects. The new constructions have significantly changed the ground cover all over the world such as urbanization, deforestation, and air pollution(Chung, Yoon, & Kim, 2004). The effect of urbanization on local climate is more important because it causes urban heat island effect (Liu, You, & Dou, 2009). Urban Heat Island effect is a worldwide issue that is disturbing the life in urban areas. (Oke, 1982) says although the topic of urban heat island effect is well documented and well-researched people still need awareness about its key factors and causes. Pavements are considered to be the major cause of the UHI effect but it can also be the solution to this problem if designed properly. This study identifies the importance of well-designed pavement for decreasing environmental temperatures in urban areas.

2 METHODOLOGY

2.1 Case study

The city of Lahore is selected as a case study. It is located in Punjab province which is the largest province concerning population. The population of Lahore that was 2.17 million in 1972 has reached 11.13 million in 2017. Lahore has been urbanizing at a very fast rate. The rapid urbanization is resulting in dramatic land cover changes. The increasing population and changes in land-cover harm the local climate of Lahore.

---

* Corresponding author

2.2 Modeling and simulation

A simple pavement model in ABAQUS software was developed to obtain the understanding of thermal interaction in the depth of pavement for a typical summer climate data of the dry and hot region of Lahore city. A typical asphalt pavement structure was chosen for simulation of infield temperature conditions. The model consists of four layers.

2.3 Model establishment

To simulate the infield temperature conditions a virtual experiment is conducted. For the simulation of temperature change in pavement during 24 hours of a day, ABAQUS software is used because of its high speed and ability to tackle complex problems. The simulation results are used to identify the enhanced cooling performance of pavement when its porosity is increased to simulate the porous pavement. The thermophysical properties of each layer used for model establishment are listed in Table 1.

Table. 1. Thermophysical Properties for Simulation.

Material	Density [kg·m³]	Specific heat [J·(kg·°C)⁻¹]	Thermal conductivity [J·(h · m ·°C)⁻¹]
Asphalt layer	2300	950	5000
Base layer	2200	920	4600
Sub-base layer	1900	942	4300
Earth subgrade	1800	1300	4100

Other parameters that are necessary for simulation of heat transfer in ABAQUS are listed in Table 2.

Table 2. Model Parameter.

Model Parameters	Value
Heat Convection Coefficient h_c* [J·(h·m²·)⁻¹]	hc=3600(3.7v$_w$+6.1) (v$_w$, wind velocity, m/s)
Absolute Zero T_Z [°C]	-273
Stefan-Boltzmann Constant σ* [J·(h· m2· K4)-1]	2.041092×10⁻⁴

The material properties and model parameters are taken from the literature review (Adkins & Merkley, 1990; Hermansson, 2004; Zachary Bean, Frederick Hunt, & Alan Bidelspach, 2007).

The environmental conditions that affect the temperature of pavement in the field are solar radiation, wind, speed, and air temperature. For this simulation, the environmental parameters of Lahore city are used. Table 3 represents the hot summer climate condition in Lahore.

These values were used to set up the boundary conditions for finite element analysis in ABAQUS. The environmental temperature was applied on the surface of the model because the heat transfer happens only between the surface of the pavement and the external environment. A uniform meshing is adopted throughout the model to improve speed and accuracy.

Table 3. Environmental Parameters.

Daily peak air temperature T_a max [°C]	Daily lowest air temperature T_a min [°C]	Daily total solar radiation volume Q [MJ/m²]	Daily effective sunlight hours c [h]	Daily average wind velocity v$_w$ [m/s]
42.7	22.77	20	14	5.39

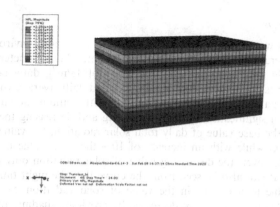

Figure 1.   Finite element model of pavement for porous asphalt with 20% porosity.

2.4   *Finite Element Analysis (FEA)*

After the completion of the model, the temperature field was studied according to the determined parameters. The analysis was performed in two stages namely the steady-state analysis and transient analysis. To study the temperature variation, three main points from top to bottom were selected on the model. The selected point located at the surface, 6 cm below the surface, and 12 cm below the surface. The developed model has been shown in Figure 1.

3   RESULTS AND ANALYSIS

3.1   *Simulation results*

Two simulation models were made to identify the cooling performance of porous asphalt pavement (20% porosity) over conventional asphalt pavement (3 % porosity). The simulation results show that the pavement surface temperature varies throughout the day which is due to the change in air temperature but the temperature change extent is falling with the increase in depth. Also, the peak temperature value approaches at a different time for different depths, the peak value for surface, 6cm, and 10cm appears at 1:30 pm, 3 pm, and around 4:30 pm respectively. The peak temperature of the pavement surface for conventional asphalt appears to be 66°C while this value is much lower in the case of porous asphalt that appears to be around 61°C but the peak values appear at the same time for both models. This difference in temperature for both models goes on decreasing with increasing depths. Figure 2 shows the simulation results of both models.

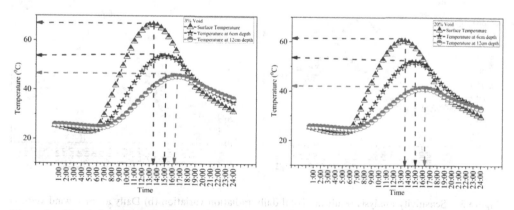

Figure 2.   Simulation results at surface, 6cm and 12cm depth (a) With 3% void (b) With 20% void.

3.2 Sensitivity analysis

A sensitivity analysis was performed to verify the effect of various environmental parameters on the surface temperature of porous asphalt pavement. The parameters used in sensitivity analysis include daily total solar radiation volume Q [MJ/m^2], daily peak air temperature $T_a$max [°C], daily lowest air temperature $T_a$min [°C], and daily average wind velocity v_w [m/s]. The effect of all four parameters is different from each other and is shown in Figure 3.

Figure 3(a) shows a significant impact of increasing and decreasing the value of daily total solar radiation. For the base value of daily total solar radiation, the value of surface temperature was around 60°C while with an increase of 10% the peak value of surface temperature increases by 11°C. However, the decreasing value of solar radiation only decreases the surface temperature by 3°C. It can also be seen from the curve that the night time low temperature is not affected much due to a change in the value of solar radiation. So, it is suggested that the day time surface temperature can be decreased by providing shadings of trees and buildings.

Figure 3 (b) shows the temperature variation curve due to wind velocity. It is obvious from the curve that daytime high temperature is reduced when wind velocity is increasing this is because the wind velocity enhances the connection between pavement and air. Alternatively, the reduced wind velocity increases the high surface temperature.

Moreover, the difference in daily peak air temperature and daily lowest air temperature has a significant impact on the surface temperature of the pavement. It is clearly shown in Figure 3 (c) & (d) when the difference of daily peak air temperature and daily lowest air temperature

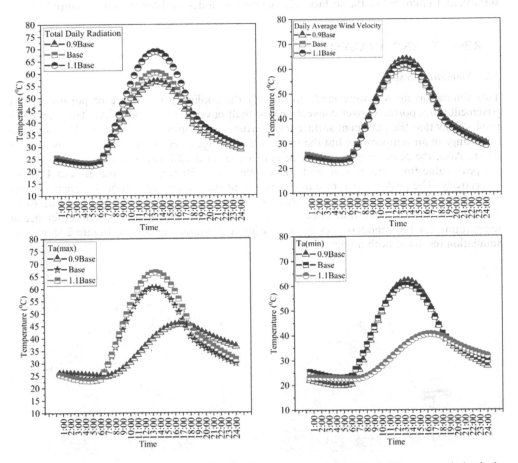

Figure 3. Sensitivity analysis result (a) Total daily radiation variation (b) Daily average wind velocity (c) Daily peak air temperature (T_a max) (d) Daily lowest air temperature (T_a min).

is high, peak surface temperature value of pavement is also very high. This difference of air temperature also controls the approaching of a peak value of pavement surface temperature, if this difference is small the approaching of peak surface temperature delays. This small difference not only decreases the surface temperature of pavement but also delays the appearance of peak temperature in the temperature variation curve.

4   CONCLUSION

This study identifies the heat island effect of pavements for Lahore. FEM model was developed to verify the cooling performance of porous asphalt pavements. Based on the developed model following conclusion was drawn from this study.

- The surface temperature of porous asphalt is 6°C less than the conventional asphalt.
- Sensitivity analysis was performed for all the factors including the wind velocity, solar radiations, daily peak air temperature, and daily lowest air temperature. Results indicate that increasing solar radiation highly affect the pavement temperature.
- Additionally, when Ta min is increased then the pavement temperature decreases more because of the difference in daily peak air temperature and daily lowest air temperature reduces. Reduction in difference introduces less heat in the simulated pavement.
- Porous pavement strategy is effective to reduce the pavement temperature and also to reduce the environment temperature.

ACKNOWLEDGMENT

The authors would like to thank the School of Transportation from Southeast University and China Road & Bridge Corporation for providing the opportunity.

REFERENCES

Adkins, D. F., & Merkley, G. P. (1990). Mathematical model of temperature changes in concrete pavements. *Journal of transportation engineering, 116*(3), 349–358.

Chung, Y.-S., Yoon, M.-B., & Kim, H.-S. (2004). On climate variations and changes observed in South Korea. *Climatic Change, 66*(1-2), 151–161.

Hermansson, Å. (2004). Mathematical model for paved surface summer and winter temperature: comparison of calculated and measured temperatures. *Cold regions science and technology, 40*(1-2), 1–17.

Liu, W., You, H., & Dou, J. (2009). Urban-rural humidity and temperature differences in the Beijing area. *Theoretical and applied climatology, 96*(3-4), 201–207.

Oke, T. R. (1982). The energetic basis of the urban heat island. *Quarterly Journal of the Royal Meteorological Society, 108*(455), 1–24.

Zachary Bean, E., Frederick Hunt, W., & Alan Bidelspach, D. (2007). Evaluation of four permeable pavement sites in eastern North Carolina for runoff reduction and water quality impacts. *Journal of Irrigation and Drainage Engineering, 133*(6), 583–592.

Functional Pavements – Chen et al (eds)
© 2021 Taylor & Francis Group, London, ISBN 978-0-367-72610-2

Evaluation on simple estimation method of slab panel's field temperature

Y.X. Wei
Nanjing Institute of Technology, Nanjing, China

P. Aela
Beijing Jiaotong University, Beijing, China

J.B. Shang
Nanjing Institute of Technology, Nanjing, China

B. Yang
China Railway Corporation, Beijing, China

ABSTRACT: Since it is impossible to accurately grasp the continuously changing field temperature of the bi-block ballastless track, a simple method of estimating the field temperature is used in engineering applications. In order to analyze and evaluate its suitability, a numerical simulation analysis of long-term field temperature changes under a large temperature difference was carried out. The results show that it is reasonably feasible to estimate the overall temperature by measuring the temperature of the point halfway through the thickness of the slab panel, however, the method of estimating the overall temperature gradient is not effective in actual application, which is obtained by calculating temperature difference between the top and the bottom sides of the cross-section. The deviation of the overall temperature estimation is less than 1 °C, and the deviation ratio of the overall temperature gradient estimation can exceed 40 %.

1 INTRODUCTION

The bi-block ballastless track, which has various structural forms such as German's Radar, Zublin and China's CRTS I, II bi-block ballastless track, is widely used in rail transit systems e.g. high-speed railway tracks [1,2]. The length of bi-block ballastless tracks has been exceeded 4000 km in China involved Lanzhou-Xinjiang High-speed Railway (1780 km), Zhengzhou-Xi'an Passenger Dedicated Line (505 km), Wuhan-Guangzhou High-speed Railway (1084 km) and Chengdu-Lanzhou High-speed Railway (458 km). The common structural form of this track is shown in Figure 1 consisted of rails, rail fastening systems, bi-block sleepers, slab panels and a continuous support layer [3,4].

The bi-block ballastless track generally has problems such as concrete cracking and warping deformation [5–7], which are mainly affected by the temperature load on the slab panel. To address this concern, scholars had carried out researches on the field temperature, and a large number of temperature tests have been carried out to obtain the temperature load. However, influenced by multi-dimensional heat penetration, the temperature of the slab panel shows a nonlinear vertical distribution that is ever-changing, making it impossible to get the temperature amplitude of each cross-section accurately [8,9]. At the same time, the temperature test methods commonly used in existing research, which are passive and limited number of point monitoring in the natural environment [9,10], can't effectively reflect the distribution characteristics and variation laws of the field temperature at different parts of the slab panel.

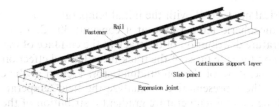

Figure 1. The structural form of the bi-block ballastless track using unit slab panels.

It can be seen that it is difficult to accurately obtain temperature-related load values, including overall temperature and overall temperature gradient.

For the convenience of structural designs and experimental works, a simple method for estimating the field temperature of the slab panel is proposed. The simple method includes the estimation of the overall temperature and the overall temperature gradient separately. The temperature of the point halfway through the thickness of the longitudinal section at the center-line of the slab panel is usually selected as the estimated value of its cross-section's overall temperature. The overall temperature gradient of each cross-section is simply estimated by calculating the temperature difference between the top and bottom of the longitudinal section at the center-line of the slab panel.

Although the simple estimation method has been applied to a large extent, its numerical accuracy and numerical deviation ratio have not yet been clearly analyzed. This article intends to analyze and evaluate the simple estimation method to provide a theoretical basis for the rationality of the method and provide safe modification suggestions for its application.

The structure of this paper is as follows: firstly, a numerical simulation of long-term field temperature changes under large temperature difference was conducted. Then the longitudinal and vertical temperature changes of the slab panel were analyzed to reflect the complexity of the field temperature of the model used. Finally, referring to the engineering practice, this paper analyzed the accuracy of the simple estimation method and put forward application suggestions.

2 NUMERICAL SIMULATION AND ANALYSIS

Modeling of a bi-block ballastless track was performed using the finite element method in the Cartesian three-dimensional coordinate system. In this regard, due to the symmetry of the structure, a 1/4 model with a scale of 1:1 was established using the commercial software ANSYS. The material of the slab panel in the model was C40 concrete with dimensions of 265 mm, 1400 mm, 5000 mm in height, width, and length, respectively. The material of the support layer was C15 concrete, with dimensions of 300 mm, 1700 mm, 5000 mm in height, width, and length [11–13], respectively. The contact faces of the two layers were bonded together, and the model surfaces not exposed to the atmospheric environment were assumed to have no heat exchange with the atmosphere. The effects of superstructure components including rails, fasteners, etc. were ignored in the model, and the bi-block sleepers were considered as a part of the slab panel.

Referring to the previous test data and the research results of relevant scientific research groups, the specific heat capacity, the thermal conductivity, the concrete density of the concrete are set to 925 J/(kg · °C), 2.2 W/(m · °C), and 2500 kg/m^3 in the calculation, respectively. Considering that the phenomenon of extreme daily temperature difference usually occurs in the period with slow air flow speed, it is assumed in the calculation that the phenomenon of extreme daily temperature difference occurs in the breeze environment with a wind speed of 3.6 km/h, and the corresponding thermal exchange coefficient of the concrete surface was 9.4 W/(m^2 · °C). In view of the rapid changes of the air temperature during the alternation time between day and night, the extreme daily temperature change of 20 °C is directly applied to the numerical model [14], with the initial temperature of the track in the model set to 10 °C and the atmospheric temperature at a constant 30 °C. Figure 2 shows the cross-sectional field temperature distribution of the track at the end face of the slab panel.

The heat exchange on the side faces of the track has a limited effect on the overall field temperature, the field temperature of the longitudinal cross-section at the center-line of the slab panel can be taken as the representative to analyze the field temperature of the slab panel. Figure 3 shows the longitudinal temperature gradient distribution of the top and bottom surfaces of the slab track at different times, wherein the abscissa l is the longitudinal distance between the cross-section and the end face. Figure 4 shows how the vertical distribution of the temperature gradient of the cross-section at different positions changes over time.

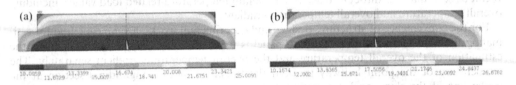

Figure 2. The field temperature distribution at the end face of the track (°C): (a) time=4 h, (b) time=8 h.

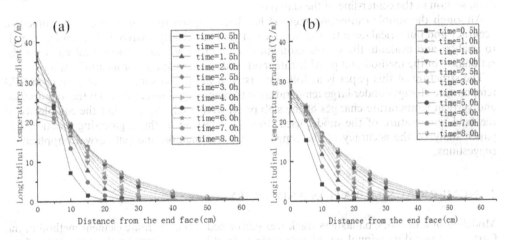

Figure 3. The longitudinal temperature gradient of the slab panel: (a) top surface, (b) bottom surface.

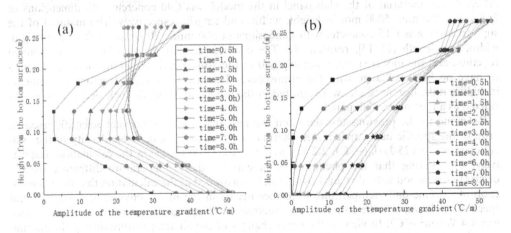

Figure 4. The vertical temperature gradient of of the slab panel: (a) l=0 cm, (b) l=60 cm.

Affected by factors such as heat penetration from the end faces, the distribution of field temperature is complex and changes with time, and the field temperature of the section will be different when the position of the section is different. When the temperature gradient of the top and bottom parts of the end face is generally larger than that of its middle part, the temperature gradient of the middle part of the slab panel decreases successively from top to bottom. It can be seen that the effect of a simple estimation method applied in different locations or at different times will be different.

3 APPLICATION ACCURACY OF SIMPLE ESTIMATION METHOD

3.1 *The overall temperature of the cross-section*

The change of the field temperature will cause the longitudinal deformation of the slab panel. Therefore the overall temperature load the slab track is subject to needs to be established in the calculation and analysis. Now the equivalent value T of the overall temperature amplitude of each cross-section is calculated by extracting the temperature values of the 7 nodes with equal vertical spacing in the cross-section. Figure 5 is a schematic diagram showing the vertical distribution of the temperature of the cross-section and the positions of its 7 nodes, Dn and Tn are respectively the numbers of nodes n and its temperature. The temperature of the point halfway through the thickness of the cross-section is usually selected as the estimated value T' of the overall temperature of the cross-section, which is the temperature parameter $T4$ in Figure 5.

Considering the interval height represented by each temperature extraction node, the mathematical relationship between the overall temperature equivalent value T of the cross-section and the node temperature values is established as such:

$$T = \frac{\frac{T_1}{2} + \frac{T_7}{2} + \sum_2^6 T_n}{6} \tag{1}$$

Take the calculation result of the above model into Formula 1, calculate the numerical differences between the equivalent value T and the estimated value T' of the overall temperature of each cross-section. Figure 6 is a graph showing the change in the numerical difference between the two with time as a function.

The numerical difference between the equivalent value T and the estimated value T' of each cross-section peaks within 2 hours after the sudden change of air temperature, and then gradually falls back. The maximum numerical difference of each cross-section in the range of $l \leq 20$ cm increases with the increase of the distance l, and the maximum differences of each cross-section in the range of $l > 20$ cm are basically the same, with the maximum difference between T and T' being just 0.71 °C. It can be seen that the method of estimating the overall temperature by measuring the temperature of the node at half thickness is reasonably feasible, If the safety of the design is considered, the estimated temperature variation of the overall temperature can be processed by + 1 °C.

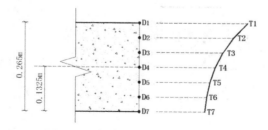

Figure 5. The position of the nodes on the cross-section.

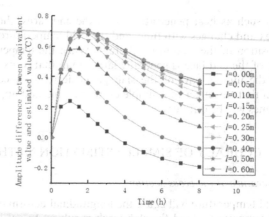

Figure 6. The numerical differences between the equivalent value and the estimated value.

3.2 The overall temperature gradient of the cross-section

The temperature gradient is the main reason for the warpage of the slab panel, which can destroy track regularity and structural stability. To control the deformation, it is necessary to calculate and analyze the overall temperature gradient load each cross-section is subject to.

Now the bending moment of each cross-section caused by the inconsistency of the vertical temperature stress is calculated to estimate the overall temperature gradient of each cross-section. First, the moment center of the warping moment will be determined by T, the equivalent value of the overall temperature. Then the cross-section is divided into 7 similar temperature segments by 7 nodes in Figure 5. Then the warping moment M of the cross-section is calculated by the temperature forces of the 7 temperature segments. Figure 7 is a schematic diagram showing the moment center of the cross-section and division of the 7 temperature segments of the cross-section.

The relationship between the warping moment M of each cross-section and the temperature amplitude T_n and the centroid depth h_n of each temperature segment of the cross-section is as such:

$$M = \alpha EBd[\frac{T_1 - T_0}{2}h_1 + \frac{T_7 - T_0}{2}h_7 + \sum_{2}^{6}(T_n - T_0)h_n] \tag{2}$$

Where α = thermal expansion coefficient of concrete, E = elastic modulus of concrete, B = width of the unit slab track, and d = vertical spacing of each temperature node.

Based on the theoretical calculation, it can be concluded that the warping moment M of each cross-section and its overall temperature gradient have the following mathematical relation:

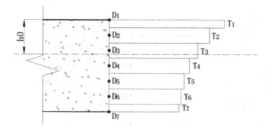

Figure 7. The moment center of the cross-section and the division of the temperature segments.

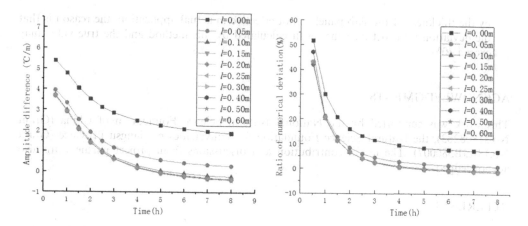

Figure 8. Comparison of the equivalent value and the estimated value: (a) amplitude difference, (b) deviation ratio.

$$M = \frac{1}{12}\alpha E \cdot S \cdot BH^3 \qquad (3)$$

Calculated by substituting numerical values, the equivalent value S of the overall temperature gradient amplitude of each cross-section will be obtained. The estimated value S' of the overall temperature gradient is the quotient of the temperature difference between the top and the bottom sides of the cross-section divided by the thickness of the slab panel.

Now the accuracy of the simple method of estimating temperature gradient will be further measured by calculating the temperature difference and deviation ratio between the equivalent value S and the estimated value S', where the deviation ratio taking the percentage of the numerical difference between the estimated value S and the equivalent value S' of each cross-section in its equivalent value as a standard, the result of the calculation is shown in Figure 8.

Although the amplitude difference and deviation ratio between the two show a rapid decline with time in the early stage of temperature change, the estimated value still deviates significantly from the true value. The maximum deviation value of the end part of the slab panel can reach 5.4 °C/m, when the deviation value of the middle part of the slab panel can reach 3.7 °C/m, and the deviation ratio of the estimated value generally exceeds 40% in the early stage of temperature change. After 4 hours, the estimated value is basically consistent with the true value. In light of the constantly changing temperature of the project site and the impossibility of ensuring enough time of temperature stability, it can be seen that the estimation method of the overall temperature gradient will have an obvious deviation in actual use, especially in the end part of the slab panel.

4 CONCLUSION

Through the numerical simulation analysis of long-term field temperature changes under large temperature difference, the evaluation of the simple estimation method on the field temperature is obtained. The following main conclusions are drawn:

(1) The method of estimating the overall temperature by measuring the temperature of the point halfway through the thickness of the longitudinal section at the center-line of the slab panel is reasonably feasible, the numerical difference between the estimated value and the true value is less than 1 °C.

(2) The method of estimating the overall temperature gradient, calculating the quotient of the temperature difference between the top and the bottom sides of the cross-section divided

by the thickness of the slab panel, is not effective in actual application, the reason is that the deviation ratio between the result calculated by this method and the true value may exceed 40%

ACKNOWLEDGMENTS

This work was supported by the National Natural Science Foundation of China (Grant No.51808283), the Natural Science Foundation for Universities of Jiangsu Province (Grant No.18KJB580007). The useful contribution and discussions from project partners are also acknowledged.

REFERENCES

[1] The National Railway Administration of China. 2014. *Code for design of high-speed railways*. Beijing: China Railway Publishing House.
[2] Zhao, G.T. 2006. *High Speed Railway Ballastless Track Structure*. Beijing: China Railway Publishing House.
[3] Liu, X.Y. & Zhao, P.R. 2010. *Theory and Method of Ballastless Track Design for Passenger Dedicated Lines*. Chengdu: Southwest Jiaotong University Publishing House.
[4] Wei, Y.X. 2014. *Study on adaptability of 6.5m bi-block ballastless track to double Lanzhou-Xinjiang railway line. Journal of the China Railway Society* 36(8): 70–74.
[5] Zhu, S.Y. & Cai, C.B. 2014. Interface damage and its effect on vibrations of slab track under temperature and vehicle dynamic loads. *International Journal of Non-linear Mechanics* 58(1): 222–232.
[6] Wu, B. & Zhang, Y. & Zeng, Z.P. 2011. Study on mechanics and crack behavior of twin-block ballastless track on subgrade under the temperature and shrinkage load. *Journal of Railway Science and Engineering* 8(1): 19–23.
[7] Zhu, S.Y. & Wang, M.Z. 2018. Mechanical property and damage evolution of concrete interface of ballastless track in high-speed railway: experiment and simulation. *Construction and Building Materials* 187: 460–473.
[8] Chen, B.J. & Qin, C.H. 2013. Analysis on temperature characteristics of large unit bi-block ballastless track. *Railway Engineering* 41(2): 88–91.
[9] Wu, C.J. & Cai, W.F. 2013. Experimental study on crack control of bi-block ballastless track bed slab on subgrade section. *High Speed Railway Technology* 4(1): 23–25.
[10] Zhao, P.R. & Yan, J.H. & Wang, K.J. 2014. Model experiment study of continuous slab track tension cracks. *Journal of Southwest Jiaotong University* 49(5): 793–798.
[11] Zhong, Y.L. & Gao, L. & Zhang, Y.R. 2018. Effect of daily changing temperature on the curling behavior and interface stress of slab track in construction stage. *Construction and Building Materials* 185: 638–647.
[12] Zhao, P.R. & Liu, X.Y. & Yang, R.S. 2016. Experimental study of temperature load determination method of bi-block ballastless track. *Journal of the China Railway Society* 38(1): 92–97.
[13] Tu, T. & Lee, S. 2017. Exact field temperature in a slab with time varying ambient temperature and time-dependent heat transfer coefficient. *International Journal of Thermal Sciences* 116: 82–90.
[14] Wei, Y.X. 2019. Influence analysis of extreme daily air temperature difference on the telescopic deformation of unit bed slab. *Railway Standard Design* 63(7): 41–45.

Functional Pavements – Chen et al (eds)
© 2021 Taylor & Francis Group, London, ISBN 978-0-367-72610-2

Comparing the environmental impacts of conventional and perpetual flexible pavements based on life cycle assessment

S. Liu, Y. Xue, G. Ni & Y. Qiao
Research Center for Digitalized Construction and Knowledge Engineering, Institute of Engineering Management, China University of Mining and Technology, Xuzhou, China

Z. Zhang & F. Giustozzi
Department of Civil and Infrastructure Engineering, RMIT University, Melbourne, Australia

J. Zhang
School of Qilu Transportation, Shandong University, Shandong University, China

ABSTRACT: This study conducted a comparative Life Cycle Assessment (LCA) between a perpetual flexible pavement and a conventional flexible pavement to quantify the associated energy consumption and carbon dioxide (CO_2) emissions. Although perpetual flexible pavements adopt thicker asphalt layers and may generate greater environmental loads in the initial production and construction phases, they can provide life cycle environmental advantages when considering material consumption for maintenance interventions. The results of the study showed that the analyzed perpetual flexible pavement can save about 14.5% energy consumption and 18.5% CO_2 emissions in the life cycle. Reduced maintenance can lead to a remarkable material decrease and is one of the main reasons for the environmental advantages associated with perpetual flexible pavements.

1 INTRODUCTION

According to the Ministry of Transport of the People's Republic of China (MOT), the total stretch of Chinese national highways reached 149,600 kilometers in 2019. Increasing pavement development has brought significant convenience but also environmental burdens. Globally, approximately 22% of carbon dioxide (CO_2) emissions and 28% of energy consumption are attributed to construction and management of road networks (Abergel et al., 2017). How to estimate pavement life cycle energy consumption and associated green-house gas (GHG) emissions becomes one of the most important choices to achieve more environmentally-balanced solutions for highway agencies (Chong & Wang, 2017).

The Life Cycle Assessment (LCA) method is defined by the International Organization for Standardization (ISO) 14040 and serves as a key evaluation tool to quantify the whole life cycle environmental influence of a specific product or system (Li et al., 2019). In recent years, it has been widely applied for pavement systems (Farina et al., 2017). A whole life cycle for pavements mainly includes raw material production, transportation, road construction and end of life disposal (Singh et al., 2020). Perpetual flexible pavements, designed with thicker structural layers than conventional pavements for preventing initiating cracks from the bottom (Romanoschi et al., 2008), are purposely constructed to achieve superior pavement performance and durability. Despite the initial higher construction costs, comparisons of the environmental impacts of perpetual and conventional flexible pavements during the life cycle stages are still needed. This can help determine environmental advantages or disadvantages for selecting perpetual or conventional pavement types in future road projects.

The main goal of this study is to compare the environmental impacts between a perpetual flexible pavement and a conventional flexible pavement using the ISO14040 LCA method, focusing on the energy consumption and CO_2 emission. The LCA results from this study could serve as suggestions or decision-making references for selecting more environmentally friendly pavements.

2 RESEARCH METHODS

ISO divides the life cycle assessment into four integrated steps: (i) goal and scope definition, (ii) inventory analysis, (iii) impact assessment and (iv) results interpretation.

2.1 Goal and scope definition

2.1.1 Goal and product definition
The goal of the LCA was to quantify and compare the energy consumption and CO_2 emissions of a perpetual flexible pavement and a conventional flexible pavement. A comparative process-based LCA was conducted. Alternative pavement structures involved in a case study are presented in Table 1, which shows the design and material usage differences between the two alternatives. They are both designed to be laid on the subbase layers (cemented treated soils) above the subgrade. The alternatives are current options being evaluated for a ring road construction in Qingdao, Shandong Province, China. The LCA considers pavement layers above the subbase, which are included in the bills of quantities of the project.

2.1.2 Functional unit
As the core of all LCA analyses, the functional unit forms the basis for the comparative LCA, which unifies different systems with the same utility and function (Santos et al., 2018). In this study, the functional unit is defined as 1 km highway with 6 lanes (two directions) which have the same lane width of 3.75 m.

2.1.3 System boundary
The system boundary is presented in Figure 1. The inputs to the system include the consumption of materials and energy. The output is CO_2 emissions. Five phases are considered in the life cycle of the two alternatives; specifically, material production, material transportation, construction, maintenance, and final disposal (end-of-life). The time boundary for the analysis is 40 years. The LCA only considers environmental loads from the agency's perspective hence user's environmental impacts from vehicle fuel consumption are excluded. The main differences in the system

Table 1. Pavement material & structure comparison.

	Conventional flexible pavement			Perpetual flexible pavement		
Layers	Material	Thickness (cm)	Amount (m^3)	Material	Thickness (cm)	Amount (m^3)
Surface	SMA-13	4	900	SMA-13	4	900
Asphalt	AC-20	6	1350	EME-16	10	2250
	AC-25	8	1800	EME-20	11	2475
	LSPM25	10	2250	LSPM25	10	2250
				AC-13F	7	1575
Base	CSM	54	12150	-	-	-
Subbase		40			80	

Note: SMA: Stone Mastic Asphalt, AC: Asphalt Concrete, EME: High Modulus Asphalt, LSPM: Large Stone Porous Asphalt Mixes, CSM: Cement Stabilized Macadam, F: anti-fatigue. The number within the material acronym is representative of the nominal maximum size of the aggregate.

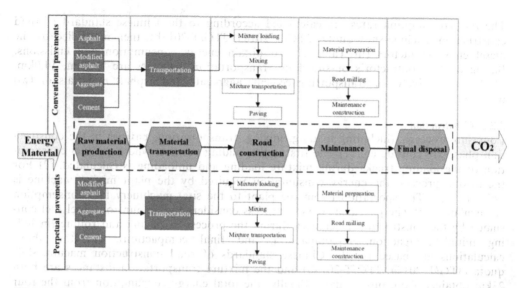

Figure 1.   System boundaries of the LCA analysis.

boundaries of the two types of pavements are the material flow (including types and quantity) in the raw production phase (see Table 1).

2.2 Inventory analysis

Inventory analysis is a technical process of collecting and collating data related to the input and output of the product system during the entire life cycle. The primary/secondary data needed in this case study was mainly provided by contractors and, partly, from literature references.

2.2.1 Material production
The calculation of energy consumption and emissions of the material production phase is based on quantifying material amounts of functional-unit of road length (Table 4), and the environmental emission factors of different materials (Table 2). The total energy consumption is a sum of energy used to produce different types of materials and transportation energy consumption based on the calorific value and fuel consumption.

2.2.2 Material transportation
Raw material transportation (from the material production site to the mixing plant or construction site) is included in this phase. Transportation occurring in the maintenance phase will be counted in the maintenance phase (Section 2.2.4). Two transportation vehicles - dump trucks and trucks - with different load capacities are mainly used.

Table 2.   Energy value and emission factors list (a) Energy value and emission factors list (b)

	Modified asphalt (Yang 2009)	Crushed stone (Timo 2011)	Cement (CLC-D2019)	Virgin asphalt (Yang 2019)		Diesel (Ding 2015)	Fuel oil (Ding 2015)	Electricity (Ding 2015)
Energy value/MJ·t-1	4583.00	37.33	3433.44	2830.73	Calorific value	42.65 (MJ/kg)	41.83 (MJ/kg)	3.60 (MJ/kwh)
CO_2 emission factors/kg·t-1	295.91	12.40	574.00	189.12	CO_2 emission factors	3.21	3.13	-

379

The fuel consumption rates are calculated according to the Chinese standard of road construction machine-shift quota (JTG/T 2018a, JTG/T-2018b), then multiplied by the diesel emission factors (in Table 2b) to obtain energy consumption and emissions. Basing on the contractor's on-site data, transportation distances are 12.7km, 14.0km, 21.7km respectively for transporting cement, aggregate and asphalt materials of two pavements.

2.2.3 *Construction*

Four sub-processes (including loading, mixing, transportation, paving) in the construction phase are considered. (1) The first is the loading process, where the fuel consumption of the loading trucks during mixture loading and unloading is calculated. (2) For the mixing process, the energy consumption required by the plant mixing machine is estimated. (3) Transportation (from the plant to the site) machinery fuel consumption is calculated. (4) The pavement paving process considers the amount of diesel fuel consumed by the construction machinery during the process of paving and rolling (including initial compaction, re-compaction, and final compaction). All the above calculations are based on the Chinese standards of road construction machine-shift quota (JTG/T 2018a, JTG/T-2018b) and the mixture transportation distances are both 23km obtained from on-site data. Finally, the total energy consumption from the four sub-processes is multiplied by the corresponding energy value and emission factor to obtain the final energy consumption and emissions during the whole construction phase.

2.2.4 *Maintenance*

During the maintenance phase, there are two pre-set plans specifically designed and indeed implemented by the road agencies for the two pavement alternatives. The maintenance scenarios are shown in Table 3. Material production, transportation, pavement construction, and pavement milling are included for maintenance activities. The calculation methods are the same as described in Section 2.2.1 (material production), 2.2.2 (transportation), and 2.2.3 (construction).

2.2.5 *Disposal*

During the disposal phase, demolition of the pavement is considered as end-of-life strategy and no asphalt recycling is considered. The quantities for demolition are decided according to Table 1 (total amounts of all materials). To obtain the energy consumption and emissions of this phase, the fuel consumption of pavement demolition machinery is calculated based on JTG/T 2018a, JTG/T-2018b standard, then the total fuel consumption is multiplied by the corresponding energy and emission factors (Table 2(b)).

2.2.6 *Inventory analysis summary*

According to the calculations for each phase of the two pavement alternatives mentioned above, the inventory analysis results are summarized in Table 4.

Table 3. Pavement maintenance scenarios.

Time (years)		8	12	16	24	32	36	40
maintenance treatment	Perpetual pavement		4 cm overlay		4 cm overlay		4 cm overlay	
	Conventional pavement	4 cm overlay		4 cm overlay	10 cm overlay	4 cm overlay		10 cm overlay

Note: 4 cm overlay = 4 cm overlay with SMA-13; 10 cm overlay = 4 cm overlay with SMA-13 + 6 cm overlay of AC-20

Table 4. Inventory analysis results summary.

Production	Material		Virgin asphalt	Modified asphalt	Crushed stone	Cement
	Quantity/FU (ton/FU)	CFP	177	434	21554	2160
		PFP	–	1126	9450	2160
Material transportation	Equipment		20 t truck		20 t dump truck	20 t truck
	Fuel consumption (kg)	CFP	6537		26461	14391
		PFP	–	12041	113553	13811

Construction	Process		Loading	Mixing		Transport	Paving
	Fuel		Diesel (t)	Fuel oil (t)	Electricity (kwh)	Diesel (t)	Diesel (t)
	Fuel consumption	CFP	7464	105115	46938	28752	5496
		PFP	4556	157556	59349	17309	6766

Maintenance	Process		Loading	Mixing		M1 Trans	M2 Trans	Milling	Paving
	Fuel		Diesel (t)	Fuel oil (t)	Electricity (kwh)	Diesel (t)		Diesel (t)	
	Fuel consumption	CFP	1084	37458	14110	6373	4022	240	1732
		PFP	433	14983	5644	2689	1609	240	899

Disposal	Demolition	CFP	Diesel (kg)	30695
		PFP		59923

Note: FU: Functional Unit, CFP: Conventional flexible pavement, PFP: Perpetual flexible pavement, M1 Trans: raw material transportation, M2 Trans: Mixture transportation

3 IMPACT ASSESSMENT AND RESULT INTERPRETATION

Figure 2(a) shows that the perpetual flexible pavement proved to generate less energy consumption (EC). About 14.5% EC reduction is obtained compared to the conventional flexible pavement alternative. The perpetual flexible pavement consumed higher energy

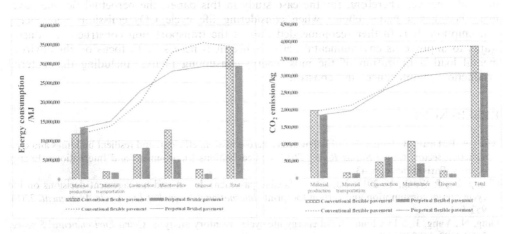

Figure 2. Comparative analysis of energy consumption (a) and CO_2 emissions (b) of two pavements. *The bar chart shows the comparative results; The lines represent the cumulative results.*

than the conventional option during the material production phase since more raw asphalt material, which has a greater amount of embedded energy, is required to be manufactured for construction. There is a significant amount of CSM base materials transported for the conventional flexible pavement, that is why in material transportation, the EC for the conventional pavement is higher. Since the need of energy for the crushed aggregate layer from the conventional pavement is significantly lower than the anti-fatigue layer AC-13F from the perpetual flexible design, the construction phase from Figure 2(a) revealed that the energy consumed by constructing the conventional pavement is about 26.5% less than the perpetual pavement. Figure 2(a) shows a significant difference in EC during the maintenance phase. The conventional flexible pavement requires double the energy than the perpetual flexible pavement. This highlighted a significant energy-saving benefit brought by the perpetual pavement at this stage. Though the solution for the disposal phase is the same for both pavement designs, the EC of perpetual flexible pavement is about 48.7% less compared with conventional one as the total volume for demolition is less.

In terms of CO_2 emissions, Figure 2(b) proves that the perpetual flexible pavement is more environmentally friendly with about 18.5% total CO_2 emission reduction. For both design strategies, the data do not show significant differences in CO_2 emissions for all phases except for the maintenance phase. In the maintenance phase, the perpetual flexible pavement will release nearly twice CO_2 compared to the conventional pavement. It is interesting to observe that the conventional pavement leads to greater CO_2 emission in the production phase. As the main emission burdens of conventional flexible pavement, 53% of CO_2 emissions are attributed to the material production phase. The use of cement in the CSM layer of the conventional pavement can be the reason, as cement was reported being much less eco-friendly (Worrell et al., 2001).

4 CONCLUSIONS

Based on the ISO14040 standardized LCA method, this paper compares the energy consumption and CO_2 emissions of conventional and perpetual flexible pavements from a life cycle perspective. It is concluded that the total energy consumption of the perpetual flexible pavement is 14.5% less than the conventional flexible pavement. The CO_2 emissions can reduce by approximately 18.5% if the perpetual flexible pavement design is adopted. The maintenance phase is primarily responsible for the greater environmental advantages of the perpetual pavement option, which generates about 61.9% and 62.1% less energy consumption and CO_2 emissions than that of the conventional flexible pavement alternative. Therefore, for the case study in this paper, the perpetual flexible pavement becomes a better choice when considering life cycle CO_2 emissions and energy consumption. It is further recommended that, if the transportation construction industry wants to improve its environmental performance, it is necessary to focus on the environmental load optimization of the most energy-consuming phases, including the material production, maintenance, and construction.

REFERENCES

Abergel, T., Dean, B., Dulac, J., 2017. Towards a zero-emission, efficient, and resilient buildings and construction sector: *Global Status Report 2017*. United Nations Environment and International Energy Agency, Paris, France, p. 43.

Chong, D., & Wang, Y. 2017. Impacts of flexible pavement design and management decisions on life cycle energy consumption and carbon footprint. *International Journal of Life Cycle Assessment*, 22(6), 952–971.

Ding, N., Yang, J. 2015 China's fossil energy life cycle inventory analysis. *China Environmental Science*, 35(5),1592–1600. (in Chinese)

Farina, A., Zanetti, M. C., Santagata, E., & Blengini, G. A. (2017). Life cycle assessment applied to bituminous mixtures containing recycled materials: Crumb rubber and reclaimed asphalt pavement. *Resources, Conservation and Recycling, 117*, 204–212.

Gao, F. 2016 Quantitative analysis of energy consumption and emissions during asphalt pavement construction based on LCA. *Chongqing Jiaotong University*. (in Chinese)

ISO 2006. International Standard ISO 14040: Environmental Management - Life Cycle Assessment: Principles and Framework, October, *International Organization for Standardization*, Geneva (Switzerland).

JTG/T 2018a. Highway Engineering Budget Quota. Compilation of fixed terminus of the Ministry of Communications, 3832-2018, China communications Press, Beijing, China. (in Chinese)

JTG/T 2018b. Highway Engineering Machinery Cost Quota. Compilation of fixed terminus of the Ministry of Communications, 3833-2018, China communications Press, Beijing, China. (in Chinese)

Li, J., Xiao, F., Zhang, L., & Amirkhanian, S. N. (2019). Life cycle assessment and life cycle cost analysis of recycled solid waste materials in highway pavement: A review. *Journal of Cleaner Production, 233*, 1182–1206.

Ministry of Transport of the People's Republic of China (MOT) 2019, *Statistical Bulletin on the Development of the Transportation Industry 2019*.

Romanoschi, S. A., Gisi, A. J., Portillo, M., & Dumitru, C. (2008). First findings from the Kansas perpetual pavements experiment. *Transportation Research Record, 2068*, 41–48.

Santos, J., Bressi, S., Cerezo, V., Lo Presti, D., & Dauvergne, M. (2018). Life cycle assessment of low temperature asphalt mixtures for road pavement surfaces: A comparative analysis. *Resources, Conservation and Recycling, 138*, 283–297.

Singh, A., Vaddy, P., & Biligiri, K. P. (2020). Quantification of embodied energy and carbon footprint of pervious concrete pavements through a methodical lifecycle assessment framework. *Resources, Conservation and Recycling, 161*(May), 104953.

Timo B, Jeff B, Frédérick ernard, et al. 2011. Life Cycle Inventory: *Bitumen. Brussels: European Bitumen Association*.

Worrell, E., Price, L., Martin, N., Hendriks, C., & Meida, L. O. (2001). Missions from * the. *Carbon, 26*, 303–329.

Yang, Q 2009 Quantitative assessment of the environmental impact of building products throughout the life cycle. Tianjin University. (in Chinese)

Functional Pavements – Chen et al (eds)
© 2021 Taylor & Francis Group, London, ISBN 978-0-367-72610-2

Opportunities and challenges for the application of asphalt supporting layer in the slab track system

S. Liu
Nanjing Forestry University, Nanjing, China

J. Yang & X. Chen
Southeast University, Nanjing, China

ABSTRACT: Dense-graded asphalt mixture was designed for asphalt supporting layer (ASL) applied in the slab track syustem. Several test sections were constructed as part of the real high-speed railway lines. Based on the field investigation on the test sections as well as the performance requirements for slab track system, both opportunities and challenges of ASL were discussed. The application of ASL can greatly improve the waterproof performance of slab track system, and the dynamic responses are comparable to that of normal roadbed. However, due to its special working environment, ASL still faces challenges in terms of local cracking risk and long-term deformation prediction.

1 INTRODUCTION

The asphalt supporting layer (ASL) is a new type of structural layer applied in the slab track system to partially substitute the upper roadbed. ASL was recently developed in China based on the successful application of asphalt layer in both ballast and ballastless track, such as asphalt sub-ballast layer, asphalt roadbed, and asphalt waterproof layer(Ramirez et al., 2014, Liu et al., 2018, Liu et al., 2019). The development of ASL aims to improve the cracking resistance of the drainage system as well as increase the carrying capacity of roadbed, to obtain good waterproof and drainage performance, enhance the stability of subgrade, and reduce the maintenance cost.

Asphalt mixture is the most commonly used pavement material due to its excellent workability, stability, and driving comfort. The asphalt pavement technology has been thoroughly studied and well documented(Motevalizadeh et al., 2018). However, the working environment of ASL in the slab track system, including loading environment and temperature conditions, is different from that of asphalt pavement(Liu et al., 2019, Liu et al., 2020). Thus, both the material and structure of ASL need to fulfill the requirements of the slab track.

Based on the research, design, and application experience of asphalt pavement in cold regions and asphalt sub-ballast layer in ballasted railway track, researchers from Southeast University and China Academy of railway science proposed guidance for material and structure design of ASL. Since 2015, ASL has been applied in the CRTS III slab track in several test sections, which were constructed as part of the real high-speed railway lines in North China(Liu et al., 2019). To investigate the performance of ASL under the actual railway environment, the monitoring systems were installed in the test sections to continually monitor the working state of ASL, a series of in-situ tests were also carried out to know the dynamic responses of ASL under high-speed train load. Also, a lot of laboratory tests, as well as numerical simulations, were conducted to evaluate the performance of ASL under extreme conditions. In this paper, the challenges and opportunities for the application of ASL in the slab track system are discussed based on laboratory test results and engineering cases.

2 MATERIAL AND STRUCTURE DESIGN OF ASL

2.1 *Material design*

The dense-graded asphalt mixture is used in ASL to secure good impermeability. The study and application experience of asphalt pavement shows that the asphalt mixture with air-void content less than 4% can provide excellent anti-permeability(Vardanega, 2014). The air-void content of the asphalt sub-ballast layer was usually controlled at the range from 1%-3%, and it had been proved to have good waterproof performance in the railway environment(Rose and Souleyrette, 2014). At the same time, it should be noted that the lower air-void content may decrease the deformation resistance of the asphalt mixture. Therefore, the target air-void content of asphalt mixture for ASL was set as 3%.

2.2 *Structure design*

Based on the classic slab track structure, ASL is applied to partially replace the existing graded gravel roadbed, as shown in Figure 1, to keep the total height of the structure unchanged. The ASL is paved not only in the area beneath the base plate but also on the shoulder, the thickness in both areas is 10 cm. The ASL beneath the base plate provides strong and stable support for the superstructure, and the ASL on the shoulder prevents the underlaying roadbed from surface water penetration.

2.3 *Field test*

To investigate the field performance of ASL in the slab track system under the actual environment, several test sections have been constructed as a part of the real high-speed railway lines. During the construction, a series of sensors, including temperature sensors, stress gauge, accelerometer, and displacement sensors, were placed inside the ASL to detect its working environment. Moisture sensors were also buried in the roadbed to record the changing of the water content of roadbed soil. The data acquisition system powered by a solar panel was installed to automatically record the data from sensors and transmit them to the remote computer server in the laboratory.

Utilizing the preburied sensors as well as the accelerometer, displacement meter installed afterward, a series of dynamic tests were conducted on the test sections to obtain the dynamic responses of ASL under high-speed train load. The dynamic test was carried out with CRH380AJ-0203 inspection train at different speed levels, and the testing system can be remote controlled and record the data automatically when the train passes through the monitoring positions.

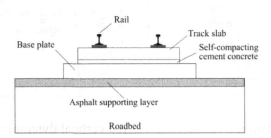

Figure 1.   The layout of ASL in the slab track system.

3 PERFORMANCE EVALUATION

3.1 *Waterproof performance*

To enhance the waterproof performance of slab track system is one of the core objectives of ASL. In the process of material design, on one hand, low air-void content was designed to secure good impermeability, on the other hand, the asphalt mixture was designed to have good anti-crack performance and fatigue resistance, to avoid water penetration through cracks in ASL. The water permeability coefficient measured in both laboratory and field were less than 10 mm/min, which is far less than the requirement for asphalt pavement.

The field monitoring data also reflects the excellent waterproof performance of ASL. Figure 2 presents the recorded subgrade water content data of the test section over two years, from December 1st 2015 to September 30th 2017. It can be found that, over two years, the subgrade water content is always kept at a stable level around the initial optimal water content. Comparing to the data from normal sections, the subgrade water content of the test section no longer fluctuates obviously with the variation of rainfall, indicating the application of ASL effectively prevents the penetration of surface water into the subgrade.

3.2 *Dynamic responses*

The application of ASL is expected to not only improve the stability and durability but also ensure the safety and riding comfort of the slab track system. Figure 3 presents the train-induced dynamic responses measured in the field test, including acceleration and dynamic displacement on the top of ASL. It can be found that the value of acceleration is linearly related

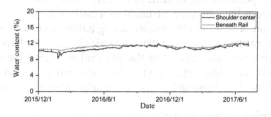

Figure 2.   The recorded subgrade water content over two years.

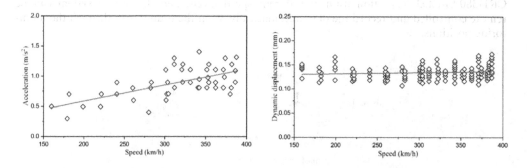

(a) vertical acceleration   (b) vertical dynamic displacement

Figure 3.   The train-induced dynamic responses of ASL. (a) vertical acceleration (b) vertical dynamic displacement.

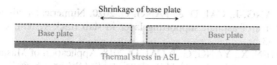

Figure 4. Schematic figure of thermal stress in ASL.

to the running speed of the train, the maximum value at the running speed of 380 Km/h is 1.35 m/s^2, which is comparable to that of conventional slab track. Meanwhile, the dynamic displacement is less affected by the train speed, for all the speed levels, the test results are scattered in the range from 0.1~0.17 mm, which well satisfies the requirements of China code for high-speed railway design.

4 THE COMING CHALLENGES

4.1 Thermal cracking

The regular field surveys show that ASL has excellent anti-crack performance during winter days, no transverse thermal crack was found on the test sections. However, a few local cracks were observed near the expansion joint of the base plate, which is deemed to be caused by the thermal shrinkage of the superstructure. As shown in Figure 4, the expansion joints were set in the cast-in-place base plate to avoid warping deformation under thermal stress, whereas the unlaying ASL was continuously paved and bonded with the base plate. As a result, during the cooling process, the shrinkage of the base plate may produce longitudinal displacement and load on ASL, which results in the local tensile stress in ASL. It should be noted that, during the cooling process, the shrinkage of ASL itself also brings thermal stress within ASL. In the area near the expansion joint, the superposition of two types of thermal stress increases the potential of cracking.

4.2 Long-term deformation

The design life of the slab track system is 60 years, which is much longer than that of asphalt pavement. Long-term deformation control of ASL is required to ensure stability during its service time. At present, the deformation resistance of ASL is evaluated by the laboratory tests used for asphalt pavement evaluation, such as the wheel tracking test and flow number test. While the differences in the working environment between asphalt pavement and ASL determine the inapplicability of these test methods. New approaches are needed to fully characterize the working environment of ASL and predict its deformation in 60 years.

5 CONCLUSIONS

The application of ASL brings new opportunities for the slab track system. The ASL was proved to have excellent waterproof performance, and its dynamic responses under high-speed railway train also satisfy the requirement. At the same time, due to the special working environment, the engineering application of ASL still faces challenges, including local cracking risk and long-term deformation prediction. Further structure optimization and evaluation methods development are needed to improve the performance of ASL in the slab track system.

REFERENCES

LIU, S., CHEN, X., MA, Y., YANG, J., CAI, D. & YANG, G. 2019. Modelling and in-situ measurement of dynamic behavior of asphalt supporting layer in slab track system. *Construction and Building Materials*, 228, 116776.

LIU, S., CHEN, X., YANG, J., CAI, D. & YANG, G. 2020. Numerical study and in-situ measurement of temperature features of asphalt supporting layer in slab track system. *Construction and Building Materials*, 233, 117343.

LIU, S., YANG, J., CHEN, X., YANG, G. & CAI, D. 2018. Application of Mastic Asphalt Waterproofing Layer in High-Speed Railway Track in Cold Regions. *Applied Sciences*, 8, 667.

MOTEVALIZADEH, S. M., AYAR, P., MOTEVALIZADEH, S. H., YEGANEH, S., AMERI, M. & BEMANA, K. 2018. Investigating the impact of different loading patterns on the permanent deformation behaviour in hot mix asphalt. *Construction and Building Materials*, 167, 707–715.

RAMIREZ, D., JAMEL, B., SOFIA, C. D. A., NICOLAS, C., ALAIN, R., DI, B. H. & CEDRIC, S. 2014. High-speed Ballasted Track Behavior With Sub-ballast Bituminous Layer. *Georail 2014*. Marne-la-Vallée, France.

ROSE, J. G. & SOULEYRETTE, R. R. 2014. Hot-Mix Asphalt (Bituminous) Railway Trackbeds—a global perspective Part III-U.S. Asphalt Trackbed Materials Evaluations and Tests. *In:* LOSA, M. & PAPAGIANNAKIS, T. (eds.) *The 3rd International Conference on Transportation Infrastructure*. Pisa. Italy: CRC Press.

VARDANEGA, P. J. 2014. State of the Art: Permeability of Asphalt Concrete. *Journal of Materials in Civil Engineering*, 26, 54–64.

Functional Pavements – Chen et al (eds)
© 2021 Taylor & Francis Group, London, ISBN 978-0-367-72610-2

Investigation into the effect of pavement albedo and environmental factors on urban heat island effect using CFD simulation

Junpeng Wang, Jun Chen, Jiantao Wu & Zheng Zhou
College of Civil and Transportation Engineering, Hohai University, Nanjing, China

ABSTRACT: The albedo of pavement is one of the main factors that can distinctly influence the urban heat island (UHI). This paper investigates the influence of pavement albedo and area, wind velocity, building albedo on UHI using Computational Fluid Dynamics (CFD) simulation. The urban microclimate model including buildings, pavement, and grassland was established in ANSYS ICEM. The radiation of the sun on the microclimate model from morning to night was simulated. The effect of pavement area and albedo (colored pavement), building albedo, wind velocity on the urban wind-heat environment are analyzed. The results showed the colored pavement has an excellent ability to reduce the temperature of the pavement surface. With the increase of pavement surface area, the air heating area and local UHI intensity increase. The building shadow and wind speed have a great influence on the urban temperature distribution. The highest temperature of buildings occurs on the roof and the temperature difference between the walls is not obvious.

1 INTRODUCTION

The urban heat island (UHI) potentially increases building energy consumption and deteriorates comfort conditions in urban areas (Santamouris, 2013). The albedo of pavement is generally considered to be one of the main factors that can distinctly influence the UHI (Chen et al., 2019). Due to the high absorption rate of solar radiation, the surface temperature of asphalt pavement can reach as high as 70°C during the summer season (Dawson et al., 2012, Jiang and Wang, 2020). Therefore, the methods to increase the albedo of asphalt pavement is the key to reduce the pavement temperature and mitigate the UHI. The color pavement is widely used in parks, which can improve the albedo of the pavement. Although the proportion of colored pavement in urban pavements increases, there are few studies on the influence of colored pavement on urban microclimate. In the past assessment of the urban microclimate, observational experiments are limited by the scale and complexity of urban environments (Toparlar et al., 2015). With the advances in computational resources, numerical simulation approaches have become increasingly popular (Toparlar et al., 2017). Computational Fluid Dynamics (CFD) has frequently used to analyze the urban microclimate (Nazarian and Kleissl, 2015).

In conclusion, the proportion of colored pavement in the urban has increased, but there are few studies on the albedo of colored pavement and its influence on UHI effects. In the analysis of the UHI effect, observational experiments are limited by the scale and complexity of urban environments. In contrast, numerical simulations are becoming increasingly popular. However, there are few analyses of the effect of pavement area and albedo, building albedo, wind velocity on the urban wind-heat environment.

The objectives of this study are to (1) establish the urban microclimate model including buildings, pavement and grassland; (2) simulate the radiation of the sun on the microclimate model from morning to night; (3) analyze the urban microclimate under conditions of pavement area and albedo, wind speed and building albedo.

2 MODEL DESCRIPTION AND VALIDATION

2.1 *Model description*

The turbulent airflow in the urban environment is computed by solving the standard κ-ε model coupled with the Enhanced wall treatment approach. The air is assumed to be an ideal incompressible gas, and the relative density is set to achieve the thermal buoyancy lift. Air is considered a non-participating medium, thus absorption, scattering, and emission of radiation by air are neglected. The solar radiation was simulated by the Discrete Ordinates model coupled with Solar Ray Tracing. Sun direction vector used direction computed from solar calculator which was set at 6 a.m. on June 21 in Nanjing, Jiangsu province. The model further assumes that all surfaces are opaque to both solar and thermal radiation and that they are diffuse. Emissivity and absorptivity are equal and independent of wavelength and direction. Figure 1 shows the dimensions of the urban microclimate model. There are pavement and three buildings surrounded by grassland on a flat wall boundary situated above the solid zone representing soil layers. The height, length, and width of the model are 30m, 200m, and 150m. The area of the target region is 50×50 m^2, and the rest of the ground areas do not participate in solar radiation. In this study, the lateral and top boundary conditions were assumed to a zero flux of all quantities across the boundaries. The climate temperature of the day is simulated by the temperature boundary conditions of the inlet and outlet. The inlet wind speed adopts gradient wind, as shown in Eq. (1).

$$U_0 = 2 \times \left(\frac{H}{10}\right)^{0.22} \tag{1}$$

Since the thickness of the building wall and pavement are small relative to the whole model, shell conduction that avoids the complex meshing to thin-walled structures was used. The temperature boundary conditions at the soil layer (exterior surfaces of the computational domain) are set to be constant (30°C). And the simulated initial temperature is 30°C. All of the walls are modeled as smooth walls.

2.2 *Validation*

In the authors' previous paper, six colors of porous Poland cement concrete (PPCC) slabs, dense Poland cement concrete (DPCC), and two types of asphalt concrete (AC) slabs are prepared (Chen et al., 2019). The internal temperature and albedo of slabs were measured using the self-developed albedometer. The internal temperature of the pavement under the same conditions as the laboratory test (no wind, 1300W irradiation intensity, irradiation for 60 min) is simulated. Figure 2 shows the comparison between test data and simulated data of the

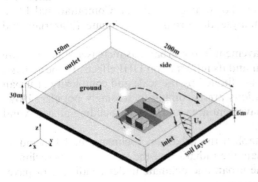

Figure 1. The dimensions of the model.

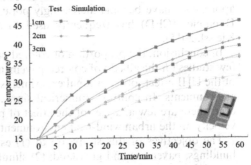

Figure 2. Temperatures inside the slab and pavement.

internal temperature of the pavement. It can be seen that the simulated internal temperature of the pavement is greater than the measured value of the laboratory test, which is mainly because the simulated pavement is solid but the laboratory slab is a porous structure. Compared with the solid slabs, the porous slab has a larger contact area with air and a higher heat transfer rate than solid. Besides, the slabs used in the laboratory test lose heat faster due to their smaller volume.

3 ANALYSIS OF THE URBAN MICROCLIMATE

3.1 Effect of the albedo and area of pavement

Table 1 shows the variables of pavement albedo in simulated cases. Four kinds of pavement materials were simulated, including porous AC (case A_1), grey PPCC (case A_2), red PPCC (case A_3), and DPCC (case A_4) with an albedo of 5.8%, 16%, 25%, and 32%, respectively. Figure 3 shows the comparison of diurnal variation of pavement surface temperature with different albedo. The temperature differences of the pavement surface are small before 9:00. With the increase of the angle and intensity of solar, the temperature differences of the pavement surface with different albedo gradually increase. In the case A_4, the DPCC with the highest albedo has the lowest temperature on the day, with the average temperature reaching a maximum of 57.5°C around 16:00. Compared to the case A_1 and case A_2, the colored PPCC with a lower absorptivity in case A_3 has a better cooling effect, with the average temperature reduced by 1.7°C and 4.2°C, respectively.

Table 2 shows the variables of the pavement area in simulated cases. In Case B_1, Case A_3, and Case B_2, the ratio of pavement area to the grassland area is 0.31, 0.725, and 1.72, respectively.

The local UHI intensity is defined as the temperature difference between the local air temperature and the inlet temperature. Area ratio is defined as the ratio between the area of the heating area and the target area at the pedestrian height (50m×50m). Figure 4 shows the area

Table 1. Variables in simulated cases.

| Case | Albedo/% | | Wind | | Pavement area/m² |
	Pavement	Building	Velocity/(m/s)	Direction	
A_1	5.8	20	2	North wind	844
A_2	16				
A_3	25				
A_4	32				

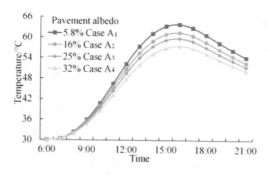

Figure 3. Diurnal variation of pavement temperature.

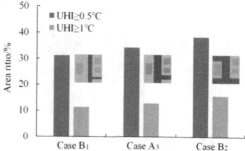

Figure 4. The local UHI intensity.

Table 2. Variables in simulated cases.

Case	Albedo/%		Wind		Pavement area/m^2
	Pavement	Building	Velocity/(m/s)	Direction	
B_1	25	20	2	North wind	472
B_2					1270

ratio of local UHI intensity of air at pedestrian height. With the increase of the pavement area, the area of heated air at the pedestrian height also increases. The areas where Case B_1, Case A_3, and Case B_2 heated by more than 0.5°C at pedestrian height were 627.2m^2, 688.7m^2, and 771.5m^2, which are 31.2%, 34.3% and 38.4% of the total area respectively. Similarly, the areas where Case E_1, Case A_3, and Case E_2 heated by more than 1°C at pedestrian height were 229.9m^2, 259.9m^2, and 317.5m^2. Meanwhile, the air temperature increases with the increase of the pavement surface area, among which the maximum temperature of Case B_2 at the pedestrian height is 0.41°C higher than that of Case B_2.

3.2 Diurnal variation of urban temperature

Figure 5 shows the representative times of the diurnal variation of urban temperature. At 9:00, large areas of building shadow slow the temperature rise rate of the pavement surface and creates the partition distribution of temperature. At noon, the solar radiation angle is the highest and the shadow area of the buildings is the smallest. However, the low-temperature area caused by shadow can still affect the distribution of temperature on the pavement. At 16:00, the urban surface temperature reaches the peak, among which the highest temperature appears on the pavement surface. Building shadows continue to have a cooling effect on the pavement surface as the position of the sun changes. At 21:00, the highest temperature of the urban environment appeared on the roof of the first windward buildings. A low-temperature area can be seen in the front and lateral of the buildings and the ground behind the building formed a local high-temperature area. Figure 6 shows the diurnal variation of air temperature and inlet temperature at pedestrian height. Compared with the temperature at the inlet, the air temperature has increased significantly. The maximum temperature of air occurs around 14:00, 0.8°C higher than the inlet temperature.

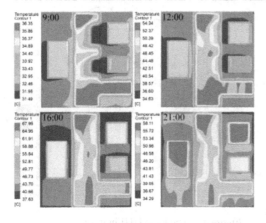

Figure 5. Diurnal variation of urban temperature.

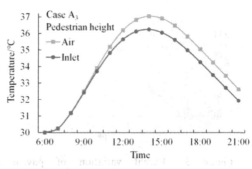

Figure 6. Diurnal variations of air temperature.

3.3 Effect of wind velocity

Table 3 presents the variables of the wind velocity in simulated cases. In this section, local air temperature and airflow speeds are analyzed. Therefore, air control volumes of 50 meters in length, 50 meters in width, and 1.8 meters in height are used for analysis at the pedestrian level. Figure 7 shows the diurnal variation of the pavement surface temperature under different wind speeds. Wind speed has a great influence on the temperature of the pavement surface, which is mainly due to the low density of buildings. Compared with the case A_3, the temperature of the pavement in case C_1 and case C_2 is reduced by 3.9°C and 6.7°C respectively.

Figure 8 shows the wall heat flux of the pavement surface. Due to the temperature difference between the pavement and the air is the largest at 16:00, the wall heat flux of the pavement is the largest. The wall heat flux increases with the increase of wind speed. As the pavement surface temperature drops, the temperature difference between the pavement surface and the air becomes smaller, and the wall heat flux decreases.

3.4 Effect of the albedo buildings

Table 4 presents the variables of the building albedo in simulated cases. Figure 9 shows the diurnal variation of the wall temperature of the buildings. As shown in Figure 9 (a), the temperature difference between 6:00 and 9:00. on the east wall of different cases is not significant. In the case D_1,

Table 3. Variables in simulated cases.

| Case | Albedo/% | | Wind | | Pavement area/m² |
	Pavement	Building	Velocity/(m/s)	Direction	
C_1	25	20	4	North wind	844
C_2			6		

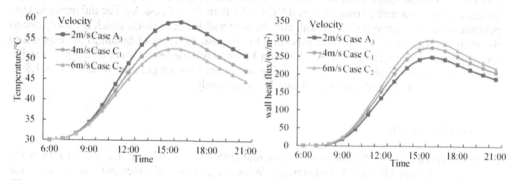

Figure 7. Diurnal variation of urban temperature.   Figure 8. Diurnal variation of wall heat flux.

Table 4. Variables in simulated cases.

| Case | Albedo/% | | Wind | | Pavement area/m² |
	Pavement	Building	Velocity/(m/s)	Direction	
D_1	25	40	2	North wind	844
D_2		30			

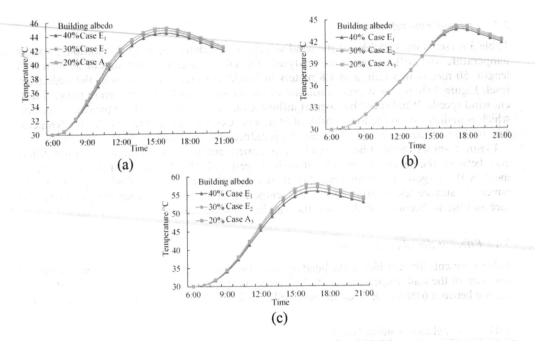

Figure 9. Diurnal temperature variations of (a) the eastern walls, (b) western walls and (c) roof of buildings.

the highest temperature on the east wall reached 44.4°C around 15:00. The maximum temperature of the east wall of case D_1 was 0.7°C cooler than that of case A_3. The temperature of the west wall of buildings was shown in Figure 9 (b). The temperature difference between the west wall of different cases was not obvious before 15:00, which is mainly because the western wall is shady. In the case D_1, the highest temperature on the west wall reached 44°C around 5 p.m. The maximum temperature of the west wall of case D_1 was 0.5°C cooler than that of case A_3. The difference in temperature difference between the east wall and the west wall is not obvious, which is mainly due to the short duration of solar radiation exposure. In contrast, the roofs of buildings are exposed to solar radiation for longer periods. As shown in Figure 9 (c), the temperature of the roof reaches a maximum of 55.9°C at 17:00 in case E_1. The maximum roof temperature of case D_1 is 1.8 °C and 1°C cooler than that of case A_3 and case D_2, respectively.

4 CONCLUSIONS

(1) The colored porous Poland cement concrete (PPCC) is 1.7°C and 4.2°C cooler than the grey PPCC and OGFC-13, respectively. With the increase of pavement surface area, the air heating area and local urban heat island intensity increase.
(2) Building shadow has a cooling effect on the local area of pavement. At the pedestrian height, the maximum temperature difference between the air and the inlet is 0.8°C.
(3) With the increase of wind speed, the pavement heat flux increases. Compared with the wind speed of 2m/s, the temperature of the pavement surface under the wind speed of 4m/s and 6m/s is reduced by 3.9°C and 6.7°C respectively.
(4) The highest temperatures of buildings with different albedo occur on the roof. The temperature difference between building walls is not obvious.

This study findings indicate that the colored pavement has an obvious mitigation effect on the urban heat island (UHI) effect due to the low albedo. It has to be noted that the UHI can be significantly affected by environmental factors such as wind.

ACKNOWLEDGMENTS

The research presented herein was sponsored by the Jiangsu Natural Science Foundation (No. BK20191300), the Fundamental Research Funds for the Central Universities (No. 2019B12714) and the Open Fund of National Engineering Laboratory of Highway Maintenance Technology (Changsha University of Science & Technology) (No. kfj180107).

REFERENCES

CHEN, J., ZHOU, Z., WU, J., HOU, S. & LIU, M. 2019. Field and laboratory measurement of albedo and heat transfer for pavement materials. *Construction and Building Materials*, 202, 46–57.

DAWSON, A. R., DEHDEZI, P. K., HALL, M. R., WANG, J. & ISOLA, R. 2012. Enhancing thermal properties of asphalt materials for heat storage and transfer applications. *Road Materials and Pavement Design*, 13, 784–803.

JIANG, L. & WANG, S. 2020. Enhancing heat release of asphalt pavement by a gradient heat conduction channel. *Construction and Building Materials*, 230, 117018.

NAZARIAN, N. & KLEISSL, J. 2015. CFD simulation of an idealized urban environment: Thermal effects of geometrical characteristics and surface materials. *Urban Climate*, 12, 141–159.

SANTAMOURIS, M. 2013. Using cool pavements as a mitigation strategy to fight urban heat island—A review of the actual developments. *Renewable and Sustainable Energy Reviews*, 26, 224–240.

TOPARLAR, Y., BLOCKEN, B., MAIHEU, B. & VAN HEIJST, G. J. F. 2017. A review on the CFD analysis of urban microclimate. *Renewable and Sustainable Energy Reviews*, 80, 1613–1640.

TOPARLAR, Y., BLOCKEN, B., VOS, P., VAN HEIJST, G. J. F., JANSSEN, W. D., VAN HOOFF, T., MONTAZERI, H. & TIMMERMANS, H. J. P. 2015. CFD simulation and validation of urban microclimate: A case study for Bergpolder Zuid, Rotterdam. *Building and Environment*, 83, 79–90.

Functional Pavements – Chen et al (eds)
© 2021 Taylor & Francis Group, London, ISBN 978-0-367-72610-2

Evaluation and optimization on the reflection and durability of reflective coatings for cool pavement

N. Xie & H. Li*

School of Transportation Engineering, Tongji University, Shanghai, China

ABSTRACT: Reflective coating is an effective way to alleviate the heat island effect and delay pavement distress under high temperature. However, the optical properties based on the full-spectrum and its aging performance need to be investigated. Based on the principle of full-spectrum reflection optical properties, this study selects near-infrared reflective fillers as the functional fillers and three types of anti-aging fillers to prepare near-infrared reflective colored cooling road coatings. The optical properties, pavement performance, and durability performance were evaluated. The results show that the total reflectance of the near-infrared reflective coating is increased from 5% to 30%-40%, and the near-infrared reflectance is increased from 3% to 50%. ZnO in nanometer benefits for the anti-aging performance of coatings. Reflective coatings applied on pavement could meet the requirement of pavement performance such as abrasion resistance and skid resistance. The research results can provide a theoretical basis for developing the reflective coatings on pavement and lead to a pavement system which is resilient to extreme urban climate.

1 INTRODUCTION

The rapid urbanization has led to the Urban Heat Island effect (Urban Heat Island) has become a catastrophic problem around the world (Santamouris *et al.*, 2017; DESA United Nation, 2018; Li *et al.*, 2019). The heat island effect can lead to increasing cooling energy consumption, decreasing air quality (accelerated surface ozone formation), increasing runoff pollution (increased owing to surface water temperature), decreased human thermal comfort, and even affected human health (H. *et al.*, 2016; Henao, Rendón and Salazar, 2020). Solar radiation is the main source of heat in cities and most of the solar radiation arriving at black asphalt pavements will absorbed and stored, which leads to 60~70°C on the pavement surface in summer. In general, road surface area occupies 30%-50% of the total area of the city (Pomerantz *et al.*, 1997; Akbari and Kolokotsa, 2016). Thus, the thermal properties of pavement influence the thermal environment in cities significantly.

Furthermore, asphalt mixture is a material with significant temperature sensitivity. Its mechanical properties such as compression resistance, bending resistance, splitting strength and stiffness modulus are all significantly affected by temperature (Guo, Nian and zhou, 2020). The traditional methods to deal with rutting diseases mainly include optimizing gradation and improving the performance of asphalt and aggregate. With the development of functional pavement technologies, cool pavement such as reflective coatings, porous pavement become the hot spots in the field of pavement engineering (Zhang *et al.*, 2018; Xie, Li, Abdelhady, *et al.*, 2019). Studies have showed that increasing the albedo of pavement is the most effective way which has the greatest effect on the thermal environment.

Reflective coatings could reduce the surface temperature up to 12°C (Qin, 2015; Xie, Li, Zhao, *et al.*, 2019). However, increasing the reflectance too much will lead to glare problem,

---

* Corresponding author

which is harmful for visual environment on road. The main goal of developing cool pavement coatings is to decrease their visible reflectance while assuring its effectiveness of decreasing temperature (Fang *et al.*, 2013; Xie, Li, Zhao, *et al.*, 2019). Comparing the traditional reflective coatings, the near-infrared coatings is able to keep high reflectance but with the relative dark color. Nevertheless, the optical properties of reflective coatings based on the full spectrum including visible reflectance and near-infrared reflectance is often neglected by the past research. The pavement performance such as wearing resistance performance and skid resistance characteristics also need further investigation. Besides, natural exposure under the effect of temperature, moisture and UV radiation will pose a threat to the durability of reflective coatings on pavement (Cao *et al.*, 2016; Morini *et al.*, 2018). It is necessary to seek for ways to perform the test in a controlled and comparable condition in laboratory to explore the durability and aging performance of reflective cool pavement coatings. This study applied near-infrared reflective fillers as functional fillers to prepare a near-infrared reflective coating. The optical properties, mechanical performance and durability performance of the coating was evaluated.

2 EXPERIMENTAL METHODS

2.1 *Materials*

The near-infrared titanium dioxide filler (NIR-TiO$_2$) is used to replace some of the traditional rutile titanium dioxide (R-TiO$_2$) as the functional filler, and iron oxide is selected as the pigment to meet the target requirements of high reflectance under low brightness conditions and auxiliary cooling. The mean particle size of NIR-TiO$_2$ is 1μm while the R-TiO$_2$ is 0.2μm. Organosilicon-modified acrylate emulsion was as the binder with other additives (dispersing agent, flatting agent, antifoaming agent and coalescing agent) to form coating samples. Three types of anti-aging fillers in this study are organic ultraviolet (UV) radiation absorber, nano-TiO$_2$ and nano-ZnO, respectively. A high-speed disperser was used in laboratory to prepare coating samples.

Firstly, in the orthogonal experimental design, NIR-TiO$_2$, Fe$_2$O$_3$ and R-TiO$_2$ were determined as the three factors. Three levels of each factors are as follows. For NIR-TiO$_2$(A), it is 5g, 8g and 16g. For red iron oxide (Fe$_2$O$_3$), it is 5g, 8g and 12g. For R-TiO$_2$, it is 8g, 16g and 25g. The design table L$_9$(3^4) was applied in the study. After determining the optimized formula, the accelerated durability test was conducted and the detailed information of samples is shown in Figure 1.

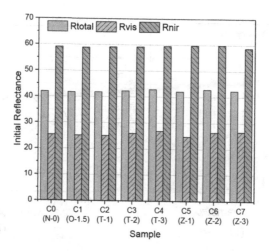

Figure 1. The initial reflectance (Rtotal, Rvisi, Rnir) of coatings with three types of anti-aging fillers. (N is none, O is the organic absorber, T is TiO$_2$, Z is ZnO and the number is the content).

2.2 Method

The optical properties, pavement performance and aging performance under the laboratory accelerated aging condition of reflective coatings were measured respectively and the detailed process is described as follows.

The optical characteristic test is carried out in accordance with the standard ASTM E903-96, using an ultraviolet/visible/near-infrared spectrophotometer (Perkin Elmer Lambda 950) and a 150mm diameter integrating sphere, with the test interval of 5nm and the test accuracy of ±0.08nm. Referring to ASTM E903-9 and ASTM G159-98, the solar spectral reflectance is based on atmospheric mass 1.5 (AM 1.5) as the standard solar radiation spectrum. Hardness of coatings refers to the ASTM D3363-00 standard through using pencil hardness (ranging from 6B to 6H) to describe the hardness of the coating film. Measurement of anti-skid performance of reflective coatings refers to standard JTGE60-2008 and JTG F80/1-2012 in China. The surface pendulum value (BPN) before and after the cooling coating application was measured. The wearing resistance performance was measured based on standard T0752-2011. The coating applied on the surface of the specimen was at an amount of $0.6kg/m^2$, which is conducted after drying of 48h at room temperature. After being completely dried, the specimens were placed in a water bath at 25°C for 1 hour. The wet wheel abrasion meter was used to comprehensively evaluate the dual effects of water damage and abrasion on the surface coatings.

Finally, considering temperature, humidity, and ultraviolet radiation, the laboratory accelerated aging test (LAAT) was conducted, which can automatically control the environmental condition in the test chamber. The device ran two scheduled cycles every day as the following selected steps: 8h of UV radiation (0.89 W/m^2·nm) at 65°C followed by 4h of high humidity (>95% RH) at 55°C. 36 experiment cycles were conducted in the study.

3 RESULTS AND DISCUSSION

3.1 Optical characterization of coatings

To obtain the influence of NIR-TiO_2 pigment on the optical characterization of cool coatings, the reflectance and its spectra of coatings with increased NIR-TiO_2 were measured. The weighted reflectance spectra results could be seen in Figure 2. It is obvious that with the increase of NIR-TiO_2 pigment, the total reflectance and the near-infrared reflectance increase firstly and then decrease, the visible reflectance decreases, the ultraviolet reflectance almost

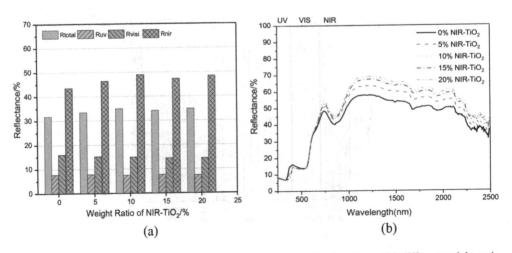

Figure 2. The reflectance variation (a) and reflectance spectra (b) of coatings with different weight ratio of NIR-TiO_2 pigment.

keep unchanged (around 7%) with some fluctuation. That's because the near-infrared reflectance of NIR-TiO$_2$ is larger but the visible reflective performance is worse. The reflectance spectra of coating samples in the orthogonal experiment #1~ #9 are shown in Figure 3. Comparing with the asphalt concrete (AC) and the cross section of asphalt concrete, the total reflectance (Rtotal) of coated samples improve from 5% to 30%-40%, the near-infrared reflectance (Rnir) increase from 3% to 50%.

3.2 Pavement performance test results

The results of BPN values of samples under dry and wet condition are shown in Figure 4 (a). Comparing to the original uncoated samples, BPN values slightly decrease but still be in the controllable range. This is owing to the fact that coatings cover the original textures of asphalt concrete and lead to a smoother surface. Under the dry condition, BPN values are all larger than 75; Under the wet condition, BPN values are larger than 50. This could satisfy the requirement of Inspection and Evaluation Quality Standards for Highway Engineering (JTG F80/1-2017), in which the BPN values of expressway and the firs-class highway should be higher than 45 in the condition of standard temperature as 20°C.

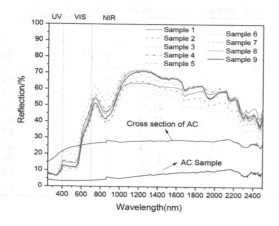

Figure 3. The reflectance spectra of coating samples and the original asphalt concrete surface.

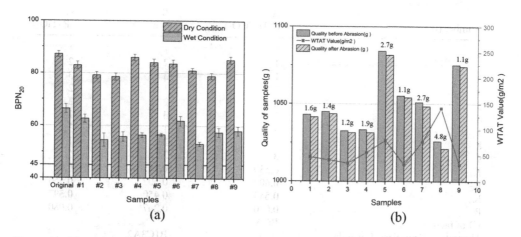

Figure 4. The BPN values and anti-abrasion test results of coating samples in the orthogonal experiment #1~ #9.

For the anti-abrasion performance considering the effect of vehicle wheels, the test results were shown in Figure 4 (b). Coatings in different formulas have different wet track abrasion test (WTAT) value. The quality losses are all less than 5 g and the WTAT values of coatings are less than 150 g/m^2. The coating samples of different formulations have different wear values, but the loss is small, which could meet the requirements for application on actual road environment.

3.3   Optimization of coatings based on the optical properties and pavement performance

The orthogonal experiment results and the comprehensive evaluation results were summarized in Table 1. For the optical performance, the total reflectance, the near-infrared reflectance and the lightness index were chosen as the indicators. For the pavement performance, the film hardness, the average BPN_{20} value, coating loss values are as the indicators. To obtain the optimized formula of coatings, the optimization calculation results were illustrated in Table 2. It could be seen that the optimized group is $B_1C_3A_2$, that is, the optimized level of $B(Fe_2O_3)$ is level 1(3%), the optimized level of $C(R-TiO_2)$ is level 3(15%) and the optimized level of $A(NIR-TiO_2)$ is level 2(5%).

Table 1.   The test results and comprehensive evaluation of coatings.

Number(i)	Test Results					
	Rtotal	Rnir	L*	Film hardness	BPN_{20}	Coating Loss
1	37.91	55.07	52.16	5	68.6	47.06
2	34.14	51.2	46.85	4	60.6	41.18
3	34.72	49.92	49.97	5	61.8	35.29
4	40.05	58.63	52.83	5	62.4	55.88
5	38.64	56.16	52.13	5	62.6	79.41
6	31.13	50.29	44.04	3	67.8	32.35
7	40.5	60.11	58.18	5	59.2	76.47
8	34.58	54.67	49.15	5	63.6	141.18
9	33.47	52.78	48.42	4	64.2	32.35

Table 2.   The optimization calculation results of coating samples.

Number (i)	A (NIR-TiO₂)	B (Fe₂O₃)	C (R-TiO₂)
	A ($NIR-TiO_2$)	B (Fe_2O_3)	C ($R-TiO_2$)
1	1	1	1
2	1	2	2
3	1	3	3
4	2	1	2
5	2	2	3
6	2	3	1
7	3	1	3
8	3	2	1
9	3	3	2
K_1	0.530	**0.717**	0.517
K_2	**0.600**	0.507	0.560
K_3	0.543	0.450	**0.597**
R	0.070	0.267	0.080
ID of factors	BCA		
Theoretical Optimized Formula		B1C3A2	

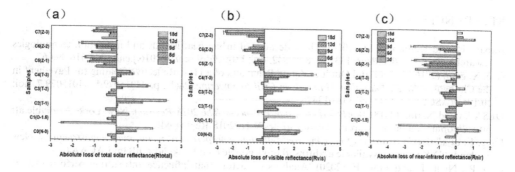

Figure 5. The absolute loss of spectral reflectance of coatings C0~C7 during the accelerated weathering process at different solar radiation region: (a) the total reflectance (Rtotal); (b) the visible reflectance (Rvis); and (c) the near-infrared reflectance.

3.4 Durability properties of coatings for cool pavement

Based on the above optimized coatings, three anti-aging fillers were added in coatings. The absolute differences of reflectance of Rtotal, Rvis and Rnir during the accelerated aging periods of 18 days are shown in Figure 5. There exists a slight increase and decrease of reflectance of coatings so that positive and negative values are both presented. The optical properties of all the reflective coatings altered no more than ±5% (absolute value), within an acceptable range. The most obvious change exists in the visible radiation region, which has the largest values. The growth of reflectance is because of chalking and whitening of coatings. During the first three days, there are reflectance decreases for all the coatings owing to the weathering. Nevertheless, the total reflectance and visible reflectance of samples C5 to C7 with ZnO as the anti-aging filler all attenuate gradually in the range of 3%. Considering the cause of the reflectance growth is chalking and whitening, it can be concluded that ZnO has better durability in the weathering environment than other fillers. Besides, there are no obvious or regular changing patterns of near-infrared reflectance, indicating that the physical composition of coatings, including chalking and powdering, influence the optical properties of reflective coatings in the visible radiation region and then changes the total reflectance. This implies that coatings with higher Rnir and lower Rvis have better durability performance.

4 CONCLUSIONS

This study selected silicone modified acrylic emulsion, deionized water and additives (dispersant, leveling agent, defoamer, film forming aid, ethanol) and near-infrared reflective fillers to develop reflective coatings for cool pavement. The reflection properties, pavement performance and the durability performance were evaluated. The main conclusions could be listed as follow.

(1) Compared with the standard new asphalt mixture sample, the total reflectance of the sample coated with the near-infrared reflective coating increases from 5% to 30%-40%, and the near-infrared reflectance increases from 3% to 50%-60%, which leading to decreased surface temperature of 9 °C. It also meets the requirement of skid resistance performance and anti-abrasion performance for applications on road.
(2) The optimized cool pavement coating materials formula is that Fe_2O_3: R-TiO_2: NIR-TiO_2=3:15:5, which is with higher reflectance, outstanding pavement performance and without glare problems.
(3) Chalking and whitening are the main reason for aging of cool coatings for pavement. Selecting ZnO as the anti-aging fillers is able to improve the durability performance of reflective coatings for cool pavement.

REFERENCES

Akbari, H. and Kolokotsa, D. (2016) 'Three decades of urban heat islands and mitigation technologies research', *Energy & Buildings*, 133, pp. 834–842. doi: https://doi.org/10.1016/j.enbuild.2016.09.067.

Cao, X. *et al.* (2016) 'Indoor and Outdoor Aging Behaviors of a Heat-Reflective Coating for Pavement in the Chongqing Area', *Journal of Materials in Civil Engineering*, 28(1), p. 04015079.1–04015079.7. doi: 10.1061/(ASCE)MT.1943-5533.0001347.

DESA United Nation (2018) *World Urbanization Prospects: The 2018 Revision.* New York. Available at: https://population.un.org/wup/Publications/Files/WUP2018-Report.pdf.

Fang, F. *et al.* (2013) *A review of near infrared reflectance properties of metal oxide nanostructures.* New Zealand: GNS Science. Available at: https://books.google.com/books?id=_hB3ngEACAAJ.

Guo, R., Nian, T. and zhou, F. (2020) 'Analysis of factors that influence anti-rutting performance of asphalt pavement', *Construction and Building Materials*, 254, p. 119237. doi: https://doi.org/10.1016/j.conbuildmat.2020.119237.

H., L. *et al.* (2016) 'Human Thermal Comfort: Modeling the Impact from Different Cool Pavement Strategies', *Transportation Research Record: Journal of the Transportation Research Board.* Washington, DC United States, 10(2), p. 2575. doi: 10.3141/2575-10.

Henao, J. J., Rendón, A. M. and Salazar, J. F. (2020) 'Trade-off between urban heat island mitigation and air quality in urban valleys', *Urban Climate*, 31, p. 100542. doi: https://doi.org/10.1016/j.uclim.2019.100542.

Li, X. *et al.* (2019) 'Urban heat island impacts on building energy consumption: A review of approaches and findings', *Energy*, 174, pp. 407–419. doi: https://doi.org/10.1016/j.energy.2019.02.183.

Morini, E. *et al.* (2018) 'Effects of aging on retro-reflective materials for building applications', *Energy and Buildings*, 179, pp. 121–132. doi: https://doi.org/10.1016/j.enbuild.2018.09.013.

Pomerantz, M. *et al.* (1997) 'Paving materials for heat island mitigation', *Proceedings of the 1998 ACEEE Summer Study on Energy Efficiency in Buildings 9;135.*

Qin, Y. (2015) 'A review on the development of cool pavements to mitigate urban heat island effect', *Renewable & Sustainable Energy Reviews*, 52, pp. 445–459. doi: https://doi.org/10.1016/j.rser.2015.07.177.

Santamouris, M. *et al.* (2017) 'Passive and active cooling for the outdoor built environment – Analysis and assessment of the cooling potential of mitigation technologies using performance data from 220 large scale projects', *Solar Energy.* Elsevier Ltd, 154, pp. 14–33. doi: 10.1016/j.solener.2016.12.006.

Xie, N., Li, H., Abdelhady, A., *et al.* (2019) 'Laboratorial investigation on optical and thermal properties of cool pavement nano-coatings for urban heat island mitigation', *Building and Environment*, 147, pp. 231–240. doi: https://doi.org/10.1016/j.buildenv.2018.10.017.

Xie, N., Li, H., Zhao, W., *et al.* (2019) 'Optical and durability performance of near-infrared reflective coatings for cool pavement: Laboratorial investigation', *Building and Environment*, 163, p. 106334. doi: https://doi.org/10.1016/j.buildenv.2019.106334.

Zhang, H. *et al.* (2018) 'Performance enhancement of porous asphalt pavement using red mud as alternative filler', *Construction & Building Materials*, 160, pp. 707–713.

Functional Pavements – Chen et al (eds)
© 2021 Taylor & Francis Group, London, ISBN 978-0-367-72610-2

Importance of pavement drainage and different approaches of modelling

S. Alber, B. Schuck & W. Ressel
University of Stuttgart, Institute for Road and Transport Science, Germany

ABSTRACT: Traffic safety is significantly influenced by sufficient drainage characteristics of the road surfaces. Surface runoff on dense asphalt or concrete pavements is influenced by several geometrical design parameters. Basic hydromechanical drainage simulation models offer a great opportunity for the evaluation of critical road sections – for both dense as well as porous surfaces. Porous surfaces show further drainage related aspects such as retention capabilities or filtering effects. These aspects also require basic adequate modelling approaches with regard to different input parameters like void characteristics or permeability.

1 INTRODUCTION

Drainage is an important capability of road pavements that affects several aspects. Traffic safety can be impaired by lower skid resistance values on wet roads or even by aquaplaning effects. Hydromechanical modelling of dense pavement surfaces with emphasis on (critical) water film thicknesses can help to improve road design in critical sections like transition zones. Different road parameters such as slope, road width or texture depth as well as further influences like rain intensity can be evaluated by simulations. Porous pavements, e.g. porous asphalt (PA), show further special aspects of drainage like water retention or filtering effects going beyond traffic safety considerations. To address these questions, an adequate basic modelling approach is also needed for porous pavements.

2 MODELLING OF PAVEMENT SURFACE DRAINAGE

Pavement drainage is a crucial requirement for traffic safety reasons. Longitudinal and cross slope have therefore always played an important role in road design – especially in the context of increasing values of the road width. A pavement surface runoff model (PSRM) has been developed by Wolff (2013) at the University of Stuttgart. It has already been described in detail in different publications, e.g. in Ressel et al. (2019). The model works with a basic hydromechanical approach, the depth-averaged shallow water equations, which makes different simulation needs possible.

Standard cases of pavement topology can be simulated and evaluated with regard to critical water film thicknesses. In particular, transition zones can be analyzed, which are known to be critical for drainage due to low cross slopes. Various parameters like longitudinal slope, cross slope, length of transition zone, relative grade, texture depth (MTD) and rain intensity can be chosen (see examples in Figure 1). A systematic study on these parameters was carried out in Lippold et al. (2019) in order to develop recommendations for road design manuals and identify critical cases. Figures 1a and 1b show two examples of water film thickness on the pavement surface. In both cases high water films occur in the transition zone and lower ones outside. Figure 1c and 1d illustrate the influence of rain intensity. Of course, rain intensity itself cannot be influenced, but the example shows that the choice of an adequate rain intensity in the simulations is quite important for drainage and safety considerations of pavements.

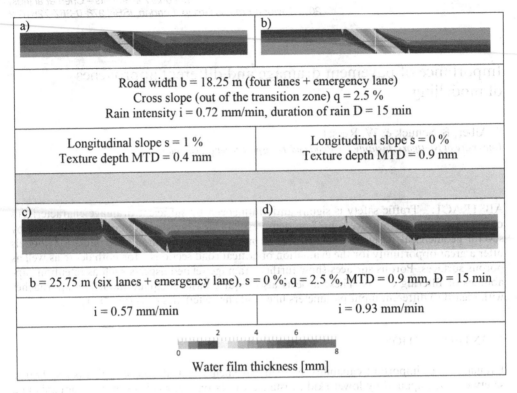

a)	b)
Road width b = 18.25 m (four lanes + emergency lane) Cross slope (out of the transition zone) q = 2.5 % Rain intensity i = 0.72 mm/min, duration of rain D = 15 min	
Longitudinal slope s = 1 % Texture depth MTD = 0.4 mm	Longitudinal slope s = 0 % Texture depth MTD = 0.9 mm

c)	d)
b = 25.75 m (six lanes + emergency lane), s = 0 %; q = 2.5 %, MTD = 0.9 mm, D = 15 min	
i = 0.57 mm/min	i = 0.93 mm/min

Water film thickness [mm]

Figure 1.   Examples of water film distribution in transition zones (calculations with PSRM).

Texture and longitudinal slope have a significant effect on drainage and the resulting water films on the surface. It would be expected that higher MTD values and higher slopes will improve drainage and thus decrease water films. But it is not that simple: One can see contradictory effects regarding both parameters in simulations, see e.g. in Lippold et al. (2019). Higher MTD values have better drainage effects in "texture valleys", but also higher flow resistances due to increased roughness. Higher longitudinal slopes result in better water runoff (and decrease water films), but they also lengthen flow paths (and increase water films). These tendencies interact or predominate in different cases and lead to differing results when simulating surface drainage with varying geometrical parameters.

The examples discussed above consider (ideal) even pavements. Real pavements often suffer from unevenness problems, e.g. rutting on asphalt pavements. Therefore, the simulation of drainage should also be able to consider these uneven surfaces, as ruts also influence drainage in a significant way – in fact, ruts are the main drainage problem in many cases. To this end, a comprehensive approach is presented in Alber et al. (2020), combining model approaches of different research partners (asphalt material model, deformation model and drainage model) to gain better prediction possibilities for drainage problems caused by rutting.

3   DRAINAGE OF POROUS PAVEMENTS

Water can infiltrate into porous pavements and thus critical water films are unlikely to appear, which is of great advantage to traffic safety. Furthermore, water retention capabilities of porous pavements can be of interest. Water runoff is delayed and several parts of the drainage systems (e.g. sewers, natural waters) can be relieved. Thus, the retention effect of porous pavements can also be part of storm water management concepts. Several authors have already dealt with this phenomenon and provide measurement results (e.g. Rodriguez-Hernandez et al. 2016).

Furthermore, the linear reservoir model is an approach widely used in hydrology, which can describe the runoff behavior of different porous pavements, e.g. porous asphalt (PA), as a whole (including infiltration, retention effect, runoff rates, time delay of runoff) with a single parameter – the storage constant K. An exemplary application of this approach using discharge measurements of different PA structures is presented in Alber et al. (in press).

4   3D DRAINAGE MODELLING

While the approach presented in section 3 is a descriptive model type, a basic hydromechanical modelling approach is currently being developed at the Institute for Road and Transport Science of the University of Stuttgart and implemented using the special software framework DuMu^X (Flemisch 2001). Such a basic approach can be applied to various questions of porous pavement drainage such as retention, danger of overflow or even filtering effects – depending on geometrical and morphological parameters of the porous asphalt structure like void content, shape of voids or permeability (Schuck et al., submitted). One of the main advantages of this model compared to other existing ones (e.g. Eck 2010) will be that it can model flow conditions and resulting characteristics with a wide variation of parameters, e.g. also concerning complex pavement topology in transition zones.

5   SUMMARY AND OUTLOOK

Different aspects of drainage as functional properties of roads and their importance for objectives like road safety or storm water management are discussed in this paper. It is shown that fundamental answers to related engineering questions require (different) basic modelling approaches – for dense pavements/surface runoff as well as for runoff through porous pavement structures. A surface drainage model for dense pavements is presented in section 2 – with exemplary results to show the influence of (some) relevant parameters of the road geometry and external variables (e.g. rain intensity). Regarding porous pavements like PA, the infiltration of water into the pavement structure is a further drainage parameter, that completely changes the drainage behavior of the pavement. Therefore, a basic model for 3D drainage (including infiltration in the vertical direction and flow through the porous structure) is required and being worked on in a basic hydromechanical approach (see section 4). Some aspects of the drainage of porous pavements – especially with regard to the overall runoff behavior (e.g. concerning runoff delay due to retention effects) – can also be investigated with existing describing models (see section 3). However, a detailed and profound analysis of the drainage of porous pavements including all relevant parameters needs basic physical modelling, as described in section 4 and is being developed in a current research project.

ACKNOWLEDGEMENT

The work underlying this study was carried out within a DFG Research Group under the research grant number FOR 2089 (sub-project RE 1620/4), on behalf of the grant sponsor, the German Research Foundation (Deutsche Forschungsgemeinschaft, DFG).

REFERENCES

Alber, S.; Schuck, B.; Ressel, W.; Behnke, R.; Canon Falla, G.; Kaliske, M.; Leischner, S.; Wellner, F. 2020. Modeling of surface drainage during the service life of asphalt pavements showing long-term rutting: A modular hydro-mechanical approach. *Advances in Materials Science and Engineering* 2020: 15 pages. Article ID 8793652. doi: 10.1155/2020/8793652

Alber, S.; Ressel, W.; Schuck, B. (in press): Explaining the drainage of porous asphalt with hydrological modelling. *International Journal of Pavement Engineering*. doi: 10.1080/10298436.2020.1811278

Eck, B. 2010. *Drainage Hydraulics of Porous Pavements: Coupling Surface and Subsurface Flow.* Technical (Online) Report- University of Texas at Austin, Center for Research in Water Resources

Flemisch, B.; Darcis, M.; Erbertseder, K.; Faigle, B.; Lauser, A.; Mosthaf, K.; Müthing, S.; Nuske, P.; Tatomir A.; Wolff, M.; Helmig, R. 2011. DuMuX: DUNE for multi- {phase, component, scale, physics, ...} flow and transport in porous media. *Advances in Water Resources* 34: 1102–1112. doi: 10.1016/j.advwatres.2011.03.007

Lippold, C.; Vetters, A.; Ressel, W.; Alber, S. 2019. Vermeidung von abflussschwachen Zonen in Verwindungsbereichen – Vergleich und Bewertung von baulichen Lösungen. *Berichte der Bundesanstalt für Straßenwesen, Reihe Verkehrstechnik* V 319. Fachverlag NW in der Carl Ed. Schünemann KG, https://bast.opus.hbz-nrw.de/frontdoor/index/index/docId/2174

Schuck, B.; Teutsch, T.; Alber, S.; Ressel, W.; Steeb, H.; Ruf, M. (submitted). Study of air void topology with focus on air void constrictions. Submitted to *Road Materials and Pavement Design*. Special Issue for EATA-Conference 2021

Ressel, W., Wolff, A., Alber, S., Rucker, I. 2019. Modelling and simulation of pavement drainage. *International Journal of Pavement Engineering* 20(7): 801–810. doi: 10.1080/10298436.2017.1347437 (available online since 07/2017)

Rodriguez-Hernandez, J.; Andrés-Valeri, V. C.; Ascorbe-Salcedo, A.; Castro-Fresno, D. 2016. Laboratory Study on the Stormwater Retention and Runoff Attenuation Capacity of Four Permeable Pavements. *Journal of Environmental Engineering* 142(2): 04015068–1-04015068-8. doi:10.1061/(ASCE)EE.1943-7870.000103

Wolff, A. 2013. Simulation of Pavement Surface Runoff using the Depth-Averaged Shallow Water Equations. *Veröffentlichungen aus dem Institut für Straßen- und Verkehrswesen* 45. PhD thesis. University of Stuttgart

Experimental and numerical investigation on the development of pore clogging in novel porous pavement based on polyurethane

Guoyang Lu
The Hong Kong Polytechnic University

Haopeng Wang
Delft University of Technology

Pengfei Liu
RWTH -Aachen University

Zhen Leng
The Hong Kong Polytechnic University

Markus Oeser
RWTH -Aachen University

ABSTRACT: Permeable pavements are often affected by pore clogging, which leads to their functional failure and reduced service life. However, the clogging mechanism and its impact on the permeability and complex pore microstructures in pervious pavement remains unclear. The aim of current study is to quantify the clogging behavior in pervious pavement materials and carry out investigations on the development of pore characteristics and permeability. Novel Polyurethane bound pervious mixture (PUPM) were adopted for comparative study in present research with conventional Porous Asphalt (PA). The Aachen Polishing Machine (APM) was selected to perfectly serve as a simulator for clogging process of pavement in the actual service condition. The development of pore characteristics in terms of clogging was experimentally illustrated. The developed experiments and analysis can further strengthen the understanding of the clogging mechanism within the porous pavement material.

1 INTRODUCTION

In contrast to a natural plant-soil system, sealed road surfaces have lower specific heat capacity and higher solar absorptivity, which results in higher temperatures in urban areas surrounded by suburbs with lower temperatures. This phenomenon is called the Urban Heat Island Effect (UHIE) [1][2].

To address these issues, urban roads with functionalities that support a healthy environment can be constructed as pervious pavement material (PPM). Such systems mostly contain different layers of materials with porous structures (e.g. porous asphalt, porous concrete). In addition to serving as a traffic-bearing system, the PPM also fulfils other functionalities such as the absorption, storage, and evaporation of rainwater [3]. Compared to conventional pavement designs, the structure of PPM facilitates fluids flow through the porous material, thus reducing and controlling surface runoff. Common types of permeable roads include: Porous Asphalt (PA), Porous Concrete (PC) and Permeable Interlocking Concrete Pavement (PICP) [4].

Polyurethane-bound pervious mixtures have demonstrated excellent functionality and more importantly good mechanical properties not exhibited by conventional PA mixtures. Additionally, PUPM has the advantage of being cleaner and more environmentally friendly. One of the reasons is that polyol is made up of organic oils. Furthermore, it's production at ambient temperature reduces energy consumption for heating and CO_2 emissions. Working conditions for laborers can be improved because the release of volatile organic compounds (VOCs) and smoke produced in mixing and placement of conventional mixtures is completely avoided [5].

However, as a porous pavement, the long-term service performance is greatly hindered by clogging characteristics [6]. Studies have been conducted by applying fine particles accumulating in the void spaces of porous pavements based on different watering methods [7]. Clogging is found to be highly correlated with the particle size and volume, the flow rate and pore characteristics of the pavement [8]. previous researches conducted were mainly focused on the observation of field measurement and the macroscopic laboratory experiments. The microscopic characterization on the clogging such as the development of pore characteristics, particles distribution and kinematic etc. are still not clarified. Additionally, most of the existing clogging experiments are based on modified permeameters by only changing the flow conditions, none of them can effectively simulate the clogging by considering the tires-road interaction. Hence, a mesoscopic study on the clogging mechanism in PPM must be carried out with a clogging test set up which can closer simulate to actual service conditions. A systematic understanding of the development hydraulic and clogging mechanisms within the PPM should be further established.

2 MATERIAL AND METHODS

2.1 Material

To investigate the pore characteristics of the PUPM, three types of the PUPM with different porosities were produced in this study: PUPM 8-H, PUPM 8, and PUPM 8-L, which denotes mixture with maximum aggregate size 8mm and high, normal, and low porosity level respectively. Conventional PPM, PA 8, was also selected as the reference material. The grain size distribution of all four chosen samples which were used in this study is illustrated in Figure 1, a and the manufacturing process can be seen in Figure 1, b:

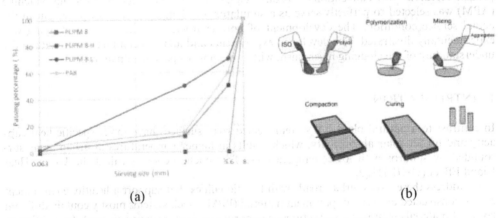

(a)   (b)

Figure 1.   Grain size distribution of PUPM 8-H, PUPM 8, PUPM 8-L and PA 8.

2.2 Methods

To simulate the clogging in the laboratory, a simulation based on the Aachen Polishing Machine (APM) was carried out, as shown in Figure 2 a. By the aid of the APM, pavement samples can be exposed to a real loading condition. For the clogging simulation in this study, sand, which has a grain size of up to 2 mm, was chosen as the clogging material. During polishing process, water was sprinkled on the surface of the specimens. In this way, the sand

| (a) | (b) | (c) |

Figure 2.   Test methods: (a) APM test; (b) Permeability; (c) CT scanning.

would enter the void of the sample with the water and the pressure of the tire. Each sample was processed for 4 different clogging periods: 20 minutes, 40 minutes, 60 minutes, and 80 minutes. These samples were then applied to the permeability tests and CT scanning respectively shown in (Figure 2 b and Figure 2 c).

3   RESULTS

3.1   Text and indenting

During the permeability test of the specimens, the data were collected in 5 hydraulic heads (from 100 to 300 mm, 50 mm per level). In each level, four measurements for each specimen were obtained at the hydraulic head from 100 to 300 mm every 50 mm. Thus, 1200 data points were recorded and were then analyzed respectively. As can be seen in Figure 3, the permeability coefficients at 300mm hydraulic head of the specimens during the clogging periods were obtained. The specimens PUPM 8-H with the highest permeability can conduct $1.x10^{-3}$ m/s of water. Followed by the PUPM 8 with the second highest permeability, and at the third position is the PA 8. In the figure, it can be recognized that the PUPM 8-L has the lowest permeability, with $3.x10^{-3}$ m/s.

For PA 8, the permeability of PA 8 decreased sharply once the clogging period started. In the first 20 minutes, the specimen's permeability sharply reduced from $1.08x10^{-3}$ m/s to $2.9x10^{-4}$ m/s, approximately 26% of its initial permeability. Conversely, during the same periods, the permeability of the PUPM 8 reduced slowly. Unlike the PA, the PU can perfectly cover the surface of

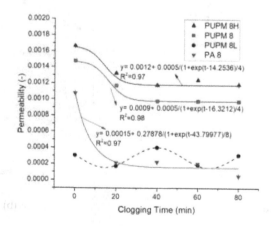

Figure 3.   Constant permeability test.

409

the aggregates and provide a smooth surface for the aggregates. Therefore, the sediments can be easier caught in the PA specimen. The characteristic of the binder of the PA is, however, another influencing factor. Because of the high viscosity of the bitumen, the sediments tend to become attached to the bitumen. The flow behavior of the PA in the experiment is consistent with the previous study. After 20 minutes of clogging simulation, the permeability of the PA 8 decreases continuously and steadily. 60 minutes later, the permeability of PA 8 reaches the 7.5x10-5 m/s which can be classified as permeable. However, the initial permeability was highly permeable. In comparison with the PUPM 8 and PUPM 8-H, the permeabilities of PA 8 are nineteen to thirteen times lower than that of PUPM 8 and PUPM 8-H.

3.2 Pore clogging characteristics

By the aid of MATLAB and Avizo software, the aggregate, polyurethane binder and clogging mixture was successfully separated. Identically, 3D visualization and quantitative analysis were performed (in Figure 4), and volume and longitudinal distribution of the polyurethane and clogging mixture are calculated. The relevant results are shown in Figure 5 (a) and (b).

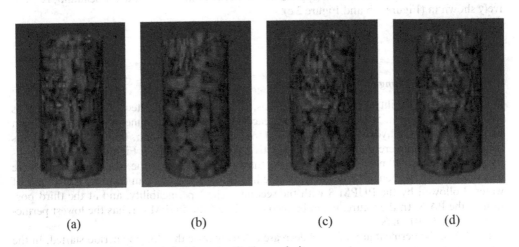

(a)	(b)	(c)	(d)

Figure 4.   3D visualization and clogging quantitative analysis.

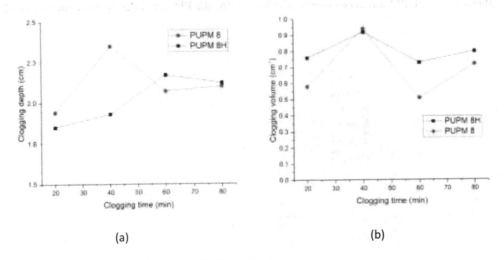

(a)	(b)

Figure 5.   (a) Depth of clogging center; (b) Clogging volume.

According to Figure 5 (a) and (b), it can be found that the clogging depths of specimen center are generally increasing, which indicates that as loading time increases, clogging such as dust tend to be pushed in deeper. At the same time, volume of clogged mixture of two groups indicates fluctuation and local increasing trend, which suggest time interval might not be long enough or surface blockage might prevent more blockage from being pushed to larger depth. The reason why these two parameters do not show a strictly increasing trend might be that the gray level of mixture and the edge of the aggregate are too close, and more badly the interference of external light during the CT scanning process, resulting in some black backgrounds being recognized as part of the mixture, which is recommended to be improved in the test technology in the future test.

4 CONCLUSION

The current research focused on the novel method to meso-evaluate the clogging characteristics in PPM. PPMs including three types of PUPM and one PA were applied for the clogging and permeability experiments. Self-developed APM and X-ray CT were firstly utilized together to comprehensively analyze the development of pore characteristics in PPM by the consideration of clogging effect.

The research further strengthened the understanding of the development of pore characteristics and clogging mechanism, which laid the foundation for future material design optimization.

As can be summarized by the current research, the clogging has a significant influence on the permeability and pore characteristics of pervious pavement materials. To further understand the clogging mechanism, experiments under more loading conditions are highly recommended. Apart from it, the theoretical models and the experimental analyses of clogging mechanism should also be enhanced.

REFERENCES

[1] Sun, W., Lu, G., Ye, C., Chen, S., Hou, Y., Wang, D., Wang, L. and Oeser, M., 2018. The state of the art: Application of green technology in sustainable pavement. Advances in Materials Science and Engineering, 2018.
[2] Wang, D., Liu, P., Leng, Z., Leng, C., Lu, G., Buch, M. and Oeser, M., 2017. Suitability of PoroElastic Road Surface (PERS) for urban roads in cold regions: Mechanical and functional performance assessment. Journal of cleaner production, 165, pp.1340–1350.
[3] Lu, G., Renken, L., Li, T., Wang, D., Li, H. and Oeser, M., 2019. Experimental study on the polyurethane-bound pervious mixtures in the application of permeable pavements. Construction and Building Materials, 202, pp.838–850.
[4] Lu, G., Liu, P., Törzs, T., Wang, D., Oeser, M. and Grabe, J., 2020. Numerical analysis for the influence of saturation on the base course of permeable pavement with a novel polyurethane binder. Construction and Building Materials, 240, p.117930.
[5] Lu, G., Törzs, T., Liu, P., Zhang, Z., Wang, D., Oeser, M., & Grabe, J. (2020). Dynamic Response of Fully Permeable Pavements: Development of Pore Pressures under Different Modes of Loading. Journal of Materials in Civil Engineering, 32(7), 04020160.
[6] Zhang, J., Jin, Q. and Cui, X., 2014. Experimental Study on Pore Clogging of a Porous Pavement under Surface Runoff. In Design, Analysis, and Asphalt Material Characterization for Road and Airfield Pavements (pp. 138–146).
[7] Zhang, J., She, R., Dai, Z., Ming, R., Ma, G., Cui, X. and Li, L., 2018. Experimental simulation study on pore clogging mechanism of porous pavement. Construction and Building Materials, 187, pp.803–818.
[8] Sansalone, J., Kuang, X., Ying, G., & Ranieri, V. (2012). Filtration and clogging of permeable pavement loaded by urban drainage. Water research, 46(20), 6763–6774.

Functional Pavements – Chen et al (eds)
© 2021 Taylor & Francis Group, London, ISBN 978-0-367-72610-2

Unconfined compressive strength and prediction model of fiber/cement modified slurry

J.B. Hu & P. Jiang*
Shaoxing University, Shaoxing, China

ABSTRACT: In order to explore the effect of fiber and cement on the modification of slurry and the influence of various factors on the unconfined compressive strength of modified slurry, an unconfined compressive strength test was carried out using slurry, fiber and cement as raw materials. The research results show that the cement and fiber content, curing age and porosity have significant effects on the strength development of modified slurry. According to 72 sets of modified slurry strength data with different ratios and different ages, an unconfined compressive strength model of the modified slurry was established. The model fitting effect is good, and it provides a certain reference basis for the prediction of the modified slurry compressive strength.

Keywords: fiber/cement modified slurry, unconfined compressive strength, prediction model

1 INTRODUCTION

In recent years, the scale of my country's basic engineering construction has gradually expanded. While engineering construction has brought economic benefits to the society, it has also brought a series of environmental problems. For example, in the process of bored pile construction, shield tunneling construction and underground continuous wall construction, a large amount of waste slurry will be generated. Due to the limited storage space and high population density in cities, improper disposal of the waste slurry from these projects will cause environmental pollution, land occupation and other problems, and even have a huge impact on the normal operation of the entire city.

In order to realize the resource utilization of waste slurry, scholars at home and abroad have conducted a lot of research on the solidification treatment of slurry (e.g. Fang et al., 2011; Su et al., 2012; Du and Wang, 2018; Yang et al., 2017; Katsioti et al., 2008; Zang et al., 2011). As a common curing agent, cement is widely used in the curing of soft soils due to its easy access to materials and high cost performance. Although the strength of cement-modified soil can generally meet the requirements of engineering, there is still the problem of brittle failure (Azadegan et al., 2012). As an auxiliary reinforcing material, fiber has the characteristics of high tensile strength, good dispersion, and strong durability. Incorporating it into the soil can effectively increase the strength of the soil and restrain the deformation of the soil. Kumar et al. (2016) have shown that fiber improves the unconfined compressive strength of cement-soil. Khattak et al. (2006) found that there is an optimal blending amount of fiber by splitting tensile test, and the optimal blending amount is related to the type of soil. Festugato et al. (2017) found that the unconfined compressive strength of fiber cement soil increases with the increase of fiber length.

At present, the research on the mechanical properties of cement and fiber-modified soil is mainly carried out around the changing law of their strength, and the

---

* Corresponding author

mathematical model of the strength and its influencing factors is established. Deng et al. (2019) studied the influence of fiber content on the parameters of the Duncan-Chang model of polypropylene fiber cement silty clay through triaxial tests. Tang et al. (2000) believed that the compressive strength of cement-solidified soil is directly proportional to the content of cement and inversely proportional to the square of the water content of the raw materials, and thus put forward a linear empirical formula between strength, volume content of cement, and water content.

In this paper, cement and fiber are used to modify the engineering waste slurry. Through the unconfined compressive strength test, the mechanical properties of the fiber cement modified slurry are analyzed, and the factors affecting the unconfined compressive strength are studied: material content, curing age, porosity, and establish related mathematical models to evaluate its unconfined compressive strength.

2 EXPERIMENTAL PREPARATION

2.1 Experimental raw material and experimental program

The slurry material used in the experiment comes from a construction site in Shaoxing City. It is a semi-colloid suspension composed of water and clay. The main physical and mechanical indexes of slurry are shown in Table 1. The cement used in the test is Conch Portland cement produced by Shaoxing Shangyu Conch Cement Co., Ltd. The cement number is P·O32.5. The fiber used in the test is a 6mm polypropylene filamentous fiber.

Using the pycnometer method to test the specific gravity of slurry, cement and fiber, the specific gravities were 2.65, 3.1 and 0.91, respectively.

The experimental program is shown in Table 2.

2.2 Sample making and testing

The sample is made according to the highway geotechnical test procedure, and the steps are as follows: calculate the amount of material according to the test plan, slowly add the weighed slurry, water, cement, and fiber into the mixer, and stir 4 times for 5 minutes each time. Put the mixed modified slurry into the test mold three times, and vibrate 50 times each time. The size of the test mold is a cylinder with a diameter of 39.1mm and a height of 80mm. Seal the test mold with plastic wrap and let it stand for 2h vertically. After the sample is solidified and stabilized as a whole, seal both ends of the sample with filter paper and place it in water for curing at a curing temperature of 20°C. According to the designed curing age, the UCS test is performed after demolding the sample after curing.

Table 1. Main physical and mechanical indexes of slurry.

Index name	Specific gravity	Liquid limit/%	Plastic limit/%	Plasticity index
Index value	2.65	43.5	23.1	20.4

Table 2. Experimental program of UCS.

Cement content/%	Fiber content/%	Moisture content/%	Curing age/d
5、10、15、20、25	0	-	7、14、28
		100	56、90、120
20	0.25、0.5、0.75、1	-	150、180

3 RESULT AND DISCUSSION

3.1 Unconfined compressive strength

A total of 9 groups of unconfined compressive strength tests were designed according to the test plan considering different cement and fiber content. The changes in the curing age in each group of tests were divided into 8 levels. The measured unconfined compressive strength of each group is shown in Table 3.

3.2 Effect of content on compressive strength

Figure 1 (a) and (b) show the original data and fitting line of the unconfined compressive strength of cement modified slurry (CMS) and fiber cement modified slurry (FCMS) samples, respectively. (a) It can be seen that the unconfined compressive strength of each age of CMS increases with the increase of cement content, and the relationship between strength and cement content is basically linear. For the specimens with a curing age of 7d, the unconfined compressive strength increases from the initial 43kPa to 122kPa, 199kPa, 298kPa and 423kPa with the increase of the cement mixing ratio. The increase rate gradually increases with the cement mixing ratio. improve. It can also be found in the figure that the slope of the fitted curve increases with the increase of age, indicating that the longer the age, the faster the unconfined compressive strength of the soil increases with the increase in cement

Table 3. Unconfined compressive strength of modified slurry.

Cement content/%	Fiber content/%	Time/d							
		7	14	28	56	90	120	150	180
5	0	43	81	113	151	184	194	195	205
10	0	122	179	221	248	288	322	331	338
15	0	199	283	339	404	448	478	489	502
20	0	298	448	553	625	701	743	752	756
20	0.25	331	508	618	699	798	847	918	902
20	0.5	352	549	650	738	839	883	952	944
20	0.75	369	563	673	777	862	920	971	951
20	1	384	572	681	751	851	864	953	902
25	0	423	518	648	755	867	899	918	953

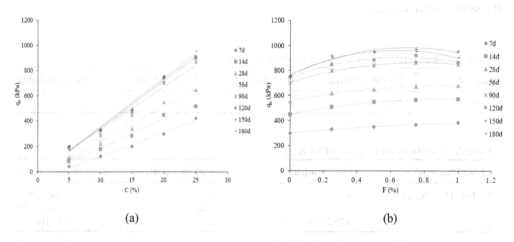

(a)　　　　　　　　　　　　　(b)

Figure 1. Variation of unconfined compression strength with cement content and fiber content.

414

mixing ratio. (b) It can be seen that after adding fibers, the unconfined compressive strength of the modified slurry has been improved. When the curing age is 7-28d, the unconfined compressive strength of FCMS increases with the increase of fiber content. After 28 days, when the fiber content is between 0.25-0.75%, the unconfined compressive strength increases with the increase of fiber content; when the fiber content is in the range of 0.75-1%, the unconfined compressive strength decreases as the fiber content increases. It can be seen that the optimal fiber content of FCMS for long age is 0.75%.

3.3 Effect of curing age on compressive strength

Figure 2(a) shows that as the curing age increases, the unconfined compressive strength of CMS increases logarithmically. The strength of CMS in the early stage has a faster growth rate. As the curing age increases, the growth rate gradually slows down, and the strength eventually tends to a fixed value. From the Figure 2(B), it can be found that when the curing age is before 150d, the intensity of FCMS increases with the increase of the curing age, but after 150d, the intensity begins to decrease. This is because as the curing age increases, the hydration reaction of the cement slowly stops, no gelling substances are produced, and the internal structure of the modified slurry is gradually stabilized, so the friction between the fibers and the modified slurry decreases.

3.4 Effect of porosity on compressive strength

The porosity η is calculated by the following formula:

$$\eta = 100 - 100\left\{\left[\frac{\gamma_d}{1 + \frac{C}{100} + \frac{F}{100}}\right]\left[\frac{1}{\gamma_{Sc}} + \frac{\frac{C}{100}}{\gamma_{Sc}} + \frac{\frac{F}{100}}{\gamma_{Sf}}\right]\right\} \tag{1}$$

where γ_d = dry density; γ_{Ss} = slurry specific gravity; γ_{Sc} = cement specific gravity; γ_{Sf} = fiber specific gravity; C = cement content; F = fiber content. Figure 3(a) shows that the unconfined compressive strength of CMS decreases with the increase of porosity, and the relationship between strength and porosity is a power function. Figure 3(b) shows that the strength of FCMS basically decreases with the increase of porosity, but the change rule is not obvious.

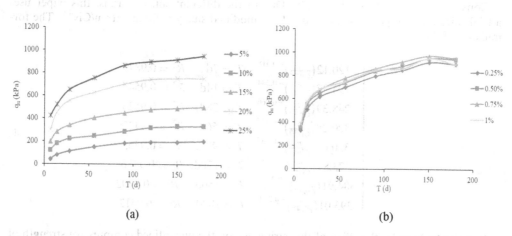

(a)   (b)

Figure 2.   Variation of unconfined compression strength with curing age.

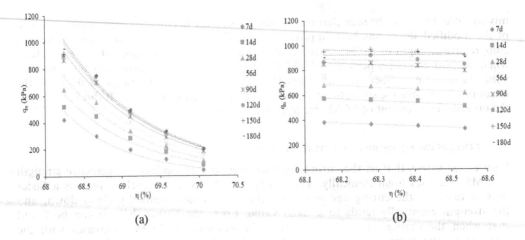

(a)   (b)

Figure 3.   Variation of unconfined compression strength with porosity.

4   PREDICTION MODEL AND VERIFICATION OF UNCONFINED COMPRESSIVE STRENGTH

4.1   Prediction model

Consoli et al. (2010) proposed the following model to evaluate unconfined compressive strength of cemented sand mixtures:

$$q_u = a\left(\frac{\eta}{Civ^b}\right)^c \tag{2}$$

where Civ = volumetric cement content, the mathematical expression is as follows:

$$Civ = \frac{C\gamma_d}{\left(1 + \frac{C}{100} + \frac{F}{100}\right)\gamma_{Sc}} \tag{3}$$

Consoli set the parameter b as 0.28. Due to the different soil materials, this paper uses a trial calculation to get b=4.9, and q_u of the modified slurry is fitted with $\eta/Civ^{4.9}$. The formula is as follows:

$$q_u = \begin{cases} 120.12\left(\frac{\eta}{Civ^{4.9}}\right)^{-0.318} & T = 7\mathrm{d}, & R^2 = 0.9849 \\ 194.37\left(\frac{\eta}{Civ^{4.9}}\right)^{-0.284} & T = 14\mathrm{d}, & R^2 = 0.9547 \\ 248.39\left(\frac{\eta}{Civ^{4.9}}\right)^{-0.266} & T = 28\mathrm{d}, & R^2 = 0.9526 \\ 300.28\left(\frac{\eta}{Civ^{4.9}}\right)^{-0.247} & T = 56\mathrm{d}, & R^2 = 0.9433 \\ 350.08\left(\frac{\eta}{Civ^{4.9}}\right)^{-0.237} & T = 90\mathrm{d}, & R^2 = 0.9353 \\ 373.84\left(\frac{\eta}{Civ^{4.9}}\right)^{-0.233} & T = 120\mathrm{d}, & R^2 = 0.9436 \\ 384.91\left(\frac{\eta}{Civ^{4.9}}\right)^{-0.24} & T = 150\mathrm{d}, & R^2 = 0.9302 \\ 393.01\left(\frac{\eta}{Civ^{4.9}}\right)^{-0.232} & T = 180\mathrm{d}, & R^2 = 0.9412 \end{cases} \tag{4}$$

In order to consider the effect of the curing age on the unconfined compressive strength of FCMS, the relationship between the parameters a, c and the curing age is analyzed. The functions of a and c obtained by function fitting are:

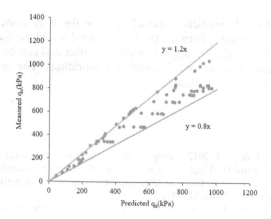

Figure 4. Comparison of predicted value and measured value.

$$a = 84.268\ln(T) - 35.48 \qquad R^2 = 0.9956$$
$$c = 0.0249\ln(T) - 0.3551 \qquad R^2 = 0.9334 \tag{5}$$

Substituting the formulas of a and c, the unconfined compressive strength formula of FCMS can be obtained:

$$q_u = [84.268\ln(T) - 35.48]\left(\frac{\eta}{Civ^{4.9}}\right)^{[0.0249\ln(T)-0.3551]} \tag{6}$$

4.2 Verification of prediction model

After the FCMS unconfined compressive strength formula is determined, the predicted value of the formula is compared with the unconfined compressive strength data obtained from the test. The result is shown in the figure below.

It can be seen from Figure 4 that the accuracy of the FCMS unconfined compressive strength model value and the actual value is basically controlled within ±20%, which has a good correlation. In general, the model is simple and easy to use, and has a good fitting effect, which can be used for reference in the engineering practice of modified slurry.

5 CONCLUSIONS

Through the unconfined compressive strength test of fiber cement modified slurry, the following conclusions can be obtained:

(1) The unconfined compressive strength of cement modified slurry increases with the increase of cement content. After fiber is added, the compressive strength of modified slurry is further improved, and within a certain range, the unconfined compressive strength increases with the increase of fiber content; the optimum content of fiber is 0.75%.
(2) The unconfined compressive strength of fiber cement modified slurry basically increases with the increase of the curing age, and the early strength increases faster, and the later increases slowly and tends to stabilize, or even decrease.
(3) The unconfined compressive strength of cement modified slurry has a power function relationship with porosity. The strength decreases with the increase of porosity, and the law is not obvious after adding fibers.

417

(4) A compressive strength prediction model suitable for fiber cement modified slurry is derived. The maximum error between the predicted value and the measured value is basically within 20%. The model has a good prediction effect and can be used to predict different cement and fiber content and different age conditions. The unconfined compressive strength of the modified slurry under.

REFERENCES

Azadegan, O., Jafari, S.H. & Li, J. 2012. Compaction characteristics and mechanical properties of lime/cement treated granular soils[J]. Electronic Journal of Geotechnical Engineering, (17):2275–2284.

Consoli, N.C. & Festugato, L. 2010. Effect of fiber-reinforcement on the strength of cemented soils. Geotextiles and Geomembranes, 28(4), 344–351.

Deng, L.F., & Ruan, B. 2019. Experimental study on triaxial compression test of polypropylene fiber reinforced cement silty clay. Journal of Railway Science and Engineering, 2019,16(05):1201–1206.

Du, Y.Q. & Wang, X.Q. 2018. Experimental study on solidification treatment of engineering waste slurry. Tianjin Construction Science and Technology, 28(01):47–50.

Fang, K., Zhang, Z.M., Liu, X.W., et al. 2011. Pollution of construction waste slurry and prevention measures. Chinese Journal of Geotechnical Engineering, 33(10): 238–241.

Festugato, L., Menger, E., Benezra, F. & Consoli N.C. 2017. Fibre-reinforced cemented soils compressive and tensile strength assessment as a function of filament length. Geotextiles and Geomembranes, 45(1), 77–82.

Katsioti, M., Katsiotis, N., Rouni, G., et al. 2008. The effect of bentonite/cement mortar for the stabilization/solidification of sewage sludge containing heavy metals. Cement and Concrete Composites, 30 (10), 1013–1019.

Khattak, M. J. & Alrashidi, M. 2006. Durability and mechanistic characteristics of fiber reinforced soil–cement mixtures. International Journal of Pavement Engineering, 7(1), 53–62.

Kumar, A. & Gupta, D. 2016. Behavior of cement-stabilized fiber-reinforced pond ash, rice husk ash–soil mixtures[J]. Geotextiles and Geomembranes, 44(3): 466–474.

Su, Q.G., Zhai, Z.G. & Deng, H.Y. 2012. Treatment of waste slurry of slurry shield machine. Tunnel Construction, 32(S2):222–226.

Tang, Y.X., Liu, H.L. & Zhu, W. 2000. Study on engineering properties of cement-stabilized soil. Chinese Journal of Geotechnical Engineering, 22(5):549–54.

Yang, A.W., Zhong, X.K., Liang, C. & Li, Y. 2017. Experiment study of solidification performance and long-term mechanical properties of dredger filled slurry. Rock and Soil Mechanics, 38(09):2589–2596.

Zhang, Z.M., Tao, T., Wang, Z.J., et al. 2011. Zero discharge treatment technology for slurry and engineering properties of separated soil. Chinese Journal of Geotechnical Engineering, 33(9):1456–1461.

Functional Pavements – Chen et al (eds)
© 2021 Taylor & Francis Group, London, ISBN 978-0-367-72610-2

Design innovation and study on pot bearings for highway bridges

M. Li
Research Institute of Highway Ministry of Transportation, Beijing, China
China Transportation Inspection & Verification Hi-Tech Co., Ltd, Beijing, China

C. Zhang
Research Institute of Highway Ministry of Transportation, Beijing, China
National Engineering Research Center of Road Maintenance Technologies/RoadMainT Co., Ltd., Beijing, China

P. Yin
China Transportation Inspection & Verification Hi-Tech Co., Ltd, Beijing, China

X.Y. Ban & J.G. Rong
National Engineering Research Center of Road Maintenance Technologies/RoadMainT Co., Ltd., Beijing, China
Research Institute of Highway Ministry of Transportation, Beijing, China

C. Wang
Highway Monitoring & Response Center, Ministry of Transport of the P.R.C, Beijing, China
National Routes Network Beijing Transportaion Technology Co., Ltd., Beijing, China

ABSTRACT: All the time, the main elements of pot bearings including pots, steel pistons and anchoring bolts are designed respectively for the three types including fixed, bilateral, and unilateral types. The optimal design idea of mechanical engineering which is to obtain the optimal design with the minimum numbers of elements is introduced in the innovative design of pot bearings for highway bridges. By the application of innovative design of elements including pots, steel pistons as well as anchoring bolts, the steel pistons can be employed universally for the two types of unilateral and bilateral pot bearings, meanwhile, pots are also interchangeable for the three types of unilateral, bilateral and fixed pot bearings, and the related elements like elastomeric discs, anchoring bolts can be employed universally, which provides great convenience for factory production and assembly. The 3D finite element analyses results indicate the pot bearings for highway bridges with innovative design can meets the design requirements.

1 INTRODUCTION

By the end of 2015, the total number of highway bridges in China had reached 779 thousand and 2 hundred kilometers. The amount of existing bridge bearings which are similar to spines to connect the upper and lower structures of bridges is huge. Because of the small proportion of bearings in the bridge construction cost, they haven't been paid enough attention. At the same time, due to the limited research level of bridge bearings enterprises in China, innovation capacity of product design is inadequate. In order to improve the core competitive power of bearings for highway bridges, design innovation and research work have been carried out on the most important type of bridge bearings, namely, pot bearings for highway bridges. Pot bearing can be classified into three types, including bilateral, unilateral and fixed bearings (China Communications Press 2009). Bilateral and unilateral bearings are mainly composed of upper bearing plates and pots, steel pistons as well as elastomeric discs. Fixed bearings are mainly composed of upper bearing plates and pots as well as elastomeric discs. It is easy to find that versatility and interchangeability haven't been taken into consideration in the existing design of pot bearings(EUROPEAN COMMITTEE FOR STANDARDIZATION 2000; EUROPEAN COMMITTEE FOR STANDARDIZATION 2005; Li 2009; Zhuang 2015). Pots and steel pistons which is lack of versatility and interchangeability need respective design according to the three different types under the same capacity, which increases the workload of mould foundry workers and go against enterprise production organization and elements inventory control.

2 INNOVATIVE DESIGNS FOR POT BEARINGS

In view of the defects of versatility and interchangeability in the design of pot bearings, a new type of universal pot bearing is developed. The key technical points are as follows.

(1) As shown in Figure 1 to Figure 3, for the same bearing capacity, the pots of pot bearings (including fixed, bilateral and unilateral types) are designed to be of the same structure and size, so that the pots become universal elements of pot bearings. As shown in Figure 4, the structure, all sizes and machining requirements of the pots are exactly the same. The production of pots can be completed with only one design drawing, one wood mould and one sand box. In this way, it can not only reduce the workload of designer as well as moulding and foundry workers, but also facilitate the exchange of the pot according to the different production tasks. At the same time, as shown in Figure 5, if the guide systems of the unilateral pot bearings are set on the two sides of pots, the pots for the bilateral and fixed pot bearings are not good in the respect of interchangeability. In addition, as shown in Figure 6, if the pots of the commonly used pot bearings are designed without considering versatility, although the pots have a certain similarity in appearance, there are still large differences in the dimensions (such as pot wall thickness and pot wall diameter), which cannot be universal and interchangeable.

(2) As shown in Figure 7, for the same bearing capacity, the steel pistons which adopt the style of two sides guide bar (including bilateral and unilateral types) are designed to be of

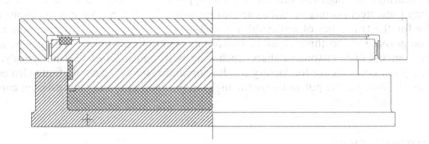

Figure 1.   Assembly drawing of universal unilateral pot bearings.

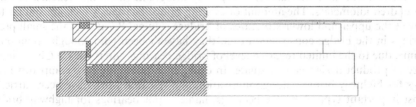

Figure 2.   Assembly drawing of universal fixed pot bearings.

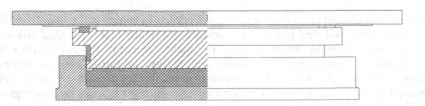

Figure 3.   Assembly drawing of universal bilateral pot bearings.

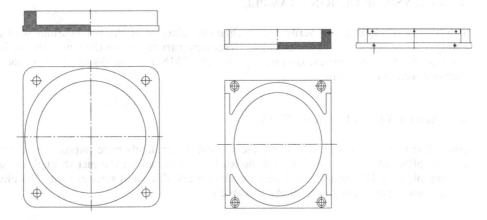

Figure 4.   Pot of universal pot bearigs.

Figure 5.   Pot of common unilateral pot bearings.

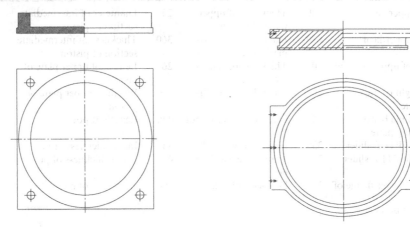

Figure 6.   Pot of common bilateral and fixed
pot bearings.

Figure 7.   Steel piston of universal pot bearigs.

the same geometry and dimensions, so that the steel pistons become interchangeable elements of the pot bearings.

(3) For the same bearing capacity, the anchoring sleeves (including fixed, bilateral and unilateral types) were designed to be of the same geometry and dimensions, so that the anchoring sleeves become a universal element of the pot bearings.

(4) For the same bearing capacity, the anchoring bolts (including fixed, bilateral and unilateral types) is designed to be of the same geometry and dimensions, so that the pot bearings
can be maintained and replaced conveniently after it had been installed.

Only four wood mould and four sand box are needed to complete the foundry of the upper bearing plates, pots and steel piston, while seven wood mould and seven sand box are needed compared with traditional design method. The main element of the pot bearings including pots, steel pistons, brass sealing rings, rubber sealing rings, anchoring bolts, anchoring sleeves can achieve the versatility and interchangeability in the factory production and assembly after employing the innovative design. Assembling of three different types of pot bearings including fixed, bilateral and unilateral types can be achieved only replacing the upper bearing plates.

3 ANALYSIS OF DESIGN EXAMPLE

In order to investigate and verify the mechanical behavior of the pot bearings with interchangeable elements, 3D finite element analyses were carried out on the innovative unilateral pot bearings with the vertical bearing capacity of 2.5MN and the displacement in the main sliding direction of ±50mm.

4 CALCULATING PARAMETERS

According to the design requirements, the vertical design loads were imposed on the upper bearing plates of the pot bearing, and the loads were passed on the pier through the upper bearing plates, PTFE sliders, steel pistons, elastomeric discs and pots in turn. The physical dimensions of pot bearing were shown in Table 1.

Table 1. Parameters of the physical dimensions of pot bearings (Units : mm).

Variable	Value	Variable	Value	Variable	Value
Width of upper bearing plate	480	Thickness of upper plate of piston	25	Diameter of intermediate section of piston	347
Length of upper bearing plate	670	Diameter of elasto-meric disc	360	Thickness of intermediate section of piston	23
Thickness of upper bearing plate	20	Thickness of elasto-meric disc	26	Length of upper plate of piston	400
Overall height of upper bearing plate	50	Inner diameter of pot	360	Width of upper plate of piston	400
Width of guide bar of upper bearing plate	34	Outer diameter of pot	440	Length of pot	450
Diameter of PTFE slider	330	Height of pot wall	45	Inner thickness of pot	·20
Thickness of PTFE slider	7	Transverse width of pot	450	Outer thickness of pot	17
Diameter of lower flange of piston	359	Diameter of pier	700	Thickness of pier	200
Thickness of lower flange of piston	11				

Material properties of the elements of the pot bearings were shown in Table2.

Table 2. Material properties of the elements of the pot bearings.

Materials	Modulus of elasticity (N/mm^2)	Poisson ratio
Structural steel	2.06E5	0.3
Rubber sheet	7.8	0.4998
Concrete	3.45E4	0.166
PTFE	1500	0.4

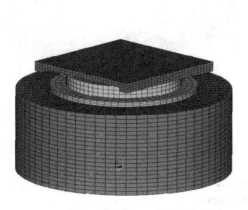

Figure 8.   Finite element model of pot bearings.   Figure 9.   1/4 finite element model of pot bearings.

The structures, loads and boundary conditions of the pot bearings are symmetrical under the vertical design load, so the 1/4 finite element model is adopted. The vertical constraint was imposed under the pier, and the symmetrical constraint was imposed on the profile. The pan bottom-shaped deformation of pier will occur under the vertical design load. In order to consider the influence of this deformation to the pot bearing, the model of a certain size of concrete pier is established to considering role of elastic supporting action.

Both two dimensional axisymmetric models and three dimensional solid models can be used to analyze the stress condition of the pot bearings under vertical design load. Because of the properties of nearly incompressible of elastomeric disc of pot bearing, the pot wall will be extruded under the vertical design load, friction between elastomeric disc and the pot will occur. Considering the simulation of contact characteristics can be helpful to model the real situation, it is necessary to establish the solid model for contact analysis. The elastomeric disc is modeled by SOLID185, and the cast steel is modeled by SOLID45. The overall finite element model is established as shown in Figure 8, and 1/4 axisymmetric model of the pot bearing is shown in Figure 9.

CONTA173 and TARGE170 were adopted for the contact element and target element respectively. Contact pairs have been established on the contact surface of the top surface of pots, pot wall and elastomeric disc. In addition, contact pairs have been established between the PTFE slider and the upper bearing plate. The units of displacement and stress are mm and MPa respectively in the following results.

5   CALCULATING RESULTS

Figure 10 shows the maximum vertical displacement of the whole model is 1.52mm. Figure 11 shows the Mises stress of the pot bearing and pier subjected to the vertical design loads was distributed uniformly. The maximum equivalent stress appears in the lower flange of the steel piston is 68.05MPa. Meanwhile, the stress of concrete pier is distributed uniformly, and bearing stress diffusion theory is verified.

It can be seen from Figure 12 to Figure 13 that the deformation of the upper bearing plate subjected to the vertical design loads is distributed uniformly. The maximum equivalent stress of upper bearings plates appears in the contact parts between the bottom surface of the upper bearing plate and PTFE slider is 38.99MPa which is smaller than the yield strength of the structural steel and meets the design requirements.

It can be seen from Figure 14 to Figure 15 that deformation of the PTFE slider is very small and the deformation distribution is uniform. The principal compressive stress is less than 30MPa except the edge of the PTFE slider, which conforms to the Transport Industry Standard of China JT/T391-2009[8].

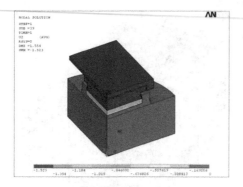

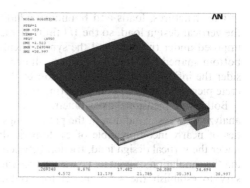

Figure 10. Vertical displacement nephogram of pot bearings.

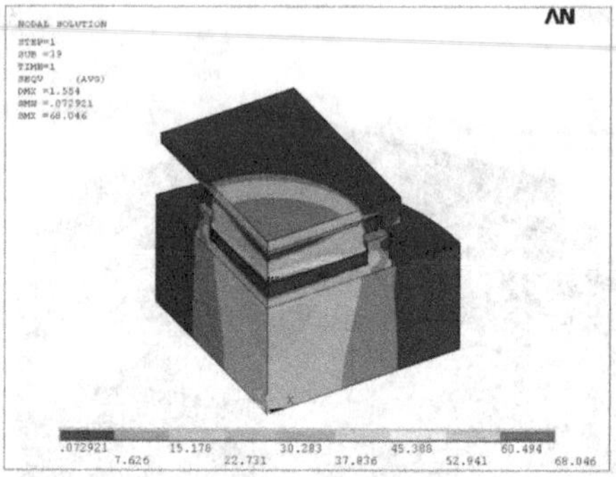

Figure 11. Mises stress nephogram of pot bearings.

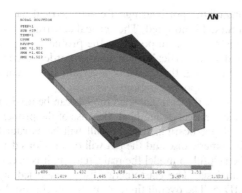

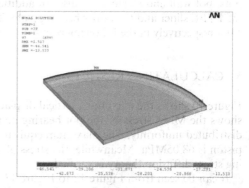

Figure 12. Global displacement nephogram.

Figure 13. Mises stress nephogram.

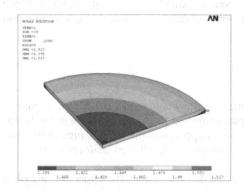

Figure 14. Global displacement nephogram.

Figure 15. Principal compressive stress nephogram.

Figure 16 and Figure 17 shows that deformation of the steel piston is very small and uniform distribution. The maximum equivalent stress and principal compressive stress appear in the periphery of the intermediate plate groove is 74.51MPa and 87.12MPa respectively. The maximum principal tensile stress appears in the contact parts between the central location of

424

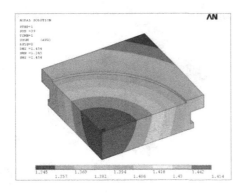

Figure 16. Global displacement nephogram.

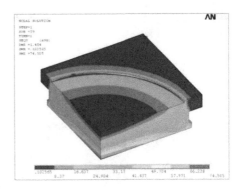

Figure 17. Mises stress nephogram.

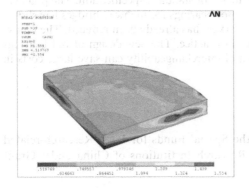

Figure 18. Global displacement nephogram.

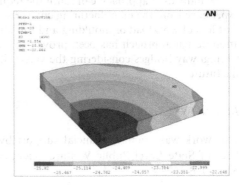

Figure 19. Principal compressive stress nephogram.

lower flange and elastomeric disc is 42.99MPa which is smaller than the yield strength of structural steel.

Figure 18 shows that compression deformation of the elastomeric disc is very small and uniform distribution, and the deformation value is smaller than 0.6mm. It can be seen from Figure 19 that stress distribution of the elastomeric disc is uniform, and the principal compression stress is between 22.65MPa and 25.82MPa, which conforms to the Transport Industry Standard of China JT/T391-2009 about the average stress of the elastomer.

Figure 20 shows that deformation of the pot is distributed uniformly and changes smoothly. Figure 21 shows that the tensile stress and compression stress are all existed under the design load, and the maximum equivalent stress is 63.37MPa.

Because of the transverse squeezing action of the compression on the pot wall, the maximum principal tensile stress which appears at the root of pot wall is 88.1MPa, but the value is still smaller than the yield strength of structural steel.

6 CONCLUSIONS

The common pot bearings for highway bridges are designed without consideration of versatility and interchangeability. The individual approach by application of optimal design idea of mechanical engineering for the design of the main element of the pot bearings for highway bridges including pots, steel pistons and so on is based on the concept that the whole optimum design of the pot bearings is achieved by minimizing the number of elements.

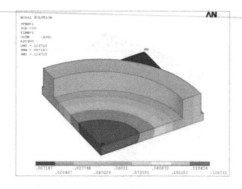
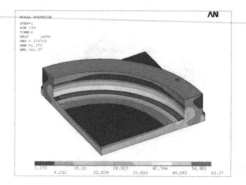

Figure 20.   Global displacement nephogram.　　Figure 21.   Mises stress nephogram.

The individual approach can fulfill the requirements of standard specification. The practical experience from the manufacturing enterprise for pot bearings for highway bridges has shown that lots of workload of moulding and foundry workers has already been reduced. The innovation design approach has been proven to work in practice. The new design of pot bearings for highway bridges considering the versatility and interchangeability can save large costs in the future.

ACKNOWLEDGEMENTS

This work was granted financial support by the Special Funds for Basic Research-related Fees of State Level Public Welfare Scientific Research Institutions of China under Grant No. 2016-9002.

REFERENCES

China Communications Press. (2009). *Pot bearings for highway bridges(JT/T 391-2009)*.
EUROPEAN COMMITTEE FOR STANDARDIZATION. (2000). *Structural Bearings—Part1: General design rules (EN 1337-1)*, Brussels.
EUROPEAN COMMITTEE FOR STANDARDIZATION. (2005). *Structural Bearings—Part3: Elastomeric Bearings (EN 1337-3)*. Brussels.
EUROPEAN COMMITTEE FOR STANDARDIZATION. (2005). *Structural Bearings—Part5: Pot Bearings (EN 1337-5)*. Brussels.
Li. Y. H. (2009). *Practical guide for highway bridge bearings*. Beijing: People Communications Press.
Zhuang, J. S. (2015). *Bridge bearings*. Beijing: China Railway Press.

Author Index

428